Post-translational modifications are now known to play a fundamental role in regulating the activity, location and function of a wide range of proteins. In plant cells work on different types of post-translational modification has progressed largely along independent lines. This book brings research workers together to allow an exchange of ideas, and reflects a diversity of interest whilst also revealing common ground. The contents provide an overview of the subject and a starting point for future research.

SOCIETY FOR EXPERIMENTAL BIOLOGY
SEMINAR SERIES: 53

POST-TRANSLATIONAL MODIFICATIONS IN PLANTS

SOCIETY FOR EXPERIMENTAL BIOLOGY SEMINAR SERIES

A series of multi-author volumes developed from seminars held by the Society for Experimental Biology. Each volume serves not only as an introductory review of a specific topic, but also introduces the reader to experimental evidence to support the theories and principles discussed, and points the way to new research.

POST-TRANSLATIONAL MODIFICATIONS IN PLANTS

Edited by

N.H. Battey

Department of Horticulture, University of Reading

H.G. Dickinson

Department of Plant Sciences, University of Oxford

A.M. Hetherington

Division of Biological Sciences, University of Lancaster

CAMBRIDGE
UNIVERSITY PRESS

Published by the Press Syndicate of the University of Cambridge
The Pitt Building, Trumpington Street, Cambridge CB2 1RP
40 West 20th Street, New York, NY 10011–4211, USA
10 Stamford Road, Oakleigh, Melbourne 3166, Australia

First published 1993

Printed in Great Britain at the University Press, Cambridge

A catalogue record for this book is available from the British Library

Library of Congress cataloguing in publication data
Post-translational modifications in plants / edited by N.H. Battey,
    H.G. Dickinson, A.M. Hetherington.
        p.   cm. – (Society for Experimental Biology seminar series;
    53)
    Papers from a symposium held at Lancaster in April 1992.
    Includes index.
    ISBN 0–521–41181–5 (hardback)
    1. Post-translational modification – Congresses.   2. Plant
proteins – Synthesis – Congresses.   I. Battey, N.H.   II. Dickinson,
Hugh G.   III. Hetherington, A.M.   IV. Series: Seminar series
(Society for Experimental Biology (Great Britain)); 53.
QK898.P8P67   1993
581.19′296 – dc20   92–28694   CIP

ISBN 0 521 41181 5 hardback

Contents

Contents

Contributors

BATTEY, N.H.
Department of Horticulture, Plant Science Laboratories, University of Reading, Whiteknights, Reading RG6 2AS, UK

BLACKBOURN, H.D.
Department of Biochemistry, University of Cambridge, Tennis Court Road, Cambridge CB2 3EA, UK

BOWLES, D.J.
Centre for Plant Biochemistry and Biotechnology, University of Leeds, Leeds LS2 9JT, UK

CALLAHAN, F.E.
Plant Molecular Biology Laboratory, USDA/ARS, Beltsville Agricultural Research Center (West), Building 006, 10300 Baltimore Avenue, Beltsville, Maryland 20705-2350, USA

CHEKKAFI, A.
LTI-CNRS URA 203, European Institute for Peptide Research, University of Rouen, 76134 Mont-St-Aignan, France

COURTNEY, S.E.
Departments of Biochemistry and Biology, Royal Holloway and Bedford New College, University of London, Huntersdale, Egham Hill, Egham, Surrey TW20 0EX, UK

DELAUNAY, A.-M.
LTI-CNRS URA 203, European Institute for Peptide Research, University of Rouen, 76134 Mont-St-Aignan, France

DICKINSON, H.G.
Department of Plant Sciences, University of Oxford, South Parks Road, Oxford OX1 3RB, UK

DIXON, R.A.
AFRC-IPSR Nitrogen Fixation Laboratory, University of Sussex, Brighton BN1 9RQ, UK

DRIOUICH, A.
University of Colorado at Boulder, DMCDB, Boulder, Colorado 80309-0347, USA

EDELMAN, M.
Plant Molecular Biology Laboratory, USDA/ARS, Beltsville
Agricultural Research Center (West), Building 006, 10300 Baltimore
Avenue, Beltsville, Maryland 20705-2350, USA
ELICH, T.D.
Plant Molecular Biology Laboratory, USDA/ARS, Beltsville
Agricultural Research Center (West), Building 006, 10300 Baltimore
Avenue, Beltsville, Maryland 20705-2350, USA
FALLON, K.M.
Institute of Cell and Molecular Biology, The King's Buildings,
University of Edinburgh, Mayfield Road, Edinburgh EH9 3JH, UK
FAYE, L.
LTI-CNRS URA 203, European Institute for Peptide Research,
University of Rouen, 76134 Mont-St-Aignan, France
FITCHETTE-LAINE, A.-C.
LTI-CNRS URA 203, European Institute for Peptide Research,
University of Rouen, 76134 Mont-St-Aignan, France
FÜHR, A.
Botanisches Institut, Universität Bonn, Venusbergweg 22, D-5300 Bonn
1, Germany
GEBAUER, G.
Institute for Cell Biology, Biochemistry and Biotechnology, University
of Bremen, D-2800 Bremen, Germany
GHIRARDI, M.L.
Plant Molecular Biology Laboratory, USDA/ARS, Beltsville
Agricultural Research Center (West), Building 006, 10300 Baltimore
Avenue, Beltsville, Maryland 20705-2350, USA
GOMORD, V.
LTI-CNRS URA 203, European Institute for Peptide Research,
University of Rouen, 76134 Mont-St-Aignan, France
HARDIE, D.G.
MRC Protein Phosphorylation Unit, Department of Biochemistry, The
University, Dundee DD1 4AN, UK
HETHERINGTON, A.M.
Division of Biological Sciences, University of Lancaster, Bailrigg,
Lancaster LA1 4YQ, UK
KALLIES, A.
Institute for Cell Biology, Biochemistry and Biotechnology, University
of Bremen, D-2800 Bremen, Germany
KNOX, J.P.
Centre for Plant Biochemistry and Biotechnology, University of Leeds,
Leeds LS2 9JT, UK

KOHLER, W.
Institute for Cell Biology, Biochemistry and Biotechnology, University of Bremen, D-2800 Bremen, Germany
LORD, J.M.
Department of Biological Sciences, University of Warwick, Coventry CV4 7AL, UK
MacKINTOSH, C.
MRC Protein Phosphorylation Unit, Department of Biochemistry, The University, Dundee DD1 4AN, UK
MacKINTOSH, R.W.
Protein Phosphorylation Group, Department of Biochemistry, The University, Dundee DD1 4AN, UK
MASTERS, A.K.
Department of Biochemistry and Genetics, Medical School, University of Newcastle upon Tyne, Newcastle upon Tyne NE2 4HH, UK
MATTOO, A.K.
Plant Molecular Biology Laboratory, USDA/ARS, Beltsville Agricultural Research Center (West), Building 006, 10300 Baltimore Avenue, Beltsville, Maryland 20705-2350, USA
MILLNER, P.A.
Department of Biochemistry and Molecular Biology, University of Leeds, Leeds LS2 9JT, UK
NIMMO, H.G.
Department of Biochemistry, University of Glasgow, Glasgow G12 8QQ, UK
RENSING, L.
Institute for Cell Biology, Biochemistry and Biotechnology, University of Bremen, D-2800 Bremen, Germany
RIDER, C.C.
Department of Biochemistry, Royal Holloway and Bedford New College, University of London, Huntersdale, Egham Hill, Egham, Surrey TW20 0EX, UK
RITCHIE, S.M.
Department of Horticulture, Plant Science Laboratories, University of Reading, Whiteknights, Reading RG6 2AS, UK
ROBERTS, L.M.
Department of Biological Sciences, University of Warwick, Coventry CV4 7AL, UK
SCHERER, G.F.E.
Botanisches Institut, Universität Bonn, Venusbergweg 22, D-5300 Bonn 1, Germany

SCHÜTTE, M.
Botanisches Institut, Universität Bonn, Venusbergweg 22, D-5300 Bonn 1, Germany

SHIRRAS, A.D.
Division of Biological Sciences, University of Lancaster, Bailrigg, Lancaster LA1 4YQ, UK

SOLL, J.
Botanisches Institut, Universität Kiel, Olshausenstrasse 40-60, D-2300 Kiel 1, Germany

SOMMERVILLE, J.
School of Biological and Medical Sciences, University of St Andrews, St Andrews, Fife KY16 9TS, UK

STEAD, A.D.
Department of Biology, Royal Holloway and Bedford New College, University of London, Huntersdale, Egham Hill, Egham, Surrey TW20 0EX, UK

SUSSMAN, M.R.
Department of Horticulture and Cell/Molecular Biology Program, University of Wisconsin, Madison, Wisconsin 53706, USA

TREWAVAS, A.J.
Institute of Cell and Molecular Biology, The King's Buildings, University of Edinburgh, Mayfield Road, Edinburgh EH9 3JH, UK

WHITE, I.R.
Department of Biochemistry and Molecular Biology, University of Leeds, Leeds LS2 9JT, UK

WISE, A.
Department of Biochemistry and Molecular Biology, University of Leeds, Leeds LS2 9JT, UK

ZAMRI, I.
Department of Biochemistry and Molecular Biology, University of Leeds, Leeds LS2 9JT, UK

List of abbreviations

AGP	arabinogalactan-protein
AHA	*Arabidopsis* H$^+$-ATPase
BSA	bovine serum albumen
Cast	castanospermine
CDPK	calcium-dependent protein kinase
4-CN	4-chloro-1-naphthol
Con A	concanavalin A
CPK	calcium-dependent, calmodulin-independent protein kinase
2,4-D	2,4-dichlorophenoxyacetic acid
DAPI	4,6-diaminido-2-phenylindole
DCMU	3(3,4-dichlorophenol)-1,1-dimethylurea
DEAE	diethylaminoethyl
DIDS	4,4'-diisothiocyanostilbene-2,2'-disulphonic acid
DMM	deoxymannojirimycin
Dol	dolichol
DTT	dithiothreitol
EDTA	ethylenediaminetetraacetic acid
EGTA	ethyleneglycol-bis(β-aminoethylether)-N,N,N',N'-tetraacetic acid
ER	endoplasmic reticulum
ERK	extracellular signal-regulated kinase
ETSH	mercaptoethanol
FMoc	fluorenylmethoxycarbonyl group
Fru6P	fructose 6-phosphate
FSBA	5'-p-fluorosulphonylbenzoyladenosine
Glc	glucose
GlcNAC	N-acetylglucosamine
G protein	GTP-binding protein
GSSG	glutathione
HPLC	high-performance liquid chromatography

HRGP	hydroxyproline-rich glycoprotein
IDA	iminodiacetic acid
IP	isoelectric point
IP_3	inositol trisphosphate
ISPK1	insulin-stimulated protein kinase 1
LHC	light-harvesting chlorophyll *a/b* binding protein
LHCP	light-harvesting chlorophyll *a/b* binding protein
Man	mannose
MAP kinase	mitogen- or messenger-activated protein kinase
M_r	relative molecular mass
NEM	N'-ethylmaleimide
NiR	reduced-ferredoxin-dependent nitrate reductase
NR	nitrate reductase
NTA	nitrilotriacetic acid
OPFP	pentafluorophenyl
PAF	platelet-activating factor
PAGE	polyacrylamide gel electrophoresis
PC	phosphatidylcholine
PEPc	phosphoenolpyruvate carboxylase
PEPCase	phosphoenolpyruvate carboxylase
PHA	phytohaemagglutinin
P_i	inorganic phosphate
PI	phosphatidylinositol
PIP	phosphatidylinositol phosphate
PIP_2	phosphatidylinositol 4,5-bisphosphate
PNA	peanut agglutinin
PP	protein phosphatase
PPD	purified protein derivative
PPDK	pyruvate P_i dikinase
PS	phosphatidylserine
PSI, PSII	photosystems I and II of chloroplast
RBCS	Rubisco small subunit
RCA	*Ricinus communis* agglutinin
RER	rough endoplasmic reticulum
RIP	ribosome-inhibiting protein
Rubisco	ribulose 1,5-bisphosphate carboxylase–oxygenase
S-allele	self-incompatibility allele
SDS	sodium dodecyl sulphate
SH2 domain	*src* homology-2 domain
SI	self-incompatibility
SRK	S receptor kinase

sSMCC	sulphosuccinimidyl-4-(N-maleimidomethyl)-cyclohex-ane-1-carboxylase
t-Boc	t-butoxycarbonyl group
TCA	trichloroacetic acid
TED	N,N,N′-tris(carboxymethyl)ethylenediamine
TGN	*trans*-Golgi network
TLE/TLC	thin layer electrophoresis/thin layer chromatography
UTase-UR	uridylyltransferase/uridylyl-removing enzyme

Preface

The papers collected in this volume are the contributions of invited speakers to a two-day meeting organised by the Plant Development and Plant Metabolism groups of the SEB and held at Lancaster in April 1992. One of the ideas of the meeting was to try and get away from the stereotyped divisions of 'hormones and receptors', 'signal transduction', 'Golgi transport', 'protein targeting', and so on, and look instead at things from the point of view of the proteins involved in these processes, and the effects of post-translational modifications on their activity. The result is a selection of papers on phosphorylation, glycosylation, acylation, ubiquitination, and protein processing, which illustrate the ways in which these modifications are brought about and their effects on cellular processes.

The first chapter, by the volume editors, is an introductory overview of the subject of post-translational modification in plants, in which key advances in the last few years are highlighted. The following chapters by Dixon and Hardie place the work on plant systems in a wider context by reviewing the roles of protein phosphorylation in bacteria and animals; Fallon & Trewavas provide a general discussion of phosphorylation in plant cells and some details on Ca^{2+}/calmodulin regulated and tyrosine kinases in plant systems. In the chapters by Mattoo *et al.*, Soll, and White *et al.*, phosphorylation is discussed with particular reference to chloroplast function; the uses of peptides in the study of G proteins and chloroplast protein kinase are outlined. Scherer *et al.* and Battey *et al.* describe the main types of Ca^{2+}-regulated protein kinases in plant cells, while the chapters by Sussman, Nimmo, Rensing *et al.*, Shirras *et al.*, and Mackintosh & Mackintosh outline some major targets of protein kinases, including H^+-ATPase, PEP carboxylase and sucrose phosphate synthase. In the next four chapters, by Faye *et al.*, Lord & Roberts, Bowles, and Knox, the roles of glycosylation in protein targeting, and the importance of post-translational processing, are discussed, along with the developmental regulation of arabinogalactan proteins of the plasma membrane. The final chapter by Courtney *et al.* provides an overview of ubiquitination in cell biology and its potential importance for plant senescence. Although this small collection of papers cannot be fully comprehensive, we hope that it is representative of work in progress, and conveys something of the excitement of this rapidly advancing area of plant science research.

The book should provide a good entry point into the literature for advanced undergraduates and postgraduates, and a useful summary of current thinking on the significance of post-translational modifications in the life of the plant. We acknowledge the financial support of the Plant Metabolism and Plant Development groups of the SEB, and of the Gatsby Foundation. We also thank the contributors for prompt delivery of their manuscripts and for taking part in a useful and stimulating meeting.

N.H. Battey
H.G. Dickinson
A.M. Hetherington

N.H. BATTEY, H.G. DICKINSON
and A.M. HETHERINGTON

Some roles of post-translational modifications in plants

Phosphorylation

Receptors and signal transduction

The first stage in an idealised signal transduction chain is binding of the ligand (hormone or other effector molecule) to its receptor. In animal cells phosphorylation on tyrosine residues has been shown to be an important consequence of ligand binding in many transmembrane receptors (for example for insulin, platelet-derived growth factor, epidermal growth factor); other receptor types, operating via the G protein/phospholipase C or adenylate cyclase routes, have their receptor activity modulated by phosphorylation (for example the β_2-adrenergic receptor). A useful summary of the roles of phosphorylation in these and other animal systems is given by Hardie (this volume).

In plants, receptors for growth regulators have been characterised to varying degrees (Napier & Venis, 1990); however, there is no evidence for a role for phosphorylation in either their regulation or their transduction mechanisms. The only evidence for a transmembrane receptor protein kinase comes from maize, in which a cDNA clone encoding a putative serine/threonine kinase structurally related to receptor tyrosine kinases has been described (Walker & Zhang, 1990). The extracellular domain of this protein is similar to that of *Brassica* S-locus glycoproteins involved in the self-incompatibility mechanism of this genus (see section on glycosylation, below). The structural similarity to tyrosine kinases like the epidermal growth factor receptor suggests that the extracellular domain of the maize protein interacts with a ligand and that this activates the kinase domain on the cytoplasmic side. These are exciting data, and demonstrate the power of molecular cloning techniques to further our understanding of phosphorylation in plants. However, there is still everything to be discovered about the function of this putative kinase.

Society for Experimental Biology Seminar Series 53: *Post-translational modifications in plants*, ed. N.H. Battey, H.G. Dickinson & A.M. Hetherington. © Cambridge University Press 1993, pp. 1–15.

The maize receptor is unusual in its greater similarity to serine/threonine kinases rather than to tyrosine kinases; it suggests that although plants do analogous things to animals – sense the environment, and communicate over long and short distances between and within cells – they have distinctly different ways of doing them. This theme of related but distinct kinase activities in plants is emphasised by the work of Lawton *et al.* (1989), who isolated bean and rice cDNAs encoding putative protein kinases with catalytic domains most similar to those of cyclic nucleotide-dependent kinases and protein kinase C, but with very distinctive regulatory domains. The recent sequencing of cDNA encoding the Ca^{2+}-dependent, calmodulin-independent protein kinase from soybean demonstrates the same point; in the predicted protein sequence a calmodulin-like domain is directly linked to the kinase catalytic region, a combination not reported for animal kinases (Harper *et al.*, 1991). The wide distribution of kinases with properties similar to the soybean enzyme, and their suggested broad specificity (see later chapter by Battey *et al.*, this volume) indicates that in plants this enzyme may play a role analogous to that of the widespread Ca^{2+}/calmodulin-dependent multiprotein kinase of animal cells (Schulman & Lou, 1989). However, the cloning of plant cDNA with deduced amino acid sequence similar to the calmodulin-binding domain of the latter enzyme (Watillon *et al.*, 1992) may indicate that the story is more complex. The reported occurrence of cAMP-regulated protein kinase in plants (Janistyn, 1986), without a clearly defined role for cAMP, also suggests the need for an open mind on the limits of plant protein kinase regulation.

Ca^{2+}-regulated protein kinases abound in plants (see Fallon & Trewavas, Scherer *et al.* and Battey *et al.*, this volume). Yet it is not so often that activation of a plant kinase has been clearly linked to an extracellular (environmental) signal. An oligosaccharide showing activity in proteinase inhibitor induction has been shown to stimulate *in vitro* phosphorylation of a 34 kDa protein from potato and tomato leaf plasma membrane (Farmer *et al.*, 1989, 1991). This example suggests a role for phosphorylation in the transmission of an attack signal in plant cells, but the consequences of phosphorylation, and its role in the defence response, remain unknown. Similarly, blue and red light have been shown to induce changes in protein phosphorylation (see Fallon & Trewavas, this volume) but the role of these changes in subsequent developmental responses is not known.

Clearly, then, there is significant recent evidence that plants use phosphorylation in their transduction mechanisms, and that these mechanisms differ in important respects from their analogues in animals. However, the exact role of phosphorylation in the process of signalling is not apparent from these studies. Better examples of the regulatory effects of protein phosphorylation in plant cells are found when enzymes involved in metabolic control are considered.

Metabolic regulation

Phosphoenolpyruvate carboxylase (PEPc) is one of the best examples of a plant enzyme whose activity is regulated by reversible phosphorylation. The novel idea is outlined by Nimmo (this volume) that day/night regulation of PEPc kinase from *Bryophyllum* is achieved not by a second messenger cascade system but by a rhythm in protein synthesis. It will be interesting to discover how widespread this type of regulatory mechanism is in plants. Sucrose phosphate synthase (SPS) is another enzyme known to be regulated by phosphorylation, activation following illumination being associated with dephosphorylation (Huber, Huber & Nielsen, 1989). In this case it is not yet known how the light signal results in dephosphorylation. The inhibition of light activation of SPS in leaves (Huber & Huber, 1990) or leaf discs (Siegl, MacKintosh & Stitt, 1990) fed with okadaic acid indicates that a light-activated phosphatase could be involved. The inhibition of this phosphatase by P_i suggests that regulation by light might be an indirect consequence of the effects of photosynthetic rate on P_i levels (Huber & Huber, 1990).

Quinate:NAD^+ oxidoreductase was described some time ago as an enzyme inactivated by dephosphorylation (Refeno, Ranjeva & Boudet, 1982), and with a complex light/dark regulation (Graziana *et al.*, 1983). This enzyme has now been shown to be a substrate for protein phosphatase 2A (MacKintosh, Coggins & Cohen, 1991). The description of this and other phosphatases (MacKintosh & Cohen, 1989) adds an important dimension to the study of reversible phosphorylation in plants (see MacKintosh & MacKintosh, this volume).

Metabolism is also regulated at the level of metabolite transport in plant cells. It is therefore of interest that malate transport in soybean symbiosomes appears to be regulated by phosphorylation, and that nodulin-26, a substrate for the soybean Ca^{2+}-dependent, calmodulin-independent kinase discussed above (see Weaver *et al.*, 1991), is a good candidate for the malate transporter (Ouyang *et al.*, 1991). This provides the first framework in which Ca^{2+}-regulated phosphorylation by a well-characterised kinase is clearly linked to a metabolic process in plants.

DNA binding proteins

The regulated expression of genes during plant development and during plant responses to the environment is of key importance. Selective transcriptional activation is achieved by the interaction of protein factors with recognition sites within the promoter regions of specific genes. Examples of these DNA binding proteins include GT-1, which binds to the gene for ribulose bisphosphate carboxylase/oxygenase small subunit (RBCS) in pea, and is involved with its light-dependent expression (Green, Kay & Chua, 1987; Lam & Chua, 1990); SBF-1 from bean is very similar to GT-1 and may control expression of chalcone synthase (Harrison *et al.*, 1991); GBF-1 is a leucine zipper protein (Schindler *et al.*, 1992) that binds to a G-box promoter sequence in RBCS genes (Giuliano *et al.*, 1988); and AT-1 is a factor from pea that binds to specific AT-rich elements in genes encoding RBCS and other proteins (Datta & Cashmore, 1989). A review of genes encoding plant transcription factors is provided by Katagiri & Chua (1992).

There is increasing evidence that phosphorylation of these DNA binding proteins influences their properties. AT-1 loses all DNA binding activity as a result of phosphorylation (Datta & Cashmore, 1989), whereas in SBF-1 phosphorylation is essential for DNA binding (Harrison *et al.*, 1991). GBF-1 has been shown to be phosphorylated by a casein kinase II activity from broccoli, and phosphorylation stimulates G-box binding activity (Klimczak, Schindler & Cashmore, 1992).

The exact role of phosphorylation of DNA binding factors *in vivo* is complex and incompletely understood. However, it is crucial that this control mechanism operates not only at the end of signal transduction chains, but also as part of a complex endogenous signalling network. An example of the former is the induction by fungal elicitors of specific, Ca^{2+}-dependent changes in the phosphorylation state of polypeptides from parsley cell suspensons (Dietrich, Mayer & Hahlbrock, 1990). Such a pathway leads to elicitor-induced gene expression and it is conceivable that some of the newly phosphorylated proteins are transcription factors. Similarly, light-induced gene expression in soybean suspension cells may involve Ca^{2+}/calmodulin in the control pathway (Lam, Benedyk & Chua, 1989). An endogenous signalling network involving phosphorylation of DNA binding proteins is suggested by the finding that the floral homeotic genes *def* from *Antirrhinum* and *ag* from *Arabidopsis* have homology with the DNA binding proteins MCM-1 from yeast and SRF from mammals (Sommer *et al.*, 1990; Yanofsky *et al.*, 1990). The fact that SRF activity can be regulated by phosphorylation (Prywes *et al.*, 1988) has led to the suggestion that the protein products of homeotic genes such as *def* and *ag* regulate flower morphogenesis by controlling specific groups of genes, and that this activity may in part be controlled by phosphorylation (Schwarz-Sommer *et al.*, 1990).

Glycosylation

Glycosylation as a targeting signal

The role of glycosylation in cellular function, and particularly in protein targeting, is explored by Faye *et al.* (this volume). The targeting of mammalian proteins to the lysosomal compartment is well documented (Sly & Fischer, 1982), the signal structure being phosphorylated mannose, which is recognised by specific receptors on the processing vesicle membrane. However, glycosylation does not seem to be involved in the targeting of plant proteins to the vacuole, a structure that has some parallel with the lysosomal compartment. In transgenic plants where the glycosylation of vacuolar proteins is modified or even absent, proteins continue to be transported to the vacuole (Voelker, Herman & Chrispeels, 1989). Glycosylation is required for the targeting of polypeptides to the extracellular compartment, for prevention of N-glycosylation of carrot suspension culture proteins by tunicamycin effectively inhibits their secretion into the medium (Driouich *et al.*, 1989). Whether glycosylation can be regarded as a specific secretion signal *per se* remains in doubt, for Faye *et al.* (this volume) have explored the role of side-chain maturation in these events using a range of processing inhibitors, and discovered that it is glycosylation itself that is needed for successful secretion, rather than a specific side-chain structure. This observation suggests that glycosylation may serve to protect the protein from proteolysis while in the cytoplasm, rather than labelling it for export.

Protein activation via glycosylation

Data presented by Bowles (this volume) provides conclusive evidence that the concanavalin A (Con A) molecule is glycosylated at its active site for the major part of its maturation, and only in the final stages of synthesis does a glycanase remove this side group and 'activate' the molecule.

Although Con A is the major constituent of the jack-bean seed, the pathway of its synthesis has proved very difficult to unravel. Early studies indicated that a glycosylated precursor, of approximately the same M_r as Con A, was first formed in the ER. This molecule was then apparently cleaved into smaller fractions, which were then religated, perhaps in the Golgi or its vesicles, to form the mature polypeptide. In the mature seed, Con A is contained within large protein bodies, originally derived from the vacuoles into which the Golgi vesicles discharge their contents. An aspect of lectin synthesis of particular interest is why a molecule such as Con A, which has intense affinity for glycosyl groups terminating in glucose or mannose, manages to pass through the ER cisternae, the *cis* and *trans* faces of the Golgi, and undergo vesicular transport to the tonoplast, without binding to the myriad glycoproteins encountered on the way. The answer for Con A was found by Bowles (this volume) through investigation of the 3D structure of the maturing lectin and, in particular, the early, large precursor molecule. A combination of N-terminal sequence data and structural predictions suggested that the apparently complex reorganisation of the precursor molecule(s) taking place during maturation could be explained by simple cleavages and a religation at amino acids 118–119 of the peptide chain on the outer surface of the molecule. Of more interest was the predicted structure of the single carbohydrate binding site, which was occupied by a glycan at all stages in the maturation of the molecule. Indeed, it was this single glycan that identified the large precursor as a glycoprotein and was responsible for its lack of affinity for other glycoproteins.

Removal of the glycan thus 'activates' the immature lectin during its maturation in the ER and the Golgi. N-glycanases of this type, which remove complete glycosyl groups from glycoproteins, have not previously been reported in plants, and it is possible that the synthesis of other potentially toxic polypeptides in plants may also involve 'glycan protection'. Nevertheless, other lectins, such as those found in pea and lentil seeds, are not glycosylated in their active sites during synthesis. Admittedly, the structure and affinities of these lectins differ from those of Con A, and thus the necessity for active site protection during synthesis may not be so great.

The synthesis of the glycosylated ribosome-inhibiting protein (RIP) ricin is also of interest in this respect, and is described in detail by Lord & Roberts (this volume). Analysis of the *Ricinus communis* agglutinin (RCA) reveals the presence of an A chain – the RIP – of 32 kDa, and a B chain of 34 kDa, which is a galactose or N-acetylgalactosamine binding lectin. There is 93% homology at the amino acid level between the A and B chains, both of which contain two potential glycosylation sites. Normally both are glycosylated, except in a 'light' form of the A chain, where only one site is occupied. The mature form of RCA consists of two A chains and one B. Three-dimensional modelling of the B chain indicates two globular domains, each binding a single galactose residue.

Both the ricin A and B chains are synthesised as a single preprotein, which is substantially modified post-translationally (Roberts & Lord, 1981). Maturation of the preprotein follows the normal route of ER, *trans* and *cis* faces of the Golgi, and then final transport to protein bodies where the glycoprotein is stored in the seed. Final processing takes place in the Golgi where the oligosaccharide is trimmed and fucose and xylose residues are added. An endopeptidase also cleaves the preprotein into A and B chains, with the elimination of 12 amino acid residues (Lord & Robinson, 1986). Glycosylation confers RNA-N-glycosidase activity on the A chain, although the B chain acts as an effective lectin in its unglycosylated form (Richardson *et al.*, 1989). Experiments with glycosylation inhibitors, such as tunicamycin, and expression of the ricin core polypeptide *in vitro*, suggest that the oligosaccharide side chains are needed neither for the transport of the molecules, nor for their stability. It would therefore seem that, in contrast to the synthesis of Con A, the glycosylation of the ricin A chain serves to activate the molecule, and thus the preprotein is retained in the inactive form to avoid toxic effects during synthesis. This assumes greater importance as more is discovered of the mode of action of ricin, which seems to involve transport to the Golgi (van Deurs *et al.*, 1988), and perhaps even the ER, before the A chain is translocated into the cytosol and ribosomes are depurinated.

The role of glycosylation in self-incompatibility (SI)

From the earliest investigations, glycoproteins have been associated with the expression of the S(incompatibility)-locus in the pistil. Nishio & Hinata (1977) reported highly charged glycoproteins to segregate with specific S-alleles in *Brassica*, while Mau *et al.* (1986) demonstrated a similar situation in *Nicotiana*, except that glycoproteins associated with particular S-alleles did not vary so markedly in charge as in *Brassica*. Subsequent investigations have shown glycoproteins to be the female product of S-locus expression in members of the Solanaceae, Cruciferae, Scrophulariaceae (A. McCubbin, unpublished data) and the Papaveraceae (F.C.H. Franklin & V.E. Franklin-Tong, personal communication). Data are not available from the male determinants of SI in these groups; certainly, a number of hypotheses, such as the dimer hypothesis of Lewis (1965), predict the presence of identical determinants in male and female organs but, despite extensive searches, such molecules have not been found.

As far as has been determined, the nature and synthesis of SI female glycoproteins is unexceptional. For example, the maturing polypeptides of *Brassica* are generally N-glycosylated with typical side chains (Takayama *et al.*, 1986), although there is some evidence that not all the potential sites are glycosylated. Thus the female product of the S_8 allele in *Brassica campestris* is composed of 405 amino acids, has a molecular mass of 46 kDa and seven oligosaccharide side chains (Takayama *et al.*, 1987).

Maturation of the side chains is also typical with the cleavage of a terminal GlcNAc, to give Man, Man, Xyl as end groups, all linked to a mannose residue. While the glycoproteins from the Solanaceae and the Cruciferae are highly glycosylated, the protein active in the SI bioassay in *Papaver* contains few potential glycosylation sites and, in one case, perhaps only one (F.C.H. Franklin & V.E. Franklin-Tong, personal communication). Interestingly, all those S-allele products sequenced feature an N-terminal signal sequence suggesting that the glycoprotein is exported from the pistillar protoplasts, either into the stylar transmitting tissue (Anderson *et al.*, 1986) or the papillar cell walls (Kandasamy *et al.*, 1990).

Since information is only available for female molecules, and since the molecular basis of SI has yet to be established for any species, it is not surprising that the role of glycosylation in SI remains unclear. Certainly differences in the peptide sequences between alleles in both *Brassica* and *Nicotiana* are sufficient to create individual glycosylation patterns for each S-allele product, but this is far from establishing that S-allele specificity resides with the organisation of the glycosyl side chains. Nevertheless, it is likely that glycosylation does play some role in either the recognition or response components of SI for treatment of the system with tunicamycin results in self-compatibility (Sarker, Elleman & Dickinson, 1988). Further, a number of recent findings have provided valuable clues to the roles played by the pistillar S-locus products. For example, the stylar glycoproteins of the Solanaceae are now known to possess RNase activity (McClure *et al.*, 1989) and it is noteworthy that the *Aspergillus* RNases, with which they share considerable homology, are also glycosylated. Glycosylation is, however, not required for RNase function itself (T.H. Kao, personal communication).

The glycoproteins encoded by a number of *Brassica* S-alleles have now been sequenced, and there is conspicuous heterogeneity in the glycosylation of these molecules. Nasrallah (Nasrallah, Yu & Nasrallah, 1988; Nasrallah, Nishio & Nasrallah, 1991) has identified two classes of allele, and has suggested that each molecule possesses variable and constant regions, such as are found in proteins of the immune system. Many of the glycosylation sites lie within the variable regions and therefore may well be involved in the determination of allelic specificity. Equally, polypeptides homologous with the S-allele products, but unlinked to the S-locus, are also heavily glycosylated. These S-locus-related glycoproteins (Lalonde *et al.*, 1989) do not differ between S-alleles and, as far as can be determined, play no direct role in SI. The most recently identified component of the *Brassica* S-locus is an 'S receptor kinase' (SRK) (Stein *et al.*, 1991) which possesses an extracellular domain with strong homology to the female S-allele glycoprotein; indeed, this homology seems to have been actively maintained during evolution. How the SRK, which is present in anthers and pistils, participates in SI is unknown, but it is interesting that it was originally cloned by homology to the ZmPK1 gene, which is expressed throughout the whole plant body of self-compatible maize (Walker & Zhang, 1990). Since the ZmPK1 product cannot itself be involved in SI, it is an attractive possibility that this class of serine/threonine kinases with highly glycosylated extracellular domains may be involved in a previously undiscovered plant-wide signalling system (Walker & Zhang, 1990). Whether glycosylation of the extracellular domain of these kinases plays a key role in their specificity remains to be determined.

Acylation

In animals, bacteria, yeasts and viruses, numerous proteins have now been shown to be modified by the covalent attachment of lipids. There are well-documented examples of the modification of proteins by the attachment of fatty acids (Schultz, Henderson & Oroszlan, 1988; McIlhinney, 1990), long-chain prenyl groups (Glomset, Gelb and Farnsworth, 1990) and glycosylphosphatidylinositol (Low, 1989). In contrast the modification of plant proteins by the covalent attachment of lipids has received much less attention, with palmitoylation of chloroplast proteins being the best studied (see Mattoo *et al.*, this volume). In this section we review the evidence from non-plant studies which suggests that the covalent modification of proteins by fatty acylation has importance in a diverse array of cell biological phenomena.

The linkage by which fatty acids are attached to proteins can be divided into two categories. In the first group, acyl proteins contain a fatty acid linked to the side chains of serine, threonine or cysteine residues via a hydroxylamine-sensitive ester or thioester linkage. These proteins are frequently localised to the cell membrane and are acylated primarily with the 16C saturated fatty acid palmitate. Interestingly, the linkage is labile and the fatty acid can turn over more rapidly than the protein. The modification (which is post-translational) is catalysed by a specific acyl transferase which uses palmitoyl CoA as acyl donor. There is evidence to suggest that this reaction can take place at a number of sites including the Golgi, ER and the plasma membrane. Membrane proteins modified in this manner include myelin proteolipoprotein, rhodopsin, the transferrin receptor, the insulin receptor, the β-adrenergic receptor, the nicotinic acetylcholine receptor, components of the Na^+ channel, and numerous viral proteins. It has been suggested that palmitoylation is involved in membrane anchorage, transport signalling and promoting coupling between receptors and their G proteins.

The second class of linkage is via a hydroxylamine-insensitive amide bond to the amino terminal glycine residue of the target protein. Acylation is with the 14C saturated fatty acid myristic acid, is stable and occurs co-translationally. Myristoylated proteins have been shown to be both membrane-bound and soluble. Examples include the retroviral oncogene product p60^{v-src}, the catalytic subunit of cAMP-dependent protein kinase, the calceneurin β subunit and the α subunits of Gi and Go. Again a specific acyl transferase (N-myristoyl transferase) is involved in the attachment to the protein substrate and this has been purified and sequenced from yeast.

A clue to the role of myristoylation may come from work on the transforming protein p60^{v-src} (a tyrosine protein kinase) of the Rous sarcoma virus. It has been observed that mutants of this protein which lack the attached myristic acid (but which still retain kinase activity) are unable to become associated with the membrane and are transformation defective. Work on G protein α subunits also indicates that membrane attachment may be an important role; however, in these proteins there are also indications that this is not the only function of myristoylation (Spiegel *et al.*, 1991). More recent studies with p60^{v-src} suggest that the myristate also serves to regulate protein–protein interactions. It would seem likely that further work will be required before the full significance of myristoylation can be assessed.

In summary, research in animal cells and viruses (reviewed in Schultz *et al.*, 1988; Towler *et al.*, 1988; Grand, 1989; McIlhinney, 1990; Spiegel *et al.*, 1991) has demonstrated that modification of proteins by the covalent attachment of lipids has important consequences for the function of the target protein. It will be of great interest to determine whether plant proteins are subject to the same range of modifications and, importantly, what is the functional significance of such covalent attachments.

References

Anderson, M.A., Cornish, E.C., Mau, S.-L., Williams, E.G., Hoggart, R., Atkinson, A., Bonig, I., Grego, B., Simpson, R., Roche, P.J., Haley, J.D., Penschow, J.D., Niall, H.D., Tregear, G.W., Coghlan, J.P., Crawford, R.J. & Clarke, A.E. (1986). Cloning of cDNA for a stylar glycoprotein associated with expression of self-incompatibility in *Nicotiana alata*. *Nature (London)* **321**, 38–44.

Datta, N. & Cashmore, A.R. (1989). Binding of a pea nuclear protein to promoters of certain photoregulated genes is modulated by phosphorylation. *Plant Cell* **1**, 1069–77.

Dietrich, A., Mayer, J.E. & Hahlbrock, K. (1990). Fungal elicitor triggers rapid, transient and specific protein phosphorylation in parsley cell suspension cultures. *Journal of Biological Chemistry* **265**, 6360–8.

Driouich, A., Gonnet, P., Makkie, M., Lainé, A.-C. & Faye, L. (1989). The role of high-mannose and complex asparagine-linked glycans in the secretion and stability of glycoproteins. *Planta* **180**, 96–104.

Farmer, E.E., Moloshok, T.D., Saxton, M.J. & Ryan, C.A. (1991). Oligosaccharide signalling in plants. *Journal of Biological Chemistry* **266**, 3140–5.

Farmer, E.E., Pearce, G. & Ryan, C.A. (1989). *In vitro* phosphorylation of plant plasma membrane proteins in response to the proteinase inhibitor inducing factor. *Proceedings of the National Academy of Sciences USA* **86**, 1539–42.

Giuliano, G., Pichersky, E., Malik, V.S., Timko, M.P., Scolnik, P.A. & Cashmore, A.R. (1988). An evolutionarily conserved protein binding sequence upstream of a plant light-regulated gene. *Proceedings of the National Academy of Sciences USA* **85**, 7089–93.

Glomset, J.A., Gelb, M.H. & Farnsworth, C.C. (1990). Prenyl proteins in eukaryotic cells: a new type of membrane anchor. *Trends in Biochemical Sciences* **15**, 139–42.

Grand, R.J.A. (1989). Acylation of viral and eukaryotic proteins. *Biochemical Journal* **258**, 625–38.

Graziana, A., Ranjeva, R., Salimath, B.P. & Boudet, A.M. (1983). The reversible association of quinate:NAD$^+$ oxidoreductase from carrot cells with a putative regulatory subunit depends on light conditions. *FEBS Letters* **163**, 306–10.

Green, P.J., Kay, S.A. & Chua, N.-H. (1987). Sequence-specific interactions of a pea nuclear factor with light responsive elements upstream of the rbcS-3A gene. *EMBO Journal* **9**, 2543–9.

Harper, J.F., Sussman, M.R., Schaller, G.E., Putnam-Evans, C., Charbonneau, H. & Harmon, A.C. (1991). A calcium-dependent protein kinase with a regulatory domain similar to calmodulin. *Science* **252**, 951–4.

Harrison, M.J., Lawton, M.A., Lamb, C.J. & Dixon, R.A. (1991). Characterization of a nuclear protein that binds to three elements within the silencer region of a bean chalcone synthase gene promoter. *Proceedings of the National Academy of Sciences USA* **88**, 2515–19.

Huber, J.L.A., Huber, S.C. & Nielsen, T.H. (1989). Protein phosphorylation as a mechanism for regulation of spinach leaf sucrose-phosphate synthase activity. *Archives of Biochemistry and Biophysics* **270**, 681–90.

Huber, S.C. & Huber, J.L. (1990). Activation of sucrose-phosphate synthase from darkened spinach leaves by an endogenous protein phosphatase. *Archives of Biochemistry and Biophysics* **282**, 421–6.

Janistyn, B. (1986). Effects of adenosine-3':5'-monophosphate (cAMP) on the activity of soluble protein kinases in maize (*Zea mays*) coleoptile homogenates. *Zeitschrift für Naturforschung* **41c**, 579–84.

Kandasamy, M.K., Dwyer, K.G., Paollilo, D.J., Doney, R.C., Nasrallah, J.B. & Nasrallah, M.E. (1990). *Brassica* S-proteins accumulate in the intercellular matrix along the path of pollen tubes in transgenic tobacco pistils. *Plant Cell* **2**, 39–49.

Katagiri, F. & Chua, N.-H. (1992). Plant transcription factors: present knowledge and future challenges. *Trends in Genetics* **8**, 22–7.

Klimczak, L.J., Schindler, U. & Cashmore, A.R. (1992). DNA binding activity of the *Arabidopsis* G-box binding factor GBF-1 is stimulated by phosphorylation by casein kinase II from broccoli. *Plant Cell* **4**, 87–98.

Lalonde, B.A., Nasrallah, M.E., Dwyer, K.G., Chen, C.-H., Barlow, B. & Nasrallah, J.B. (1989). A highly conserved *Brassica* gene with homology to the S-locus specific glycoprotein structural gene. *Plant Cell* **1**, 249–58.

Lam, E., Benedyk, M. & Chua, N.-H. (1989). Characterization of phytochrome-regulated gene expression in a photoautotrophic cell suspension: possible role for calmodulin. *Molecular and Cellular Biology* **9**, 4819–23.

Lam, E. & Chua, N.-H. (1990). GT-1 binding site confers light responsive expression in transgenic tobacco. *Science* **248**, 471–4.

Lawton, M.A., Yamamoto, R.T., Hanks, S.K. & Lamb, C.J. (1989). Molecular cloning of plant transcripts encoding protein kinase homologs. *Proceedings of the National Academy of Sciences USA* **86**, 3140–4.

Lewis, D. (1965). A protein dimer hypothesis on incompatibility. In *Genetics Today* (ed. S.J. Geerts) (*Proc. XI Int. Congress Genet. 1963*), pp. 657–63. Oxford: Pergamon.

Lord, J.M. & Robinson, C. (1986). Role of proteolytic enzymes in the post-translational modification of proteins. In *Plant Proteolytic Enzymes* (ed. M.J. Dalling), pp. 69–80. Boca Raton, Florida: CRC Press.

Low, M.G. (1989). Glycosyl-phosphatidylinositol: a versatile anchor for cell surface proteins. *FASEB Journal* **3**, 1600–8.

MacKintosh, C., Coggins, J. & Cohen, P. (1991). Plant protein phosphatases. *Biochemical Journal* **273**, 733–8.

MacKintosh, C. & Cohen, P. (1989). Identification of high levels of type 1 and type 2A protein phosphatases in higher plants. *Biochemical Journal* **262**, 335–9.

Mau, S.-L., Williams, E.G., Atkinson, A., Anderson, M.A., Cornish, E.C., Grego, B., Simpson, R.J., Kheyr-Pour, A. & Clarke, A.E. (1986). Style proteins of a wild tomato (*Lycopersicon peruvianum*) associated with expression of self-incompatibility. *Planta* **169**, 184–91.

McClure, B.A., Haring, V., Ebert, P.R., Anderson, M.A., Simpson, R.J., Sakiyama, F. & Clarke, A.E. (1989). Style self-incompatibility gene products of *Nicotiana alata* are ribonucleases. *Nature (London)* **342**, 955–7.

McIlhinney, R.A.J. (1990). The fats of life: the importance and function of protein acylation. *Trends in Biochemical Sciences* **15**, 387–91.

Napier, R.M. & Venis, M.A. (1990). Receptors for plant growth regulators: recent advances. *Journal of Plant Growth Regulation* **9**, 113–26.

Nasrallah, J.B., Nishio, T. & Nasrallah, M.E. (1991). The self-incompatibility genes of *Brassica*: expression and use in genetic ablation of floral tissues. *Annual Review of Plant Physiology and Plant Molecular Biology* **42**, 393–422.

Nasrallah, J.B., Yu, S.M. & Nasrallah, M.E. (1988). Self-incompatibility genes of *Brassica oleracea*: expression, isolation and structure. *Proceedings of the National Academy of Sciences USA* **85**, 5551–5.

Nishio, T. & Hinata, K. (1977). Analysis of S-specific proteins in stigma of *Brassica oleracea* L. by isoelectric focusing. *Heredity* **38**, 391–6.

Ouyang, L.-J., Whelan, J., Weaver, C.D., Roberts, D.M. & Day, D.A. (1991). Protein phosphorylation stimulates the rate of malate uptake across the peribacteroid membrane of soybean nodules. *FEBS Letters* **293**, 188–90.

Prywes, R., Dutta, A., Cromlish, J.A. & Roeder, R.G. (1988). Phosphorylation of serum response factor, a factor that binds to the serum response element of the c-*FOS* enhancer. *Proceedings of the National Academy of Sciences USA* **85**, 7206–10.

Refeno, G., Ranjeva, R. & Boudet, A.M. (1982). Modulation of quinate:NAD$^+$ oxidoreductase activity through reversible phosphorylation in carrot cell suspensions. *Planta* **154**, 193–8.

Richardson, P.T., Westby, M., Roberts, L.M., Gould, J.H., Colman, A. & Lord, J.M. (1989). Recombinant proricin binds galactose but does not depurinate 28S ribosomal RNA. *FEBS Letters* **255**, 15–20.

Roberts, L.M. & Lord, J.M. (1981). The synthesis of *Ricinus communis* agglutinin. Co-translational and post-translational modification of agglutinin polypeptides. *European Journal of Biochemistry* **119**, 31–41.

Sarker, R.H., Elleman, C.J. & Dickinson, H.G. (1988). The control of pollen hydration in *Brassica* requires continued protein synthesis whilst glycosylation is necessary for intraspecific incompatibility. *Proceedings of the National Academy of Sciences USA* **85**, 4340–4.

Schindler, U., Menkens, A.E., Beckmann, H., Ecker, J.R. & Cashmore, A.R. (1992). Heterodimerization between light-regulated and ubiquitously expressed *Arabidopsis* GBF bZIP proteins. *EMBO Journal* **11**, 1261–73.

Schulman, H. & Lou, L.L. (1989). Multifunctional Ca^{2+}/calmodulin-dependent protein kinase: domain structure and regulation. *Trends in Biochemical Sciences* **14**, 62–6.

Schultz, A.M., Henderson, L.E. & Oroszlan, S. (1988). Fatty acylation of proteins. *Annual Review of Cell Biology* **4**, 611–47.

Schwarz-Sommer, Z., Huijser, P., Nacken, W., Saedler, H. & Sommer,

H. (1990). Genetic control of flower development by homeotic genes in *Antirrhinum majus*. *Science* **250**, 931–6.

Siegl, G., MacKintosh, C. & Stitt, M. (1990). Sucrose-phosphate synthase is dephosphorylated by protein phosphatase 2A in spinach leaves. *FEBS Letters* **270**, 198–202.

Sly, W.S. & Fischer, M.D. (1982). The phosphomannosyl recognition system for intracellular and intercellular transport of lysosomal enzymes. *Journal of Cell Biochemistry* **18**, 67–85.

Sommer, H., Beltrán, J.-P., Huijser, P., Pape, H., Lönning, W.-E., Saedler, H. & Schwarz-Sommer, Z. (1990). *Deficiens*, a homeotic gene involved in the control of flower morphogenesis in *Antirrhinum majus*: the protein shows homology to transcription factors. *EMBO Journal* **9**, 605–13.

Spiegel, A.M., Backlund, P.S., Butrynski, J.E., Jones, T.L.Z. & Simonds, W.F. (1991). The G protein connection: molecular basis of membrane association. *Trends in Biochemical Sciences* **16**, 338–41.

Stein, J.C., Howlett, B., Boyes, D.C., Nasrallah, M.E. & Nasrallah, J.B. (1991). Molecular cloning of a putative receptor protein kinase gene encoded as the self-incompatibility locus of *Brassica oleracea*. *Proceedings of the National Academy of Sciences USA* **88**, 8816–20.

Takayama, S., Isogai, A., Tsukamoto, C., Ueda, Y., Hinata, K., Okazaki, K., Koseki, K. & Suzuki, A. (1986). Structure of carbohydrate chains of S-glycoproteins in *Brassica campestris* associated with self-incompatibility. *Agricultural and Biological Chemistry* **50**, 1673–6.

Takayama, S., Isogai, A., Tsukamoto, C., Ueda, Y., Hinata, K., Okazaki, K. & Suzuki, A. (1987). Sequences of S-glycoproteins, products of the *Brassica campestris* self-incompatibility locus. *Nature (London)* **326**, 102–5.

Towler, D.A., Gordon, J.I., Adams, S.P. & Glaser, L. (1988). The biology and enzymology of eukaryotic protein acylation. *Annual Review of Biochemistry* **57**, 69–99.

van Deurs, B., Sandvig, K., Peterson, O.W., Olsnes, S., Simons, K. & Griffiths, G. (1988). Estimation of the amount of internalized ricin that reaches the trans-Golgi network. *Journal of Cell Biology* **106**, 253–67.

Voelker, T.A., Herman, E.M. & Chrispeels, M.J. (1989). *In vitro* mutated phytohemagglutinin genes expressed in tobacco seeds: role of glycans in protein targeting and stability. *Plant Cell* **1**, 95–104.

Walker, J.C. & Zhang, R. (1990). Relationship of a putative receptor protein kinase from maize to the S-locus glycoproteins of *Brassica*. *Nature (London)* **345**, 743–6.

Watillon, B., Kettmann, R., Boxus, Ph. & Burny, A. (1992). Cloning and characterization of an apple (*Malus domestica* (L.) Borkh) cDNA encoding a calmodulin-binding protein domain similar to the calmodulin-binding region of type II mammalian Ca^{2+}/calmodulin-dependent protein kinase. *Plant Science* **81**, 227–35.

Weaver, C.D., Crombie, B., Stacey, B. & Roberts, D.M. (1991). Calcium-dependent phosphorylation of symbiosome membrane proteins from nitrogen-fixing soybean nodules. *Plant Physiology* **95**, 222–7.

Yanofsky, M.F., Ma, H., Bowman, J.L., Drews, G.N., Feldmann, K.A. & Meyerowitz, E.M. (1990). The protein encoded by the *Arabidopsis* homeotic gene *agamous* resembles transcription factors. *Nature (London)* **346**, 35–9.

R.A. DIXON

Signal transduction and protein phosphorylation in bacteria

Introduction

Prior to 1986 it was assumed that protein phosphorylation played a relatively minor role in controlling the response of prokaryotes to environmental stimuli. Although many phosphoproteins had been detected by gel electrophoresis and a number of protein kinases had been purified, the physiological relevance of phosphorylation was not evident (Cozzone, 1984). The discovery of a role for phosphorylation in the response of bacteria to nitrogen status provided the first example of a defined stimulus–response pathway in which protein activity is controlled reversibly by phosphorylation and dephosphorylation (Ninfa & Magasanik, 1986). In the nitrogen regulation system two protein components are involved; one protein is a 'sensor' which has protein kinase activity and the other protein is a 'regulator' whose activity is responsive to phosphorylation. This pattern of 'two-component regulation' is surprisingly well conserved and serves to regulate a range of diverse processes in prokaryotes.

Bacteria are responsive to a wide variety of environmental signals including changes in nutrient concentration, osmolarity, temperature and the presence of other organisms. When DNA sequences of several genes involved in different signal transduction pathways were determined, it became obvious that many examples were representative of two families of proteins sharing amino acid similarities in common with the 'sensor' and 'regulator' proteins identified in the nitrogen regulation system (Nixon, Ronson & Ausubel, 1986; Winans et al., 1986; Ronson, Nixon & Ausubel, 1987b; Drummond & Wootton, 1987). Further biochemical studies established that members of each protein pair communicate with each other by a conserved mechanism of protein phosphorylation as predicted from the sequence data. The signal transduction event involves a phosphotransfer reaction in which a phosphoryl group is donated from a

Society for Experimental Biology Seminar Series 53: *Post-translational modifications in plants*, ed. N.H. Battey, H.G. Dickinson & A.M. Hetherington. © Cambridge University Press 1993, pp. 17–37.

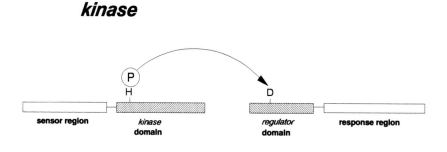

Fig. 1. Model for signal transduction in two-component regulatory systems. Cross-hatching indicates the conserved domains. H and D represent conserved histidine and aspartate residues on *kinases* and *regulators*, respectively. The direction of phosphotransfer is indicated by the arrow.

histidine residue on the 'sensor' to an aspartate residue on the 'regulator'. The 'sensor' family are characterised by a domain of around 250 amino acids near the carboxy-terminus whereas members of the 'regulator' family are defined by a domain of approximately 120 amino acids near the amino-terminus (Fig. 1). Although these protein pairs are commonly referred to as 'two-component regulatory systems' it must be emphasised that they often form part of a complex signal transduction pathway comprised of several additional protein components. Furthermore, in some cases both types of domain are located on the same protein. In this article I shall refer to the 'sensor' proteins as *kinases* with the caveat that this property has not yet been demonstrated for all members of the family. It should also be noted that the term 'kinase' only partly describes their function. Other terms in common use include 'sensor', 'histidine protein kinase' and 'modulator'. The partner proteins will be termed *regulators* although the alternatives 'effector' and 'response regulator' are commonly used.

Two-component regulatory systems of this type are found in both Gram-negative and Gram-positive bacteria, controlling adaptive responses such as chemotaxis, outer membrane protein expression and dicarboxylic acid transport as well as developmental processes such as sporulation and bacterial pathogenicity. This article firstly describes common properties of these kinases and regulators and then briefly explores the features of three well-defined systems. Most of this material has been covered in recent detailed reviews (Albright, Huala & Ausubel, 1989;

Stock, Ninfa & Stock, 1989*b*; Bourret, Borkovich & Simon, 1991). Protein phosphorylation also plays an important role in controlling bacteriophage infection, carbohydrate transport and metabolism in bacteria, using components which are distinct from the kinase and regulator homologues discussed here. Such systems have been reviewed elsewhere (Cozzone, 1988; Saier, Wu & Reizer, 1990).

Kinases

This family is defined by a domain containing a region of conserved sequence, which is normally located at the C-terminus of the protein; most members also have a highly conserved histidine residue, which is located at the N-terminus of this domain. Eight of these proteins have been examined for autophosphorylation activity *in vitro*; the relative lability of the phosphorylated residue suggests that it is phosphohistidine. In the case of CheA (Hess, Bourret & Simon, 1988*a*) and NTRB (Ninfa & Bennett, 1991) the histidine residue has been unambiguously identified as the phosphorylation site. Mutations in the conserved histidine residue in the EnvZ and CheA proteins completely inactivate kinase activity (Forst, Delgardo & Inouye. 1989; Hess *et al.*, 1988*a*). The N-terminal non-conserved domains of the kinases are quite variable in length and are postulated to have a sensor role, controlling the activity of the C-terminal domain in response to environmental signals through protein–protein or protein–ligand interactions. The majority of the kinases are membrane-associated with putative transmembrane sequences spanning a region of the N-terminus which is predicted to lie on the outer surface of the cytoplasmic membrane. This extracytoplasmic region therefore modulates the activity of the intracellular kinase domain via transmembrane signalling. As a variant on this theme, the FixL kinase is a haemoprotein containing transmembrane spanning regions in the N-terminal region, a cytoplasmic central region predicted to bind haem and a C-terminal kinase domain. The membrane location of this protein may facilitate its response to oxygen via the haem moiety (Gilles-Gonzales, Ditta & Helinski, 1991).

Regulators

Members of this family are characterised by a conserved N-terminal domain of approximately 125 amino acids which contains the target site for phosphorylation by their kinase partners. An aspartate residue around position 55 is invariant in these proteins and several other residues are very highly conserved. The structure of one regulator, CheY, has been solved initially to a resolution of 0.27 nm (Stock *et al.*, 1989*a*) and more recently refined to 0.17 nm (Volz & Matsumura, 1991). The protein is composed of a single domain forming a core of five parallel β-strands surrounded by five α-helices. Sequence alignments of CheY with its regulator homologues reveal a conserved pattern of hydrophobic residues corresponding to the core of CheY protein, strongly suggesting that all members of this family have N-terminal domains of similar structure (Stock, Stock & Mottonen, 1990). Three aspartate residues (Asp12, Asp13 and Asp57) cluster together on one face of the CheY molecule, forming an acidic pocket which is located close to another highly conserved residue, Lys109. It seems most probable that this pocket is conserved among regulator proteins and in each case forms the phosphoacceptor site.

CheY is unusual among the regulators in that most members of this family (with the exception of SpoOF) are considerably larger than CheY and possess additional C-terminal domains which define several distinct sub-families (Table 1). In most cases these domains function to regulate transcription. Members of the NTRC sub-class contain two further domains: firstly, a central domain required for positive control of promoters recognised by RNA polymerase holoenzyme containing the alternative sigma factor, σ^{54}, and secondly, a C-terminal domain which in all cases is most probably required for site-specific DNA recognition (Kustu et al., 1989). A different C-terminal region is shared by the OmpR sub-class, which is also postulated to have a role in transcriptional activation and DNA-binding (Mizuno & Mizushima, 1990). The FixJ sub-class have a third type of C-terminal domain, which shows some homology with the proposed DNA recognition regions of bacterial sigma factors (Kahn & Ditta, 1991). The final group of unclassified regulators include those which have only a single domain as described above, as well as CheB, which has a C-terminal domain with methylesterase activity.

Common phosphotransfer chemistry

Signal transduction in two component systems involves phosphotransfer between the conserved kinase and regulator domains, with the non-conserved domains serving to provide an appropriate response to specific stimuli. The kinase domains contain determinants for autophosphorylation and phosphotransfer activities; as expected, truncated forms of these proteins lacking N-terminal non-conserved regions retain their kinase function. The autophosphorylation reaction exclusively utilises the γ-phosphate of ATP; as mentioned above, the phosphorylation site appears to be a histidine residue in all cases. The relative acid lability of phosphohistidine groups clearly distinguishes them from phosphorylated serine,

Table 1. *Two-component regulatory systems in bacteria*

Sub-family	Kinase	Regulator	Function
NTRC regulator sub-class			
	NTRB	NTRC	Nitrogen regulation
	PgtB	PgtA	Phosphoglycerate transport
	DctB	DctD	Dicarboxylic acid transport
OmpR regulator sub-class			
	EnvZ	OmpR	Osmoregulation
	VirA	VirG	*Agrobacterium* virulence
	PhoQ	PhoP	*Salmonella* virulence
	PhoR	PhoB	Phosphate regulation
	PhoM	PhoMORF2	
	ArcB	ArcA	Oxygen regulation
	CpxA		
FixJ regulator sub-class			
	FixL	FixJ	Symbiotic N_2-fixation
	NodU	NodW	*Rhizobium* nodulation
	NarX	NarL	Nitrate reductase regulation
	DegS	DegU	Degradative enzymes
	UhpB	UhpA	Sugar phosphate transport
	RscC	RcsB	Capsule synthesis
	BvgS	BvgA	*Bordetella* virulence
Unclassified			
	CheA	CheY	Chemotaxis
		CheB	
	KinA	SpoOA	Sporulation
	KinB	SpoOF	
	FrzE	FrzG	Motility/development

threonine, aspartate or tyrosine, which tend to be more stable in acid than in alkali. Some of these proteins (e.g. NTRB and EnvZ) exhibit phosphatase activity in the presence of phosphorylated regulators. This reaction requires ATP although it may not require ATP hydrolysis (Keener & Kustu, 1988; Igo *et al.*, 1989*b*).

The regulator domain provides the phosphoacceptor site, which appears to be a functionally discrete property of this domain since regulators lacking their C-terminal non-conserved regions are phosphorylated as are those proteins which exclusively consist of this domain. The chemical stability of the phosphoryl group on a number of regulators studied *in vitro* suggests that it is located on an acidic residue. The phosphorylation site on CheY has been identified as Asp57, a conserved residue in all regulator homologues (Sanders *et al.*, 1989). Mutational analysis of this residue in several members of this family suggests that an aspartyl carboxylate side chain at this position is essential for the phosphotransfer reaction. A divalent cation such as Mg^{2+} is also required for phosphotransfer to both NTRC (Weiss & Magasanik, 1988) and CheY, and fluorescence studies with CheY indicate that the acidic pocket forms a single metal-ion binding site (Lukat, Stock & Stock, 1990). The phosphoaspartyl group on each regulator is readily hydrolysed to inorganic phosphate although the rate of hydrolysis varies considerably from a half-life of around 10 seconds for CheY and CheB phosphates (Hess *et al.*, 1988*b*) to around one hour for OmpR and VirG (Igo *et al.*, 1989*b*; Jin *et al.*, 1990). Since acyl phosphates such as acetyl phosphate have an intrinsic half-life on the order of hours and similar half-lives for phosphorylated regulators are observed under denaturing conditions, it appears that stability of the phosphorylated residue is determined by the local environment within each regulator. Active destabilisation of the phosphate group is referred to as autodephosphorylation or autophosphatase activity and in the case of NTRC and CheY this reaction requires Mg^{2+} ions (Weiss & Magasanik, 1988; Lukat *et al.*, 1990). As mentioned above, some kinases can accelerate regulator dephosphorylation but in other cases, for example, the chemotaxis system, a separate *trans*-acting phosphatase, CheZ, is required (Stock, Lukat & Stock, 1991). Phosphatase activity can therefore ensure that a given response is rapidly shut off. Moreover, the wide ranging stability of the phosphoryl group on different regulators provides an appropriate time scale for each individual transcriptional response.

Although in many two component systems the phosphotransfer event involves communication between a single kinase and its partner regulator there are cases when more than one kinase can act on a single regulator. For example, two kinases, CpxA and ArcB, respond to different signals but act on the same regulator ArcA (see Table 1). In the chemotaxis system, the opposite situation occurs and a single kinase CheA controls the activities of two separate regulators, CheY and CheB. Finally, there are five examples (ArcB, BvgS, FrzE, RscC and VirA) where kinase and regulator domains appear to be fused in the same protein although each of these systems includes a separate regulator protein. The function of these additional regulator domains is unknown although they appear to be essential for activity. In the case of FrzE, intramolecular phosphotransfer occurs directly between the fused kinase and regulator domains (McCleary & Zusman, 1990).

Given the homology among members of the kinase and regulator families and the conserved mechanism of phosphotransfer between partners it seems likely that 'cross-talk' could occur between components of different signal transduction pathways. Evidence for cross-talk is strong both *in vivo* and *in vitro*. Mutant strains lacking a particular kinase often display a low level of constitutive regulator activity which is only eliminated when the phosphoacceptor site is inactivated. Moreover, *in vitro* studies with purified proteins show that kinases can act as phosphodonors to several different regulators (Ninfa *et al.*, 1988; Igo *et al.*, 1988*b*; Olmedo *et al.*, 1990). The physiological relevance of cross-talk is, however, unclear since in most cases it only appears to provide a background level of activity which may be of little significance compared with the high signal-to-noise ratio of the cognate system.

Well-defined systems

Nitrogen regulation

Bacteria can utilise a wide variety of nitrogen sources, but the ammonium ion is the preferred source of nitrogen for growth. When cells are grown in ammonia-limiting conditions there is a considerable increase in transcription of a number of genes whose products facilitate the utilisation of alternative nitrogen sources such as proline, histidine and, in the case of nitrogen-fixing organisms, atmospheric dinitrogen. In Gram-negative bacteria, nitrogen regulation is mediated by a two-component system in which NTRB (also termed GlnL or NR_{II}) is the kinase and NTRC (also called GlnG or NR_I is the regulator. NTRC activates transcription at promoters recognised by RNA polymerase holoenzyme containing the alternative sigma factor σ^{54} (Kustu *et al.*, 1989). Under nitrogen-limiting conditions NTRC is phosphorylated by NTRB and is then competent to activate transcription of those genes whose expression allows growth on alternative nitrogen sources. Under nitrogen-excess conditions NTRB dephosphorylates NTRC so that transcriptional activation is impaired and ammonia is utilised as the nitrogen source.

In vitro studies have characterised the phosphotransfer reaction and the role of phosphorylation in activating NTRC. NTRB autophosphorylates on His139 (Ninfa & Bennett, 1991) and the phosphate is transferred to an aspartate residue on NTRC (Keener & Kustu, 1988; Weiss & Magasanik, 1988) which as suggested by mutational analysis is most probably Asp54 (Dixon *et al.*, 1991). An isolated 12.5 kDa N-terminal fragment of NTRC released by partial tryptic digestion can be phosphorylated by NTRB just as well as the complete protein (Keener & Kustu, 1988). NTRC-phosphate is labile and undergoes rapid auto-dephosphorylation with a half-life of 3–4 min at 37 °C. The level of phosphorylation of NTRC correlates with its ability to activate transcription (Ninfa & Magasanik, 1986) and efficient levels of activation are only obtained *in vitro* when NTRB and ATP are also present (Keener *et al.*, 1987; Ninfa, Reitzer & Magasanik, 1987; Austin, Henderson & Dixon, 1987). In accordance with the three-domain model for the structure of σ^{54}-dependent transcriptional activators (Drummond, Whitty & Wootton, 1986), phosphorylation of NTRC could influence the positive control function of its central domain and/or its ability to recognise specific DNA binding sites, a property determined by a helix–turn–helix motif in the C-terminal domain (Contreras & Drummond, 1988). Deletion of the N-terminal domain inactivates NTRC suggesting that phosphorylation has a positive function rather than suppressing a negative influence of this domain (Drummond, Contreras & Mitchenall, 1990). However, deletion of the regulator domain from DCTD, another member of this sub-class, gives derivatives which are active in transcriptional activation and DNA binding suggesting that there may be important structural differences between members of this activator family (Ledebur *et al.*, 1990). Phosphorylation of NTRC does increase its affinity for DNA binding sites and in particular may promote cooperative interactions between bound NTRC dimers (Minchin, Austin & Dixon, 1988). However, the magnitude of this effect is probably insufficient to account for the influence of phosphorylation on transcriptional activation.

In common with other members of the σ^{54}-dependent family of transcriptional activators (Buck *et al.*, 1986), NTC-phosphate activates transcription at a distance by binding to upstream sequences which can function when moved more than 1 kb from the transcriptional start site (Reitzer & Magasanik, 1986). Since NTRC-binding sites can function in transcriptional activation when positioned downstream of the promoter, they share properties in common with eukaryotic enhancers. Activation of transcription by NTRC is face-of-the-helix-dependent (Minchin, Austin & Dixon, 1989; Reitzer, Movsas & Magasanik, 1989), suggesting that the enhancer-bound activator may contact σ^{54}-RNA polymerase via the formation of a DNA loop; such loops have been visualised directly by electron microscopy (Su *et al.*, 1990).

NTRC-phosphate catalyses the isomerisation of closed promoter complexes formed with σ^{54} RNA polymerase to open promoter complexes in which the sequence surrounding the transcription start site is locally melted (Ninfa *et al.*, 1987; Sasse-Dwight and Gralla, 1988; Minchin, Austin & Dixon, 1989; Popham *et al.*, 1989). This reaction requires ATP (Popham *et al.*, 1989) consistent with the presence of a putative nucleotide binding pocket in the central domain of σ^{54}-dependent activators (Ronson *et al.*, 1987*a*). A mutation in the phosphate-binding loop of this proposed ATP binding site inactivates the positive control function of NTRC (Drummond *et al.*, 1990) and prevents isomerisation of closed promoter complexes to open complexes (Austin, Kundrot & Dixon, 1991). NTRC possesses an ATPase activity which is strongly phosphorylation-dependent (Weiss *et al.*, 1991) and is also stimulated by DNA binding (Austin & Dixon, 1992). The influence of DNA on the ATPase activity is apparently cooperative and may be consequent upon the binding of higher-order NTRC oligomers to multiple DNA binding sites.

Mutations that allow NTRC to function as an activator in the absence of phosphorylation have been isolated in several laboratories (Weglenski *et al.*, 1989; Popham *et al.*, 1989; Dixon *et al.*, 1991). The best-characterised mutations are located at position 160, close to the putative phosphate binding loop of the ATP binding site. The Ser160 to Phe substitution (S160F) promotes ATP hydrolysis in the absence of phosphorylation (Weiss *et al.*, 1991) but the ATPase activity of this protein, like that of the wild-type, is strongly stimulated by the presence of plasmid DNA or double-stranded oligonucleotides containing tandem NTRC-binding sites (Austin & Dixon, 1992). The mutant protein also has increased affinity for these sites in the absence of phosphorylation. It is possible that the S160F mutation alters the configuration of the phosphate binding loop such that the non-phosphorylated form hydrolyses ATP, and this conformational change may also influence DNA binding. A tyrosine substitution at position 160 gives properties similar to the S160F mutant. Interestingly, tryptophan and cysteine substitutions at this position do not give rise to phosphorylation-independent activity but they apparently increase the response of NTRC to cross-talk by kinases other than NTRB (Dixon *et al.*, 1991). This region of the protein therefore influences communication between the N-terminal and central domains and the latter mutations may alter the conformation of the regulator domain.

A complex signal transduction cascade determines the phosphorylation state of NTRC *in vivo* (Fig. 2). The availability of fixed nitrogen influences the ratio of glutamine to 2-ketoglutarate, which in turn controls the activity of a uridylyltransferase/uridylyl-removing enzyme (UTase-UR). This enzyme catalyses the uridylylation and deuridylylation of tyrosine residues in another regulatory protein called P_{II}. Under conditions of nitrogen excess when the ratio of glutamine to 2-ketoglutarate is relatively high, UTase-UR acts primarily to remove uridylyl groups from P_{II}. Under nitrogen-limiting conditions the opposing activity of UTase-UR is favoured and P_{II} is predominantly in its uridylylated form (Chock, Rhee & Stadtman, 1980). When non-uridylylated P_{II} is added *in vitro* to mixtures containing NTRB, ATP and NTRC-phosphate, a rapid dephos-

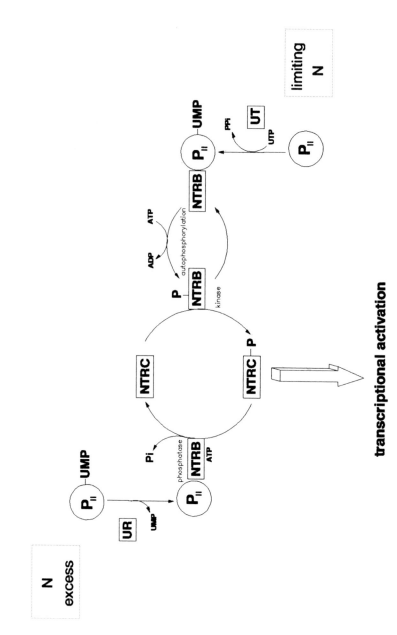

Fig. 2. The nitrogen regulation system in enteric bacteria. UR and UT indicate the uridylyl-removing and uridylyl transferase activities of UTase-UR, respectively.

phorylation of NTRC occurs which is significantly faster than the first order rate constant for autodephosphorylation by NTRC alone (Ninfa & Magasanik, 1986; Keener & Kustu, 1988). However, this phosphatase activity is not demonstrated by a mutant form of NTRB which is incapable of responding to nitrogen regulation *in vivo* and presumably has lost the ability to respond to P_{II} (Ninfa & Magasanik, 1986). Other NTRB mutants may be locked in a 'phosphatase mode' since the *in vivo* phenotype of such mutants suggests that cross-talk to NTRC is inhibited (MacFarlane & Merrick, 1987). The rapid dephosphorylation of NTRC by wild-type NTRB and P_{II} *in vitro* correlates with the loss of ability of NTRC to activate transcription (Ninfa & Magasanik, 1986). The phosphatase activity of NTRB therefore ensures that the positive control function of NTRC is rapidly inactivated in response to conditions of nitrogen excess.

Osmoregulation and porin expression

Bacteria modify the porin composition of their outer membranes in response to changes in osmolarity. The relative levels of the major porins OmpF and OmpC are reciprocally regulated, OmpC being preferentially expressed at high osmolarity whereas OmpF is predominantly expressed under conditions of low osmolarity (Csonka & Hanson, 1991). This reciprocal relationship is maintained by a two-component system comprising the kinase EnvZ and its partner regulator OmpR (Fig. 3).

EnvZ is an inner membrane protein comprised of a periplasmic domain thought to function as an osmosensor, which is linked to a cytoplasmic kinase domain via two transmembrane regions (Forst *et al.*, 1987). When incubated with ATP, EnvZ autophosphorylates on a histidine residue (Aiba, Mizuno & Mizushima, 1989*a*; Forst *et al.*, 1989; Igo & Silhavy, 1988) and the phosphoryl group is transferable to the N-terminal region of OmpR (Kato *et al.*, 1989; Igo, Ninfa & Silhavy, 1989*a*). The half-life of OmpR-phosphate is relatively long (about 1.5 h); this has facilitated the identification of OmpR-phosphate *in vivo*. The level of phosphorylation increases considerably at high osmolarity. However, this effect is also observed with *envZ* null mutants suggesting that OmpR can be readily phosphorylated by cross-talk with other kinases (Forst *et al.*, 1990). Mutagenesis of the putative phosphorylation site in OmpR (Asp55 to Gln) completely abolishes transcriptional activation of *ompC* and *ompF* *in vivo*. However, second-site suppressor mutations located in the N-terminal domain (T83A and G94S) restore this activity. It seems likely that these mutations alter the conformation of OmpR so that it is active in the absence of phosphorylation, although these mutants still respond to

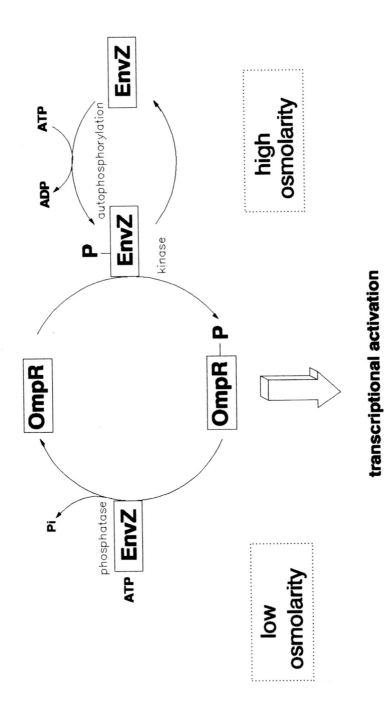

Fig. 3. Regulation of transcription of major porin genes by the EnvZ–OmpR signal transduction system.

changes in osmolarity, suggesting that the osmotic response can be mediated by mechanisms other than EnvZ-mediated phosphorylation (Brissette, Tsung & Inouye, 1991).

Phosphorylation stimulates both DNA binding (Aiba *et al.*, 1989*b*) and transcriptional activation by OmpR at the *ompC* and *ompF* promoters *in vitro* (Igo *et al.*, 1989*b*; Aiba & Mizuno, 1990). Cross-linking studies suggest that phosphorylation may promote oligomerisation of OmpR. Specific mutations in the regulator domain (E96A and R115S) appear to prevent oligomerisation and phosphorylation-dependent enhancement of DNA-binding (Nakashima *et al.*, 1991). Titration experiments indicate that *ompF* transcription requires a relatively low level of OmpR-phosphate *in vitro* and is inhibited at high levels, whereas *ompC* transcription is prefentially stimulated (Aiba & Mizuno, 1990). The level of phosphorylated OmpR may determine differential expressions of *ompF* and *ompC* (see Stock *et al.*, 1989*b*) in accord with *in vivo* observations that mutations that stabilise high levels of OmpR-phosphate give rise to an $OmpF^-OmpC^+$ phenotype (Aiba *et al.*, 1989*c*). Genetic studies suggest that OmpR may activate transcription through an interaction with the α-subunit of RNA polymerase (Slauch, Russo & Silhavy, 1991) and may also stabilise the binding of RNA polymerase at promoters (Tsung, Brissette & Inouye, 1990).

Like NTRB, EnvZ has a phosphatase activity which catalyses the dephosphorylation of OmpR-phosphate in the presence of ATP. This property is intrinsic to the cytoplasmic domain and does not appear to require ATP hydrolysis since non-hydrolysible analogues of ATP will substitute (Aiba *et al.*, 1989*a*; Igo *et al.*, 1989*b*). Mutant forms of EnvZ which appear to be locked in the 'phosphatase mode' give rise to an $OmpC^-OmpF^-$ phenotype, confirming that EnvZ mediates osmoregulation by controlling the concentration of OmpR-phosphate (Russo & Silhavy, 1991).

Chemotaxis

Chemotaxis in motile bacteria is effected by biasing the direction of flagella rotation in response to various attractants and repellants. When flagella rotation is counterclockwise bacteria adopt a smooth swimming behaviour, whereas clockwise rotation favours a tumbling motion in which cells have the opportunity to reorient themselves. Chemosensory signals are transduced by a set of transmembrane receptors consisting of a periplasmic ligand binding domain connected via the inner membrane to a cytoplasmic signalling domain. The activities of these transducers are further modulated by methylation of glutamine residues within the

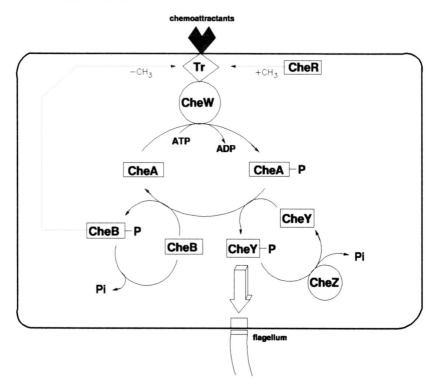

Fig. 4. Central role of phosphorylation in chemotaxis. Tr represents the transmembrane receptors; $-CH_3$ and $+CH_3$ indicate the methylesterase and methyltransferase activities of CheB and CheR, respectively.

cytoplasmic domain, catalysed by the methyltransferase CheR and antagonised by the methylesterase CheB (reviewed by Stock et al., 1991).

We focus on the central role of phosphotransfer in chemotaxis, involving the kinase CheA and the regulators CheZ and CheB (Fig. 4). CheA is rather an unusual member of the kinase family in that the site of autophosphorylation, His48, is located in the N-terminal domain rather than in the kinase domain itself (Hess et al., 1988a). Substitutions at this position, as expected, abolish autophosphorylation in vitro and chemotaxis in vivo (Hess et al., 1988a; Stewart, Roth & Dahlquist, 1990). The rate of autophosphorylation by CheA alone is predicted to be too slow to provide sufficient control of flagella rotation. Experiments with reconstituted in vitro systems show, however, that CheA forms a ternary complex with transmembrane receptors and another cytoplasmic protein, CheW (Bourret et al., 1991; McNally & Matsumara, 1991; Ninfa et al.,

1991). This complex has more than 100-fold greater kinase activity than CheA alone.

Phosphorylation of CheY by CheA most probably induces a change in flagella rotation and hence tumbling via an interaction between CheY-phosphate and the flagella motor, although there is as yet no direct evidence for this. Phosphotransfer occurs between the imidazole side chain of His48 in CheA to the carboxylate side chain of Asp57 in CheY (Hess *et al.*, 1988*a*; Sanders *et al.*, 1989). A partial tryptic cleavage product containing the isolated phosphorylated N-terminal domain of CheA is equally proficient in this reaction although, as expected, this fragment does not have autophosphorylation activity (Hess *et al.*, 1988*a*). Interestingly, low-molecular-mass phosphodonors including acetyl phosphate, carbamoyl phosphate and phosphoamidates can also phosphorylate CheY (Lukat *et al.*, 1992). CheY-phosphate rapidly autodephosphorylates with a half-life in the range of 6–15 s. This reaction is Mg^{2+}-dependent and is most probably promoted by binding of the metal to the acidic pocket since a D13N mutation reduces both the autophosphatase activity and the affinity for Mg^{2+} (Lukat *et al.*, 1990). Unlike other systems in which the kinase component has phosphatase activity, a separate protein, CheZ, increases the rate of dephosphorylation of CheY-phosphate (Hess *et al.*, 1988*b*). It is not clear how the activity of CheZ is regulated or how it induces dephosphorylation although this reaction, like autophosphatase activity, is Mg^{2+}-dependent (Lukat *et al.*, 1990). The phosphatase activity of CheZ appears to be specific to CheY-phosphate so CheZ may recognise a particular conformation which is relatively unique to CheY.

The kinase CheA also phosphorylates the regulator CheB in order to control the rate of receptor methylation. Phosphorylation of the N-terminal domain of CheB increases methylesterase activity of the C-terminal catalytic domain at least 10-fold (Lupas & Stock, 1989; Stewart *et al.*, 1990). CheB-phosphate is extremely unstable and possesses greater autophosphatase activity than CheY although it is not subject to the phosphatase activity of CheZ (Hess *et al.*, 1988*b*). Removal of the regulator domain stimulates the esterase activity of the catalytic domain (Simms, Keane & Stock, 1985). Thus phosphorylation apparently suppresses a negative function of the N-terminal domain.

References

Aiba, H. & Mizuno, T. (1990). Phosphorylation of a bacterial activator OmpR by a protein kinase, EnvZ, stimulation transcription of the *ompF* and *ampC* genes in *Escherichia coli*. *FEBS Letters* **261**, 19–22.

Aiba, H., Mizuno, T. & Mizushima, S. (1989*a*). Transfer of phosphoryl group between two regulatory proteins involved in osmoregulatory expression of the *ompF* and *ampC* genes in *Escherichia coli*. *Journal of Biological Chemistry* **264**, 8563–7.

Aiba, H., Nakasai, F., Mizushima, S. & Mizuno, T. (1989*b*). Phosphorylation of a bacterial activator protein OmpR by a protein kinase EnvZ, results in a stimulation of its DNA binding ability. *Journal of Biochemistry (Tokyo)* **106**, 5–7.

Aiba, H., Nakasai, F., Mizushima, S. & Mizuno, T. (1989*c*). Evidence for physiological importance of the phosphotransfer osmoregulation of *Escherichia coli*. *Journal of Biological Chemistry* **264**, 14090–4.

Albright, L.M., Huala, E. & Ausubel, F.M. (1989). Prokaryotic signal transduction mediated by sensor and regulator protein pairs. *Annual Review of Genetics* **23**, 311–36.

Austin, S. & Dixon, R. (1992). The ATPase activity of the bacterial enhancer-binding protein NTRC is phosphorylation and DNA-dependent. *EMBO Journal* **11**, 2219–28.

Austin, S., Henderson, N. & Dixon, R. (1987). Requirements for transcriptional activation *in vitro* of the nitrogen regulated *glnA* and *nifLA* promoters from *Klebsiella pneumoniae*: dependence on activator concentration. *Molecular Microbiology* **1**, 92–100.

Austin, S., Kundrot, C. & Dixon, R. (1991). Influence of a mutation in the putative nucleotide binding site of the nitrogen regulatory protein NTRC on its positive control function. *Nucleic Acids Research* **19**, 2281–7.

Bourret, R.B., Borkovich, K.A. & Simon, M.I. (1991). Signal transduction pathways involving protein phosphorylation in prokaryotes. *Annual Review of Biochemistry* **60**, 401–41.

Brissette, R.E., Tsung, K. & Inouye, M. (1991). Intramolecular second-site revertants to the phosphorylation site mutation in OmpR, a kinase-dependent transcriptional activator in *Escherichia coli*. *Journal of Bacteriology* **173**, 3749–55.

Buck, M., Miller, S., Drummond, M. & Dixon, R. (1986). Upstream activator sequences are present in the promoters of nitrogen fixation genes. *Nature (London)* **320**, 374–8.

Chock, P.B., Rhee, S.G. & Stadtman, E.R. (1980). Interconvertible enzyme cascades in cellular recognition. *Annual Review of Biochemistry* **49**, 813–43.

Contreras, A. & Drummond, M. (1988). The effect on the function of the transcriptional activator NtrC from *Klebsiella pneumoniae* of mutations in the DNA recognition helix. *Nucleic Acids Research* **16**, 4025–39.

Cozzone, A.J. (1984). Protein phosphorylation in bacteria. *Trends in Biochemical Sciences* **9**, 400–3.

Cozzone, A.J. (1988). Protein phosphorylation in prokaryotes. *Annual Review of Microbiology* **42**, 97–125.

Csonka, L.N. & Hanson, A.D. (1991). Prokaryotic osmoregulation – genetics and physiology. *Annual Review of Microbiology* **45**, 569–606.

Dixon, R., Eydmann, T., Henderson, N. & Austin, S. (1991). Substitutions at a single amino acid residue in the nitrogen-regulated activator protein NTRC differentially influence its activity in response to phosphorylation. *Molecular Microbiology* **5**, 1657–67.

Drummond, M.H., Contreras, A. & Mitchenall, L.A. (1990). The function of isolated domains and chimaeric proteins constructed from the transcriptional activators NifA and NtrC of *Klebsiella pneumoniae*. *Molecular Microbiology* **4**, 29–37.

Drummond, M., Whitty, P. & Wootton, J. (1986). Sequence and domain relationships of *ntrC* and *nifA* from *Klebsiella pneumoniae*: homologies to other regulatory proteins. *EMBO Journal* **5**, 441–7.

Drummond, M. & Wootton, J. (1987). Sequence of *nifL* from *Klebsiella pneumoniae*: mode of action and relationship to two families of regulatory proteins. *Molecular Microbiology* **1**, 37–44.

Forst, S., Comeau, D.E., Norioka, S. & Inouye, M. (1987). Localisation and membrane topology of EnvZ, a protein involved in osmoregulation of OmpF and OmpC in *Escherichia coli*. *Journal of Biological Chemistry* **262**, 16433–8.

Forst, S., Delgardo, J. & Inouye, M. (1989). Phosphorylation of OmpR by the osmosensor EnvZ modulates the expression of the *ompF* and *ompC* genes in *Escherichia coli*. *Proceedings of the National Academy of Sciences USA* **86**, 6052–6.

Forst, S., Delgardo, J., Rampersaud, A. & Inouye, M. (1990). *In vivo* phosphorylation of OmpR, the transcriptional activator of *ompF* and *ampC* genes in *Escherichia coli*. *Journal of Bacteriology* **172**, 3473–7.

Gilles-Gonzalez, A., Ditta, G.S. & Helinski, D.R. (1991). A haemoprotein with kinase activity encoded by the oxygen sensor of *Rhizobium meliloti*. *Nature (London)* **350**, 170–2.

Hess, J.F., Bourret, R.B. & Simon, M.I. (1988*a*). Histidine phosphorylation and phosphoryl group transfer in bacterial chemotaxis. *Nature (London)* **336**, 139–43.

Hess, J.F., Oosawa, K., Kaplan, N. & Simon, M.I. (1988*b*). Phosphorylation of three proteins in the signalling pathway of bacterial chemotaxis. *Cell* **53**, 79–87.

Igo, M.M., Ninfa, A.J. & Silhavy, T.J. (1989*a*). A bacterial environmental sensor that functions as a protein kinase and stimulates transcriptional activation. *Genes and Development* **3**, 598–605.

Igo, M.M., Ninfa, A.J., Stock, J.B. & Silhavy, T.J. (1989*b*). Phosphorylation and dephosphorylation of a bacterial activator by a transmembrane sensor. *Genes and Development* **3**, 1725–34.

Igo, M.M. & Silhavy, T.J. (1988). EnvZ, a transmembrane environmental sensor in *Escherichia coli* K12 is phosphorylated *in vitro*. *Journal of Bacteriology* **170**, 5971–3.

Jin, S., Prusti, R.K., Roitsch, T., Ankenbauer, R.G. & Nester, E.W.

(1990). Phosphorylation of the VirG protein of *Agrobacterium tumefaciens* by the autophosphorylated VirA protein: essential role in biological activity of VirG. *Journal of Bacteriology* **172**, 4945–50.

Kahn, D. & Ditta, G. (1991). Modular structure of FixJ: homology of the transcriptional activator domain with the -35 domain of sigma factors. *Molecular Microbiology* **5**, 987–97.

Kato, M., Aiba, H., Tate, S., Nshimura, Y. & Mizuno, T. (1989). Location of phosphorylation site and DNA-binding site of a positive regulator, OmpR, involved in activation of the osmoregulatory genes of *Escherichia coli*. *FEBS Letters* **249**, 168–72.

Keener, J. & Kustu, S. (1988). Protein kinase and phosphoprotein phosphatase activities of nitrogen regulatory proteins NTRB and NTRC of enteric bacteria: roles of the conserved amino-terminal domain of NTRC. *Proceedings of the National Academy of Sciences USA* **85**, 4976–80.

Keener, J., Wong, P., Popham, D., Wallis, J. & Kustu, S. (1987). A sigma factor and auxiliary proteins required for nitrogen-regulated transcription in enteric bacteria. In *RNA polymerase and the regulation of transcription* (ed. W.S. Reznikoff, R.R. Burgess, J.E. Dahlberg, C.A. Gross & M.P. Wickens), pp. 159–75. New York: Elsevier.

Kustu, S., Santero, E., Keener, J., Popham, D. & Weiss, D. (1989). Expression of σ^{54} (*ntrA*)-dependent genes is probably united by a common mechanism. *Microbiological Reviews* **53**, 367–76.

Ledebur, H., Gu, B., Sojda, J. & Nixon, B.T. (1990). *Rhizobium meliloti* and *Rhizobium leguminosarum dctD* gene products bind to tandem sites in an activation sequence located upstream of σ54-dependent *dctA* promoters. *Journal of Bacteriology* **172**, 3888–97.

Lukat, G.S., McCleary, W.R., Stock, A.M. & Stock, J.B. (1992). Phosphorylation of bacterial response regulator proteins by low molecular weight phospho-donors. *Proceedings of the National Academy of Sciences USA* **89**, 718–22.

Lukat, G.S., Stock, A.M. & Stock, J.B. (1990). Divalent metal ion binding to the CheY protein and its significance to phosphotransfer in bacterial chemotaxis. *Biochemistry* **29**, 5436–42.

Lupas, A. & Stock, J. (1989). Phosphorylation of an N-terminal regulatory domain activates the CheB methylesterase in bacterial chemotaxis. *Journal of Biological Chemistry* **264**, 17337–42.

MacFarlane, S. & Merrick, M.J. (1987). Analysis of the *Klebsiella pneumoniae ntrB* gene by site-directed *in vitro* mutagenesis. *Molecular Microbiology* **1**, 133–42.

McCleary, W. & Zusman, D.R. (1990). Purification and characterisation of the *Myxococcus xanthus* FrzE protein shows that it has autophosphorylation activity. *Journal of Bacteriology* **172**, 6661–8.

McNally, D.F. & Matsumara, P. (1991). Bacterial chemotaxis signalling complexes – formation of CheA/CheW complex enhances autophosphorylation and affinity for CheY. *Proceedings of the National Academy of Sciences USA* **88**, 6269–73.

Minchin, S.D., Austin, S. & Dixon, R.A. (1988). The role of activator binding sites in transcriptional control of the divergently transcribed *nifF* and *NifLA* promoters from *Klebsiella pneumoniae*. *Molecular Microbiology* **2**, 433–42.

Minchin, S.D., Austin, S. & Dixon, R.A. (1989). Transcriptional activation of the *Klebsiella pneumoniae nifL* promoter by NTRC is face-of-the-helix dependent and the activator stabilizes the interaction of sigma 54-RNA polymerase with the promoter. *EMBO Journal* **8**, 3491–9.

Mizuno, T. & Mizushima, S. (1990). Signal transduction and gene regulation through the phosphorylation of two regulatory components: the molecular basis for the osmotic regulation of the porin genes. *Molecular Microbiology* **4**, 1077–82.

Nakashima, K., Kanamaru, K., Aiba, H. & Mizuno, T. (1991). Signal transduction and osmoregulation in *Escherichia coli*: a novel type of mutation in the phosphorylation domain of the activator protein, OmpR, results in a defect in its phosphorylation-dependent DNA binding. *Journal of Biological Chemistry* **266**, 10775–80.

Ninfa, A.J. & Bennett, R.L. (1991). Identification of the site of autophosphorylation of the bacterial protein kinase/phosphatase NR_{II}. *Journal of Biological Chemistry* **266**, 6888–93.

Ninfa, A.J., Gottlin-Ninfa, E., Lupas, A.N., Stock, A., Magasanik, B. & Stock, J. (1988). Crosstalk between bacterial chemotaxis signal transduction proteins and regulators of transcription of the Ntr regulon: Evidence that nitrogen assimilation and chemotaxis are controlled by a common phosphotransfer mechanism. *Proceedings of the National Academy of Sciences USA* **85**, 5492–6.

Ninfa, A.J. & Magasanik, B. (1986). Covalent modification of the *glnG* product, NR_I, by the *glnL* product, NR_{II}, regulates the transcription of the *glnALG* operon in *Escherichia coli*. *Proceedings of the National Academy of Sciences USA* **83**, 5909–13.

Ninfa, A.J., Reitzer, L.J. & Magasanik, B. (1987). Initiation of transcription at the bacterial glnAp2 promoter by purified *E. coli* components is facilitated by enhancers. *Cell* **50**, 1039–46.

Ninfa, E.G., Stock, A., Mowbray, S. & Stock, J. (1991). Reconstitution of the bacterial chemotaxis signal transduction system from purified components. *Journal of Biological Chemistry* **266**, 9764–70.

Nixon, B.T., Ronson, C.W. & Ausubel, F.M. (1986). Two-component regulatory systems responsive to environmental stimuli share strongly conserved domains with the nitrogen assimilation regulatory genes *ntrB* and *ntrC*. *Proceedings of the National Academy of Sciences USA* **83**, 7850–4.

Olmedo, G., Ninfa, E.G., Stock, J. & Yangman, P. (1990). Novel mutations that alter the regulation of sporulation in *Bacillus subtilis* –

evidence that phosphorylation of regulatory protein SpoOA controls the initiation of sporulation. *Journal of Molecular Biology* **215**, 359–72.

Popham, D.L., Szeto, D., Keener, J. & Kustu, S. (1989). Function of a bacterial activator protein that binds to transcriptional enhancers. *Science* **243**, 629–35.

Reitzer, L.J. & Magasanik, B. (1986). Transcription of *glnA* in *E. coli* is stimulated by activator bound to sites far from the promoter. *Cell* **45**, 785–92.

Reitzer, L.J., Movsas, B. & Magasanik, B. (1989). Activation of *glnA* transcription by Nitrogen Regulator I (NR_I)-phosphate in *Escherichia coli*: Evidence for a long-range physical interaction between NR_I-phosphate and RNA polymerase. *Journal of Bacteriology* **171**, 5512–22.

Ronson, C.W., Astwood, P.M., Nixon, B.T. & Ausubel, F.M. (1987a). Reduced products of C-4 dicarboxylate transport regulatory genes of *Rhizobium leguminosarum* are homologous to nitrogen regulatory gene products. *Nucleic Acids Research* **15**, 7921–34.

Ronson, C.W., Nixon, B.T. & Ausubel, F.M. (1987b). Conserved domains in bacterial regulatory proteins that respond to environmental stimuli. *Cell* **49**, 579–81.

Russo, F.D. & Silhavy, T.J. (1991). EnvZ controls the concentration of phosphorylated OmpR to mediate osmoregulation of the porin genes. *Journal of Molecular Biology* **222**, 567–80.

Saier, M.H., Wu, F.-L. & Reizer, J. (1990). Regulation of bacterial physiological processes by three types of protein phosphorylating systems. *Trends in Biochemical Sciences* **15**, 391–5.

Sanders, D.A., Gillece-Castro, B.L., Stock, A.M., Burlingame, A.L. & Koshland, D.E. (1989). Identification of the site of phosphorylation of the chemotaxis response regulator protein CheY. *Journal of Biological Chemistry* **264**, 21770–8.

Sasse-Dwight, S. & Gralla, J.D. (1988). Probing the *Escherichia coli glnALG* upstream activation mechanism *in vivo*. *Proceedings of the National Academy of Sciences USA* **85**, 8934–8.

Simms, S.A., Keane, M. & Stock, J. (1985). Multiple forms of the CheB methylesterase in bacterial chemosensing. *Journal of Biological Chemistry* **260**, 10161–8.

Slauch, J.M., Russo, F.D. & Silhavy, T.J. (1991). Suppressor mutations in *rpoA* suggests that OmpR controls transcription by direct interaction with the alpha-subunit of RNA polymerase. *Journal of Bacteriology* **173**, 7501–10.

Stewart, R.C., Roth, A.F. & Dahlquist, F.W. (1990). Mutations that affect control of the methylesterase activity of CheB, a component of the chemotaxis adaptation system in *Escherichia coli*. *Journal of Bacteriology* **172**, 3388–99.

Stock, J.B., Lukat, G.S. & Stock, A.M. (1991). Bacterial chemotaxis and the molecular logic of intracellular signal transduction networks. *Annual Review of Biophysics and Biophysical Chemistry* **20**, 109–36.

Stock, A.M., Mottonen, J.M., Stock, J.B. & Schutt, C.E. (1989*a*). Three-dimensional structure of CheY the response regulator of bacterial chemotaxis. *Nature (London)* **337**, 745–9.

Stock, J.B., Ninfa, A.J. & Stock, A.M. (1989*b*). Protein phosphorylation and regulation of adaptive responses in bacteria. *Microbiological Reviews* **53**, 450–90.

Stock, J.B., Stock, A.M. & Mottonen, J.M. (1990). Signal transduction in bacteria. *Nature (London)* **344**, 395–400.

Su, W., Porter, S., Kustu, S. & Echols, H. (1990). DNA-looping and enhancer activity: association between DNA-bound NtrC activator and RNA polymerase at the bacterial *glnA* promoter. *Proceedings of the National Academy of Sciences USA* **87**, 5504–8.

Tsung, K., Brisette, R.E. & Inouye, M. (1990). Enhancement of the RNA polymerase binding to the promoters by a transcriptional activator, OmpR, in *Escherichia coli*: its positive and negative effects on transcription. *Proceedings of the National Academy of Sciences USA* **87**, 5940–4.

Volz, K. & Matsumura, P. (1991). Crystal structure of CheY refined at 1.7Å resolution. *Journal of Biological Chemistry* **266**, 15511–17.

Weglenski, P., Ninfa, A.J., Ueno-Nishio, S. & Magasanik, B. (1989). Mutations in the *glnG* gene of *Escherichia coli* that result in increased activity of nitrogen regulator I. *Journal of Bacteriology* **171**, 4479–85.

Weiss, D.S., Batut, J., Klose, K.E., Keener, J. & Kustu, S. (1991). The phosphorylated form of the enhancer-binding protein NTRC has an ATPase activity that is essential for activation of transcription. *Cell* **67**, 155–67.

Weiss, V. & Magasanik, B. (1988). Phosphorylation of nitrogen regulator I (NR$_1$) of *Escherichia coli*. *Proceedings of the National Academy of Sciences USA* **85**, 8919–23.

Winans, S.C., Ebert, P.R., Stachel, S.E., Gordon, M.P. & Nester, E.W. (1986). A gene essential for *Agrobacterium* virulence is homologous to a family of positive regulatory loci. *Proceedings of the National Academy of Sciences USA* **83**, 8278–82.

D.G. HARDIE

Roles of protein phosphorylation in animal cells

Introduction

Protein phosphorylation is a reversible covalent modification of great regulatory significance in animal cells. It is catalysed by protein kinases, which transfer the terminal phosphate of ATP onto the target protein, and is reversed by protein phosphatases, which catalyse a distinct reaction involving hydrolysis and releasing inorganic phosphate (Fig. 1). This contrasts with the use of covalently bound phosphate as a reaction intermediate in many metabolic enzymes, where the phosphate is both inserted and removed by the enzyme itself, and turns over during each cycle of catalysis.

The fact that different reactions are used for phosphorylation and dephosphorylation is significant. Given the normal prevailing concentrations of ATP, ADP and phosphate in the cell, the equilibrium ratio of protein : phosphoprotein lies well in favour of phosphorylation for a protein kinase reaction, and dephosphorylation for a protein phosphatase reaction. Therefore by controlling the *rate* of the protein kinase and/or the protein phosphatase reactions, the system can be switched completely from protein to phosphoprotein and back again, irrespective of

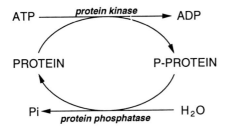

Fig. 1. A protein kinase–phosphatase cycle.

Society for Experimental Biology Seminar Series 53: *Post-translational modifications in plants*, ed. N.H. Battey, H.G. Dickinson & A.M. Hetherington. © Cambridge University Press 1993, pp. 39–51.

equilibrium constraints. A protein kinase/phosphatase cycle can therefore act as a molecular *switch* or *trigger*. The only switching mechanism of comparable importance in animal cells is the GTP ↔ GDP cycle of GTP-binding or *G* proteins (Kaziro *et al.*, 1991). However, a protein kinase–phosphatase cycle has an additional built-in advantage because both of its components are enzymes which act catalytically. It is therefore capable of achieving a large amplification of a small initial signal. A particularly large amplification is possible if two or more protein kinase phosphatase cycles are arranged in series, a system known as a *protein kinase cascade*. This property of signal amplification is particularly important in the mechanism of action of hormones, which typically vary in concentration in blood in the nanomolar range, but affect the function of protein targets which may be at concentrations of up to 100 μM in the target cells.

Proteins regulated by phosphorylation

Since the discovery that glycogen phosphorylase was regulated by phosphorylation (Krebs & Fischer, 1956), the number of proteins shown to be modified in this way in animal cells has continued to grow dramatically. However, demonstration of the modification of a protein by a protein kinase in a cell-free assay is not sufficient proof that this occurs in intact cells. Similarly, some protein phosphorylation events that do occur in intact cells do not have any known functional significance. Nevertheless, enough examples are known to enable us to state that almost every aspect of cellular function is controlled by this phenomenon in animal cells. Examples of physiologically significant phosphorylation events are found in proteins of almost every functional class, as follows:

Enzymes

The key regulatory enzyme in many of the most important metabolic pathways is regulated by phosphorylation. Some examples are given in Table 1.

Cytoskeletal proteins

Many cytoskeletal proteins are known to be phosphorylated in intact cells, although in most cases the functional significance of this is not yet clear. However, phosphorylation of the nuclear lamins may underlie the disaggregation of the nuclear envelope during mitosis (Peter *et al.*, 1992).

Table 1. *Some enzymes that are regulated by phosphorylation–dephosphorylation*

Enzyme	Pathway	Protein kinase	Effect
Glycogen phosphorylase	glycogen breakdown	phosphorylase kinase	↑
Glycogen synthase	glycogen synthesis	various	↓
L-pyruvate kinase	glycolysis	cAMP-dependent	↓
Acetyl-CoA carboxylase	fatty acid synthesis	AMP-activated	↓
HMG-CoA reductase	cholesterol synthesis	AMP-activated	↓
Hormone-sensitive lipase	triglyceride/cholesterol	cAMP-dependent	↑
	ester breakdown	AMP-activated	↓
Phenylalanine hydroxylase	phenylalanine breakdown	cAMP-dependent	↑
Tyrosine hydroxylase	catecholamine synthesis	various	↑

Ion channels and pumps

The functions of many ion channels have been shown to be modified by addition of purified protein kinases and phosphatases, or of inhibitors of these enzymes, in patch clamp experiments. In many cases biochemical analysis has suggested that this is due to direct phosphorylation, e.g. the phosphorylation of dihydropyridine-sensitive Ca^{2+} channels in muscle (Chang *et al.*, 1991). Similarly the activity of the Ca^{2+} pump of cardiac sarcoplasmic reticulum is increased by phosphorylation of the accessory protein phospholamban (Chiesi *et al.*, 1991).

Receptors

Many receptors for extracellular messengers are now known to be regulated by phosphorylation (e.g. the β_2-adrenergic receptor). In cases where this phenomenon has been well studied it seems to underlie the phenomenon of *desensitisation* by which the activated receptor is switched off (Collins, Caron & Lefkowitz, 1992).

Transcription factors

Almost as rapidly as transcription factors are being discovered, so is the importance of their regulation by reversible phosphorylation. Protein phosphorylation appears to underlie the regulation of gene expression by cyclic AMP (Lamph *et al.*, 1990), which activates cyclic AMP-dependent protein kinase, and by phorbol esters (Boyle *et al.*, 1991), which activate protein kinase C.

Identification of protein kinases

From the discovery of the first protein kinase (phosphorylase kinase) (Krebs & Fischer, 1956), the number of protein kinases grew relatively slowly until the advent of DNA cloning, and since then has increased dramatically with no sign of any slackening. Around 120 mammalian protein kinases are now known, not counting probable homologues in different species. These protein kinases all belong to the same gene family, and it has been claimed that this is the largest known protein family (Lindberg, Quinn & Hunter, 1992). Its members fall into three functional classes:

1   those specific for modification of serine or threonine residues;
2   those specific for tyrosine residues;
3   a recently discovered class which can phosphorylate serine, threonine *and* tyrosine residues (e.g. *wee1* protein kinase) (Featherstone & Russell, 1991).

In addition there are protein kinases that phosphorylate histidine residues, but it is not yet known whether these represent a distinct gene family (Wei & Matthews, 1991). Although about one third of the known protein kinases are in the tyrosine-specific class, this modification is rather rare, accounting for less than 0.1% of total protein phosphorylation in cultured cells (Cooper *et al.*, 1982). This appears to be because the tyrosine kinases are highly localised in the cell, particularly at the plasma membrane, with many being transmembrane proteins and others associating with the inner surface of the membrane via lipid modifications (Cross *et al.*, 1984) and protein–protein interactions (Mayer, Jackson & Baltimore, 1991).

How are the protein kinases regulated? One can distinguish several general mechanisms.

Constitutively active protein kinases

Some protein kinases appear to be intrinsically active and have no known form of regulation. This may of course merely indicate that the mechanism of regulation has not been found. However, in some cases the target sites also appear to be constitutively phosphorylated in cells, e.g. those phosphorylated by casein kinase II (Poulter *et al.*, 1988). Phosphate at these sites may be inserted during or immediately after synthesis of the target protein, and may not turn over rapidly. Regulation of these phosphorylation events, if it occurs, must be via modulation of protein phosphatases.

Second-messenger-dependent protein kinases

An important class of protein kinases is activated by the binding of second messengers, which are produced inside the cell in response to an extracellular (first) messenger (Hardie, 1991). Examples include cyclic AMP-dependent protein kinase, cyclic GMP-dependent protein kinase and protein kinase C (which is activated by the second messenger diacylglycerol). There are also several protein kinases which are activated by the second messenger Ca^{2+} via the small Ca^{2+}-binding protein calmodulin. These include calmodulin-dependent protein kinase II, phosphorylase kinase, myosin light chain kinase, and elongation factor-2 kinase. The second-messenger-dependent protein kinases mediate most, if not all, intracellular responses to these second messengers.

Extracellular-messenger-activated protein kinases (receptor protein kinases)

In these cases, the protein kinase catalytic domain is present as the cytoplasmic domain of a transmembrane receptor in which the external domain binds an extracellular messenger. This binding activates the internal protein kinase, possibly by causing aggregation of receptors. Most well-studied examples are receptors for polypeptide hormones or growth factors in which the protein kinase is tyrosine-specific, for example the receptors for insulin, epidermal growth factor and platelet-derived growth factor (Hardie, 1991). However, a few recent examples have been found where the protein kinase domain is more closely related to serine/threonine-specific kinases, e.g. the activin receptor (Mathews & Vale, 1991).

Extracellular-messenger-activated protein kinases (indirect, via protein–protein interactions)

A number of tyrosine-specific protein kinases are not transmembrane proteins, but there is increasing evidence that they associate with cell surface proteins, including some known receptors, via protein–protein interactions. A large proportion of these non-receptor tyrosine kinases contain so-called SH2 (*src* homology-2) domains, named after the archetypal member of this class, the *c-src* protein kinase. There is now evidence that the SH2 domains are required for interaction with phosphotyrosine-containing regions of other proteins (Mayer *et al.*, 1991), for example the autophosphorylated regions on receptor tyrosine kinases. Recently a serine/threonine-specific protein kinase with a putative SH2 domain has also been described (Bellacosa *et al.*, 1991).

Protein-kinase-regulated protein kinases

Some protein kinases are themselves regulated by phosphorylation by other protein kinases. A protein kinase cascade is a sequence of protein kinases in which one protein kinase phosphorylates and activates the next. Although few well-characterised examples of this phenomenon have been described to date, the number is increasing. Examples include activation of phosphorylase kinase by cyclic AMP-dependent protein kinase (Heilmeyer, 1991), activation of the AMP-activated protein kinase by its kinase kinase (Hardie, 1992), and the triple component cascade MAP kinase kinase → MAP kinase → S6 kinase (Nakielny *et al.*, 1992). There are also examples in which phosphorylation inactivates the protein kinase, e.g. phosphorylation of *c-src* protein kinase on tyrosine-527 by *c-src* kinase (Nada *et al.*, 1991), and phosphorylation of p34^{cdc2} on threonine-14 and tyrosine-15, probably by the *weel* protein kinase (Clarke & Karsenti, 1991).

Identification of protein phosphatases

Like the protein kinases, the protein phosphatases (PP) exist in two main functional classes, serine/threonine-specific and tyrosine-specific, but in this case they belong to distinct gene families (Fig. 2). The serine/threonine-specific protein phosphatases themselves contain two distinct gene families, a large class containing numerous representatives (PP1, PP2A, PP2B, and many others defined by cDNA cloning but not yet biochemically), and a second class currently only comprising the two

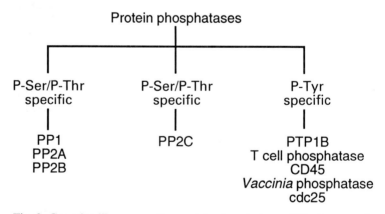

Fig. 2. Gene families among the protein phosphatases. PP1 etc., protein phosphatase-1, -2A, -2B, -2C; see Fischer *et al.* (1991) for a discussion of the phosphotyrosine-specific phosphatases.

isoforms of PP2C (Cohen, 1989; Cohen *et al.*, 1990). The analysis of tyrosine-specific protein phosphatases has lagged behind, but with the application of DNA cloning the number of representatives is now increasing dramatically (Fischer, Charbonneau & Tonks, 1991). A subclass of this group includes the protein phosphatase from *Vaccinia* virus, which can dephosphorylate both phosphotyrosine *and* phosphoserine residues (Guan, Broyles & Dixon, 1991), and which shows sequence similarity with the cdc25 gene product of *Schizosaccharomyces pombe* (Clarke & Karsenti, 1991). Genetic evidence (Nurse, 1990) suggests that the latter protein is a protein phosphatase which reverses the effect of the *wee1* protein kinase, and the latter can phosphorylate both serine/threonine and tyrosine residues (Featherstone & Russell, 1991).

Functions of protein phosphorylation

After over 30 years of study of protein phosphorylation in animal cells, it is becoming clear that protein phosphorylation has at least three general functions:

1   It represents the major mechanism by which extracellular messengers (for example hormones, local mediators, growth factors) which bind to cell surface receptors exert their ultimate intracellular effects.
2   It represents the major mechanism through which events that occur discontinuously during the cell cycle (e.g. DNA replication, mitosis) are timed.
3   In at least one case (the AMP-activated protein kinase) it represents a system which prevents a loss of essential metabolites or build-up of toxic metabolites, a form of stress response.

Each of these three functions will now be briefly reviewed.

Response to extracellular messengers

As discussed above, the second-messenger-dependent protein kinases account for most, if not all, of the effects of extracellular messengers which act via second messengers such as cyclic AMP, cyclic GMP, diacylglycerol and Ca^{2+} (Hardie, 1991). Ligand-gated ion channel receptors (e.g. the nicotinic acetylcholine receptor) also often cause increases in cytosolic Ca^{2+} due to opening of voltage-gated Ca^{2+} channels, and hence exert many of their effects via Ca^{2+}/calmodulin-dependent protein kinases. The third major type of signal transduction mechanism at cell surface receptors is the receptor tyrosine kinases. How these exert their

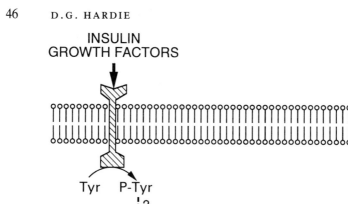

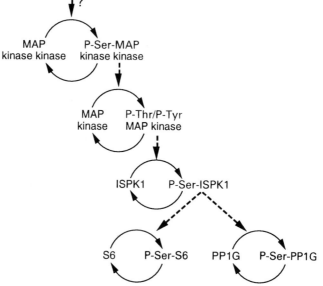

Fig. 3. The protein kinase cascade initiated by insulin and polypeptide growth factors (e.g. nerve growth factor). S6 is ribosomal protein S6; PP1G is the glycogen-binding subunit of protein phosphatase-1. See text for further details.

downstream effects remains unclear, but recent work suggests that they activate a cascade of serine/threonine protein kinases (Fig. 3). Central components of this cascade are the MAP kinases (mitogen- or messenger-activated protein kinases, also known as extracellular signal-regulated kinases or ERKs (Boulton *et al.*, 1991)). MAP kinase is activated by phosphorylation on tyrosine and threonine residues by a single enzyme, MAP kinase kinase (Nakielny *et al.*, 1992). Phosphorylation of both residues is required for activation; one explanation for this unusually stringent requirement is that it may prevent non-specific or accidental activation of MAP kinase by other protein kinases. MAP kinase itself phosphorylates a downstream kinase known as insulin-stimulated protein kinase-1 (ISPK1, also known as p90^{rsk}). As well as phosphorylating ribosomal protein S6, which may be involved in activation of protein synthesis by growth factors, ISPK1 also phosphorylates the G subunit of protein phosphatase-1, increasing its glycogen synthase phosphatase activity. This explains how an extracellular messenger like insulin can activate both phosphorylation *and* dephosphorylation (Dent *et al.*, 1990).

MAP kinase kinase is itself activated by insulin or growth factors, and this is reversed by protein (serine/threonine) phosphatases, but not protein (tyrosine) phosphatases. An enigma therefore remains as to how MAP kinase kinase is coupled to the receptor tyrosine kinases. An attractive idea is that it, or an upstream serine/threonine specific protein kinase, may have an SH2 domain which allows it to bind to a phosphotyrosine protein such as an autophosphorylated receptor, bringing about its autophosphorylation and activation. Serine/threonine-specific protein kinases containing putative SH2 domains have recently been described (Bellacosa *et al.*, 1991).

Cell cycle regulation

Exciting progress in this field came from the realisation that maturation promoting factor, a preparation which initiates mitosis in *Xenopus* oocyte extracts, is a complex between the p34^{cdc2} protein kinase and a cyclin protein. The *cdc2* gene was originally identified in *Schizosaccharomyces pombe* via mutations which cause delayed entry into mitosis, while cyclins are proteins that are synthesised during defined stages of the cell cycle. It is now clear that this mechanism is almost universal in eukaryotes (Nurse, 1990). The current model for how mitosis is initiated in mammals is shown in Fig. 4 (Clarke & Karsenti, 1991). The B form of cyclin, which is synthesised during the G2 phase of the cell cycle after DNA replication is complete, forms a complex with the *cdc2* protein kinase. As this happens, *cdc2* is phosphorylated on neighbouring tyrosine and threonine residues, probably *via* the *wee1* protein kinase. Mitosis is then triggered by dephosphorylation of *cdc2* by the *cdc25* protein phosphatase, which probably dephosphorylates both the tyrosine and the threonine. Once again this requirement for both residues to be dephosphorylated may ensure that mitosis is not triggered partly or accidentally by some other protein phosphatase. Although few of the substrates for *cdc2* have yet been identified, one may be the nuclear lamins, triggering breakdown of the nuclear envelope prior to chromosome separation (Peter *et al.*, 1992).

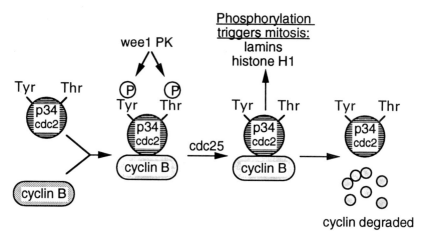

Fig. 4. Current model for regulation of p34^{cdc2} protein kinase during the G2 phase of the cell cycle (Clarke & Karsenti, 1991).

Response to potentially harmful changes in intracellular metabolites

The only well-established example of this is the system involving the AMP-activated protein kinase (Hardie, 1992). As its name suggests, this kinase is allosterically activated by AMP, which also promotes its phosphorylation and further activation by a kinase kinase (Fig. 5). The com-

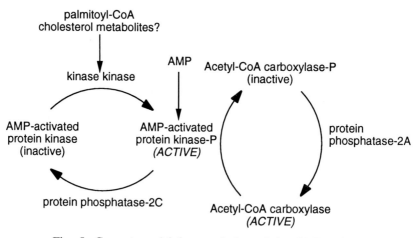

Fig. 5. Current model for regulation of the AMP-activated protein kinase (Hardie, 1992).

bination of the two effects can give rise to 100-fold activation of the kinase. The kinase phosphorylates and inactivates the rate-limiting enzymes in the pathways of fatty acid and cholesterol synthesis, and the release of fatty acids and cholesterol from intracellular stores (triglyceride and cholesterol esters).

What is the physiological significance of this system? AMP rises in animal cells whenever there is a net conversion of ATP to ADP, due to the action of the enzyme adenylate cyclase: $2ADP \leftrightarrow ATP + AMP$. The system therefore switches off the energy-requiring lipid biosynthetic pathways whenever the cell is compromised for energy, a form of stress response (Moore, Weekes & Hardie, 1991). There are also indications that it may exert an additional feedback control to prevent a toxic build-up of the free lipids, fatty acid and cholesterol.

Conclusion

Protein phosphorylation can be justifiably claimed to be the most important mechanism for cellular regulation in animal cells. Hunter (1987) has drawn an interesting analogy between protein kinase/phosphatase cycles and the transistors in electronic circuits. Both can switch the system from one state to another, and can also amplify signals, but require a small input of energy in order to do this. If this analogy is reasonable, the protein kinase/phosphatase cycles are where the decisions are taken, and everything else can be regarded as merely the wiring!

References

Bellacosa, A., Testa, J.R., Staal, S.P. & Tsichlis, P.N. (1991). A retroviral oncogene, *akt*, encoding a serine-threonine kinase containing an SH2-like region. *Science* **254**, 274–7.

Boulton, T.G., Nye, S.H., Robbins, D.J., Ip, N.Y., Radziejewska, E., Morgenbesser, S.D., DePinho, R.A., Panayotatos, N., Cobb, M.H. & Yancopoulos, G.D. (1991). ERKs: a family of protein-serine/threonine kinases that are activated and tyrosine phosphorylated in response to insulin and NGF. *Cell* **65**, 663–75.

Boyle, W.J., Smeal, T., Defize, L.H., Angel, P., Woodgett, J.R., Karin, M. & Hunter, T. (1991). Activation of protein kinase C decreases phosphorylation of *c-jun* at sites that negatively regulate its DNA-binding activity. *Cell* **64**, 573–84.

Chang, C.F., Gutierrez, L.M., Mundina-Weilenmann, C. & Hosey, M.M. (1991). Dihydropyridine-sensitive calcium channels from skeletal muscle. II. Functional effects of differential phosphorylation of channel subunits. *Journal of Biological Chemistry* **266**, 16395–400.

Chiesi, M., Vorherr, T., Falchetto, R., Waelchli, C. & Carafoli, E.

(1991). Phospholamban is related to the autoinhibitory domain of the plasma membrane Ca^{2+}-pumping ATPase. *Biochemistry* **30**, 7978–83.

Clarke, P.R. & Karsenti, E. (1991) Regulation of p34^{cdc2} protein kinase: new insights into protein phosphorylation and the cell cycle. *Journal of Cell Science* **100**, 409–14.

Cohen, P. (1989). The structure and regulation of protein phosphatases. *Annual Review of Biochemistry* **58**, 453–508.

Cohen, P.T.W., Brewis, N.D., Hughes, V. & Mann, D.J. (1990). Protein serine/threonine phosphatases; an expanding family. *FEBS Letters* **268**, 355–9.

Collins, S., Caron, M.G. & Lefkowitz, R.J. (1992). From ligand binding to gene expression: new insights into the regulation of G-protein-coupled receptors. *Trends in Biochemical Sciences* **17**, 37–9.

Cooper, J.A., Bowen-Pope, D.F., Raines, E., Ross, R. & Hunter, T. (1982). Similar effects of platelet-derived growth factor and epidermal growth factor on the phosphorylation of tyrosine in cellular proteins. *Cell* **31**, 263–73.

Cross, F.R., Garber, E.A., Pellman, D. & Hanafusa, H. (1984). A short sequence in the p60^{src} N terminus is required for p60^{src} myristylation and membrane association and for cell transformation. *Molecular and Cellular Biology* **4**, 1834–42.

Dent, P., Lavoinne, A., Nakielny, S., Caudwell, F.B., Watt, P. & Cohen, P. (1990). The molecular mechanism by which insulin stimulates glycogen synthesis in mammalian skeletal muscle. *Nature (London)* **348**, 302–8.

Featherstone, C. & Russell, P. (1991). Fission yeast p107^{wee1} mitotic inhibitor is a tyrosine/serine kinase. *Nature (London)* **349**, 808–11.

Fischer, E.H., Charbonneau, H. & Tonks, N.K. (1991). Protein tyrosine phosphatases: a diverse family of intracellular and transmembrane enzymes. *Science* **253**, 401–6.

Guan, K., Broyles, S.S. & Dixon, J.E. (1991). A tyr/ser protein phosphatase encoded by *Vaccinia* virus. *Nature (London)* **350**, 359–60.

Hardie, D.G. (1991). *Biochemical Messengers*. London: Chapman & Hall.

Hardie, D.G. (1992). Regulation of fatty acid and cholesterol metabolism by the AMP-activated protein kinase. *Biochimica et Biophysica Acta* **1123**, 231–8.

Heilmeyer, L.M.G. (1991). Molecular basis of signal integration in phosphorylase kinase. *Biochimica et Biophysica Acta* **1094**, 168–74.

Hunter, T. (1987). A thousand and one protein kinases. *Cell* **50**, 823–9.

Kaziro, Y., Itoh, H., Kozasa, T., Nakafuku, M. & Satoh, T. (1991). Structure and function of signal-transducing GTP-binding proteins. *Annual Review of Biochemistry* **60**, 349–400.

Krebs, E.G. & Fischer, E.H. (1956). The phosphorylase *b* to *a* converting enzyme of rabbit skeletal muscle. *Biochimica et Biophysica Acta* **20**, 150–7.

Lamph, W.W., Dwarki, V.J., Ofir, R., Montminy, M. & Verma, I.M. (1990). Negative and positive regulation by transcription factor cAMP response element-binding protein is modulated by phosphorylation. *Proceedings of the National Academy of Sciences USA* **87**, 4320–4.

Lindberg, R.A., Quinn, A.M. & Hunter, T. (1992). Dual specificity protein kinases: will any hydroxyl do? *Trends in Biochemical Sciences* **17**, 114–19.

Mathews, L.S. & Vale, W.W. (1991). Expression cloning of an activin receptor, a predicted transmembrane serine kinase. *Cell* **65**, 973–82.

Mayer, B.J., Jackson, P.K. & Baltimore, D. (1991). The noncatalytic *src* homology region 2 segment of *abl* tyrosine kinase binds to tyrosine-phosphorylated cellular proteins with high affinity. *Proceedings of the National Academy of Sciences USA* **88**, 627–31.

Moore, F., Weekes, J. & Hardie, D.G. (1991). Evidence that AMP triggers phosphorylation as well as direct allosteric activation of rat liver AMP-activated protein kinase. A sensitive mechanism to protect the cell against ATP depletion. *European Journal of Biochemistry* **199**, 691–7.

Nada, S., Okada, M., MacAuley, A., Cooper, J.A. & Nakagawa, H. (1991). Cloning of a complementary DNA for a protein-tyrosine kinase that specifically phosphorylates a negative regulatory site of p60$^{c\text{-}src}$. *Nature (London)* **351**, 69–72.

Nakielny, S., Cohen, P., Wu, J. & Sturgill, T. (1992). MAP kinase activator from insulin-stimulated skeletal muscle is a protein threonine/tyrosine kinase. *EMBO Journal*, **11**, 2123–9.

Nurse, P. (1990). Universal control mechanism regulating onset of M-phase. *Nature (London)* **344**, 503–8.

Peter, M., Sanghera, J.S., Pelech, S.L. & Nigg, E.A. (1992). Mitogen-activated protein kinases phosphorylate nuclear lamins and display sequence specificity overlapping that of mitotic protein kinase p34^{cdc2}. *European Journal of Biochemistry* **205**, 287–94.

Poulter, L., Ang, S.G., Gibson, B.W., Williams, D.H., Holmes, C.F.B., Caudwell, F.B., Pitcher, J. & Cohen, P. (1988). Analysis of the *in vivo* phosphorylation state of rabbit skeletal muscle glycogen synthase by fast-atom-bombardment mass spectrometry. *European Journal of Biochemistry* **175**, 497–510.

Wei, Y.F. & Matthews, H.R. (1991) Identification of phosphohistidine in proteins and purification of protein-histidine kinases. *Methods in Enzymology* **200**, 388–414.

K.M. FALLON and A.J. TREWAVAS

The significance of post-translational modification of proteins by phosphorylation in the regulation of plant development and metabolism

Background

The transfer of intercellular and intracellular information is essential for the control of plant growth and development. In the form of cellular proteins, a framework is present for the reception, amplification and transmission of such information through the plant. The most widely studied mechanism for the transfer of information on this network is phosphorylation. Studies of signal transduction via phosphorylation–dephosphorylation began more than three decades ago in animal cells (Sutherland & Wosilait, 1955), and have reached new intensities in the last decade. As a result, a complex pattern of interacting elements has been identified which, beginning with the receipt of a defined stimulus, results in the modulation of cellular function. In plant cells, where clearly defined agonist–receptor–stimulus models are not so readily available for study, progress has not been so rapid, and investigation has concentrated on finding systems to match those in animals. Significant progress has been achieved, from the identification of a number of plant protein kinases in the early 1970s (Ralph *et al.*, 1972; Keates & Trewavas, 1974), to the identification of several plant GTP-binding proteins (G proteins) (see, for example, Drobak *et al.*, 1988; Blum *et al.*, 1988), and the recent cloning of a calcium-dependent protein kinase (CDPK) (Harper *et al.*, 1991). Progress has been aided by the use of tools developed to elucidate more clearly defined animal pathways, such as anti-G-protein antibodies and specific promoters and inhibitors of phosphorylation. Systems paralleling those of animals have been sought particularly in the stimulus–response coupling involved in the responses of plants to environmental stimuli, especially since analysis of such stimulus-triggered protein modification should shed light on the nature of the agonist–receptor interac-

Society for Experimental Biology Seminar Series 53: *Post-translational modifications in plants*, ed. N.H. Battey, H.G. Dickinson & A.M. Hetherington. © Cambridge University Press 1993, pp. 53–64.

tion involved in the response. In this overview of the field our knowledge of the kinases themselves is considered together with investigations of their reaction with their protein substrates and developments in enzyme phosphorylation.

The significance of phosphorylation/dephosphorylation cascades

A key role of protein phosphorylation in cell function is through enzyme regulation and hence stimulus amplification and integration, achieved through 'cascade systems'. The phosphorylation cascade originated in the discovery of phosphorylation/dephosphorylation regulated enzyme activity (Sutherland & Wosilait, 1955). Studies have since shown their effectiveness and efficiency (Stadtman & Chock, 1977). These covalent modification systems have great regulatory capacities and respond to changes in metabolite levels while providing signal amplification through increases in the activity of enzymes, which in turn act on further substrate enzymes (see Hardie, this volume, for further discussion). Fig. 1 shows how the coupling of two cascades produces a steady state with the inter-convertible enzyme E being cycled between two states via activation of enzymes A and B and concomitant hydrolysis and reconstitution of ATP. With their importance well established in animal cells (Cohen, 1982), a role in plants is indicated, but without supporting data as yet, although we expect such information to be published soon.

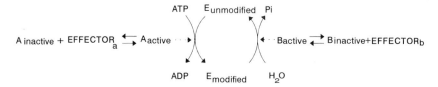

Fig. 1. Monocyclic cascade system. A, B, converter enzymes; E, convertible enzyme. (See text for further details.)

Protein phosphorylation in plant cells

Protein phosphorylation in different cell organelles

Protein phosphorylation in plants has been surveyed in recent years, largely with the aid of sodium dodecyl sulphate (SDS) gel electrophoresis together with γ-^{32}P ATP *in vitro* or ^{32}P$_i$ *in vivo* to label and examine the range of different phosphoproteins present (Ranjeva & Boudet, 1987). In this way large numbers of tissues and fractions have been assayed under different conditions. Generally, *in vitro* studies have been more successful owing to the closer control of conditions attainable and the less complicated generation of acceptable autoradiographs of separated phosphoproteins.

Early studies revealed numerous phosphorylated proteins in plant tissues (Trewavas, 1976), and these have now been found throughout the plant cell: in nuclei (Trewavas, 1979), plastids (see, for example, Laing & Christeller, 1984; see also Bennett (1984) for review of older literature), mitochondria (Danko & Markwell, 1985) and membrane preparations (Hetherington & Trewavas, 1984).

Chloroplast phosphorylation has long been of interest (Bennett, 1977), and the phosphoprotein distribution within the plastid has been studied (Laing & Christeller, 1984; Soll, 1985). Protein kinases have been identified on the outer chloroplast membrane (Soll & Buchanan, 1983), in the stroma (Guitton, Dorne & Mache, 1984) and bound to the thylakoid (Lin, Lucero & Racker, 1982). The stromal enzymes pyruvate P_i dikinase (PPDK) and ribulose 1,5 bisphosphate carboxylase–oxygenase (Rubisco) are phosphorylated *in vivo* (Budde & Chollet, 1988; Roeske & Chollet, 1989), though the effect of this on the activity of Rubisco is unclear (Guitton & Mache, 1987) and the mediator of changes in PPDK activity through phosphorylation is not established (Roeske & Chollet, 1989). Similarly, though the cytosolic enzymes phosphoenolpyruvate carboxylase (PEPCase) and sucrose phosphate synthase (SPS) are reversibly phosphorylated, the mechanism whereby PEPCase phosphorylation is controlled is not yet fully established (Chollet, 1990; but see Nimmo, this volume, for recent data on PEPCase regulation in CAM plants), though SPS phosphorylation may be regulated through changes in P_i (Huber & Huber, 1990).

Though studies have shown that some phosphoproteins occur persistently in plant extracts, the role of the majority remains unclear. What has been achieved is some understanding of the events to which these phosphoproteins are related. Phosphorylation patterns have been shown to change during germination and on treatment with plant growth substances (Chapman, Trewavas & Van Loon, 1975), and in parallel with increased transcription (Wielgat & Kleczkowski, 1981).

Regulation of protein kinases by calcium

Plasma membrane calcium-dependent protein kinases in plants

As with phosphoproteins, protein kinases have been identified throughout the plant cell, in nuclei, cytosol and subcellular organelles (for review, see Hetherington *et al.*, 1990). Recent years have seen attempts to link protein phosphorylation to particular kinases and some progress has been made. A novel technique has been developed to study phosphorylation in polypeptides of interest after separation (Blowers, Hetherington & Trewavas, 1985). Once separated, the protein of interest is reconstituted and incubated with γ-^{32}P ATP. Using this method, three rapidly labelled peptides have been identified localised to the plasma membrane, in pea bud tissue, whose phosphorylation is light-dependent *in vitro* (Trewavas & Blowers, 1990). Of these one of 18 kDa was found to autophosphorylate. It is the approximate site of a protein kinase (pp[18]) which appears to be calcium and calmodulin-dependent; pp[18] autophosphorylates rapidly (15–30 s) compared with other membrane proteins, and converts phosphatidylinositol (PI) to phosphatidylinositol phosphate (PIP). Its high affinity for ATP and rapid cycling suggest a role in cell signalling. Unfortunately, however, low protein yields in the chosen system have so far made it impossible to investigate this further through sequence data (Blowers & Trewavas, 1989). The available data suggest a 67 kDa located active site with an associated regulatory pp[18], which can modulate the response of the kinase to calcium and calmodulin. Sequence data would allow the design of probes to investigate homology with other pea protein kinases. In this way other investigators have made significant progress in understanding plant protein kinase action.

Calcium-dependent protein kinase in soybean and other plants

A well-characterised calcium-dependent protein kinase (CDPK) has been purified from soybean cells (Harmon, Putnam-Evans & Cormier, 1987). Consisting of two main peptide bands of 52 and 55 kDa, it is stimulated 50- to 100-fold by calcium; like pp[18], renatured CDPK autophosphorylates in a calcium-dependent manner. Identification and sequencing of cDNA clones encoding CDPK and comparison with the kinase gave incomplete homology, suggesting the existence of CDPK isozymes (Harper *et al.*, 1991). Like pp[18], CDPK has characteristics which indicate a possible role in signal transduction, but without a clearly defined substrate *in vivo* to aid identification of its function.

Other CDPKs have been purified from a wide range of plant tissues (Polya & Chandra, 1990; see also Battey *et al.*, this volume), suggesting that these kinases serve a fundamental and widespread role. Many reports show CDPKs, both soluble and membrane-associated, to be activated by phospholipids, although this sensitivity is not universal. Such heterogeneity is similar to the situation in animal cells, where a variety of CDPK isozymes vary in their response to phospholipids and calcium (Hardie, 1990, and references therein). Together with reports that inositol trisphosphate (IP$_3$) triggers the release of vacuolar calcium, elements of the signal transduction chain appear to be taking shape.

The role of tyrosine kinase in the regulation of development

Early studies related modifications in phosphorylation status to discrete periods in development. Thus, seed germination and tuber cell division have long been correlated with phosphorylation changes (Melanson & Trewavas, 1982). In recent years the role of phosphorylation in the cell cycle has become clearer, and progress has been rapid.

In mammalian cells, it has been known for some time that proteins are phosphorylated on tyrosine in addition to serine and threonine residues, and that the protein–tyrosine kinases responsible are associated with the plasma membrane (see, for example, Eckhart, Hutchinson & Hunter, 1979). Most revealing has been the discovery that a protein–serine kinase essential for cell-cycle progression (cdc2) is phosphorylated on tyrosine (Draetta *et al.*, 1988). The phosphorylation is thought to prevent ATP binding and thus inhibit phosphorylation by cdc2. This key role of tyrosine kinase has prompted a search for phosphotyrosine in plant tissues. Obviously the prime targets for such a search are rapidly dividing tissues, and evidence has now been produced showing that traces of phosphotyrosine are present in the nuclei of germinating pea seedlings (P. Brusa & A.J. Trewavas, 1992, unpublished). When such nuclei are run on SDS gels and the proteins blotted onto nitrocellulose, the resulting blot may be probed with the phosphotyrosine antibodies now available. Such probing reveals a high-molecular-mass tyrosine phosphopeptide, which has been partly purified by column and thin-layer chromatography.

It would have been surprising if tyrosine kinases had not been present in plants in the light of their crucial role in the cell cycle of other eukaryotic cells. This finding promises new insights into plant development, although the very low levels of phosphotyrosine so far noted may prove a complication in surveying its occurrence in plant tissues. Future work to localise phosphotyrosine with the cell and between tissues of different developmental status should prove revealing.

Regulation of plant protein kinase by light

Phosphorylation in relation to photosynthesis
A number of studies have been carried out to probe the link between phosphorylation and the response to light. Results have suggested that photosynthesis is controlled in part through the light-dependent phos-

phorylation status of chlorophyll *a/b* protein complex (LHCP). LHCP controls the input of light quanta to photosystems I and II (PSI & PSII), and its phosphorylation is mediated by a membrane-bound protein kinase and phosphatase (Allen *et al.*, 1981). On phosphorylation LHCP separates from PSII, decreasing its area, and becomes attached to PSI. The area of PSI is thereby increased in line with the decrease in PSII area, so restoring the system to equilibrium (Coughlan, 1990).

Red-light-dependent phosphorylation

Studies of plant responses to non-photosynthetic light have focused primarily on the photoreceptor phytochrome. Purified phytochrome from *Avena* has been shown to have polycation-stimulated protein kinase activity (Wong *et al.*, 1986), and there is some evidence that phytochrome may autophosphorylate (McMichael & Lagarias, 1990). This has led to speculation that phytochrome is a protein kinase, although other reports showing no phytochrome kinase activity do not support this (Grimm, Gast & Rudiger, 1989; Kim, Bai & Song, 1989).

Other studies have taken a more physiological approach in attempting to link light-dependent changes in protein phosphorylation to phytochrome. Early studies had shown a range of polypeptides to be phosphorylated in plant nuclei (Trewavas, 1979). Later investigation found a number of proteins in pea nuclei which were phosphorylated in a calcium- and red-light-dependent fashion (Datta, Chen & Roux, 1985). Since the red-light effect was far-red reversible, the action of phytochrome is indicated in the response. Although a direct calcium activation was not demonstrated, the phosphorylation was blocked by EGTA and calmodulin antagonists. This report would thus suggest the presence of a calcium/calmodulin-dependent protein kinase active in the nucleus with implications for control of gene expression. Interestingly, recent reports have indicated that the binding of a nuclear protein to promoters of photoregulated genes may be regulated by phosphorylation (Datta & Cashmore, 1989).

Red-light-dependent phosphorylation in a wheat protoplast model system

In order to investigate further the operation of this signal–response pathway, we selected a system in which the red-light response was immediately apparent, and also measurable with accuracy and reproducibility. Previous work had indicated that protoplasts of wheat respond to red-light irradiation by an increase in volume (Blakeley *et al.*, 1983), and that the effect is far-red reversible, suggesting a phytochrome-mediated response. Using this system, swelling of protoplasts in response to red light was first confirmed; experiments were then designed to determine whether a change in phosphorylation could be detected in parallel with the swelling response. Protoplasts were labelled *in vivo* with $^{32}P_i$ and irradiated; phosphoproteins were then separated on SDS gels. The kinetics of labelling of the phosphoproteins varied markedly, with a 70 kDa polypeptide showing extremely rapid labelling (less than 1 min) while others labelled more slowly during the remaining 60 min. Since very rapid phosphate turnover is an indication of a possible role in signalling (Garrison, 1983), the most rapidly labelled bands were monitored under various irradiation regimes. The 70 kDa phosphoprotein was found to label more rapidly in 15 s under red-light irradiation, but was unaffected by blue-light irradiation. The labelling of the phosphoprotein was dependent on $[Ca^{2+}]$ in the medium ($[Ca^{2+}]_m$) with an optimum $[Ca^{2+}]_m$ of 10 mM, and labelling time was reduced with increasing $[Ca^{2+}]$ within the protoplast ($[Ca^{2+}]_i$) released from caged-calcium, and with release of caged inositol trisphosphate. Furthermore, parallel studies using laser scanning confocal microscopy together with the release of caged-calcium and caged-IP$_3$ within the protoplast have demonstrated both that protoplast swelling is induced by increases in $[Ca^{2+}]_i$, and that $[Ca^{2+}]_i$ increases with red light irradiation and caged-IP$_3$-release (Shacklock, Read & Trewavas, 1992).

In an earlier study using oat protoplasts, light-dependent phosphorylation was investigated by other workers (Park & Chae, 1989). The kinetics of labelling were not investigated, and labelling was carried out at 4 °C, over a much longer time period than in the present study. Changes in phosphorylation were not visible after red-light irradiation, but a less convincing densitometric analysis of phosphorylated bands appeared to show some increase in phosphorylation of two phosphopeptide bands after red-light illumination. The detected increase was banished by calcium chelation, although no experiments were performed to determine the optimal $[Ca^{2+}]_m$ for the response, or to determine $[Ca^{2+}]_i$. Studies similar to those of Park & Chae (1989) have been carried out in our laboratory with the wheat protoplast system, but have proved inconclusive. Only with the rapid labelling conditions previously described have changes in phosphorylation under red-light conditions proved convincing.

Blue-light-dependent phosphorylation

Other studies have identified phosphorylation changes dependent on other light spectra. One report has localised light-dependent phosphorylation events in pea plasma membrane preparations (Gallagher *et al.*, 1988). In this paper, it was noted that a strongly phosphorylated polypeptide band was present on SDS gels of *in vitro* phosphorylated membrane preparations from dark-grown peas. When similar preparations were examined after previous exposure of the seedlings to white light for a few hours, phosphorylation of the band was comparatively much less. Further investigation revealed that while blue light eliminated this 120 kDa band completely, red light had little effect on the phosphorylation density. Curiously, when isolated membranes were irradiated *in vitro* with blue light prior to phosphorylation, the 120 kDa band became more densely phosphorylated, the reverse of the *in vivo* effect (Short & Briggs, 1990). A suggested explanation is that changes in conformation after irradiation allow sites to be phosphorylated *in vivo* and hence prevent ^{32}P-labelling of those sites *in vitro*. This phenomenon suggests that complementary studies using *in vivo* and *in vitro* systems may provide insights into the mechanism of the phosphorylation reaction.

A blue-light receptor
A GTP-binding protein (G protein) has recently been identified in the plasma membrane of etiolated peas, which is recognised by a polyclonal anti-G-protein antibody, and binds to a GTP analogue only after blue-light illumination (Warhepa *et al.*, 1991). Its GTPase activity is induced by blue light administered *in vitro*, and is not responsive to red-light illumination. In the above system, then, we have a situation where a precise stimulus (blue light) and clearly defined end response (de-etiolation) are well established, and where a possible receptor and effector are beginning to be revealed. With identification of any protein kinase that may be involved, a definite pathway may begin to take shape.

Conclusions

Research into protein phosphorylation in plants has reached an exciting stage, with studies providing insights into the control of cellular activity. Key elements of the information transfer network, first described in animal cells, have now been identified in plants, and the post-translational modification of proteins through phosphorylation has been located at the heart of this network. It now seems certain that the understanding gained in mammalian systems will aid us in gaining the same depth of knowledge of the regulation of growth and development in plants.

References

Allen, J., Bennett, J., Steinback, K.E. & Arntzen, C.J. (1981). Chloroplast protein phosphorylation couples plastoquinone redox state to distribution of excitation energy between photosystems. *Nature (London)* **291**, 1–15.

Bennett, J. (1977). Phosphorylation of chloroplast membrane phosphoproteins. *Nature (London)* **269**, 344–6.

Bennett, J. (1984). Regulation of photosynthesis by protein phosphorylation. In *Enzyme Regulation by Reversible Phosphorylation – Further Advances* (ed. P. Cohen), pp. 227–46. Amsterdam: Elsevier.

Blakeley, S.D., Thomas, B., Hall, J.L. & Vince-Prue, D. (1983). Regulation of swelling of etiolated-wheat-leaf protoplasts by phytochrome and gibberellic acid. *Planta* **158**, 416–21.

Blowers, D.P., Hetherington, A.M. & Trewavas, A.J. (1985). Isolation of plasma membrane bound calcium/calmodulin regulated protein kinase from pea using Western blotting. *Planta* **166**, 208–15.

Blowers, D.P. & Trewavas, A.J. (1989). Autophosphorylation of pp[18], a calcium and calmodulin regulated plasma membrane located protein from pea, is a very rapid cycling of phosphate on serine residues. *Plant Physiology* **90**, 1279–85.

Blum, W., Hinsch, K.-D., Schultz, G. & Weiler, E.W. (1988). Identification of GTP-binding proteins in the plasma membrane of higher plants. *Biochemical and Biophysical Research Communications* **156**, 954–9.

Budde, R.J.A. & Chollet, R. (1988). Regulation of enzyme activity in plants by reversible phosphorylation. *Physiologia Plantarum* **72**, 435–9.

Chapman, K.S.R., Trewavas, A. & Van Loon, L.C. (1975). Regulation of the phosphorylation of chromatin-associated proteins in *Lemna* and *Hordeum*. *Plant Physiology* **55**, 293–6.

Chollet, R. (1990). Light/dark modulation of C_4-photosynthesis enzymes by regulatory phosphorylation. In *Current Topics in Plant Biochemistry and Physiology*, vol. 9 (ed. D.D. Randall & D.G. Blevins), pp. 344–56. Columbia, Missouri: Interdisciplinary Plant Group, University of Missouri.

Cohen, P. (1982). The role of protein phosphorylation in the neural and hormonal control of cellular activity. *Nature (London)* **296**, 613–20.

Coughlan, S.J. (1990). Thylakoid protein phosphorylation; regulation of light energy distribution in photosynthesis. In *Current Topics in Plant Biochemistry and Physiology*, vol. 9 (ed. D.D. Randall & D.G. Blevins), pp. 271–281. Columbia, Missouri: Interdisciplinary Plant Group, University of Missouri.

Danko, J.S. & Markwell, J.P. (1985). Protein phosphorylation in plant mitochondria. *Plant Physiology* **79**, 311–14.

Datta, N. & Cashmore, A.R. (1989) Binding of a pea nuclear protein to promoters of certain photoregulated genes is modulated by phosphorylation. *Plant Cell* **1**, 1069–77.

Datta, N., Chen, Y.-R. & Roux, S.J. (1985). Phytochrome and calcium stimulation of protein phosphorylation in isolated pea nuclei. *Biochemical and Biophysical Research Communications* **128**, 1403–8.

Draetta, G., Piwnica-Worms, H., Morrison, D., Druker, B., Roberts, T. & Beach, D. (1988). Human cdc2 protein kinase is a major cell cycle regulated tyrosine kinase substrate. *Nature (London)* **336**, 738–44.

Drobak, B.K., Allan, E.F., Comerford, J.G., Roberts, K. & Dawson, A.P. (1988). Presence of guanine nucleotide-binding proteins in a plant hypocotyl microsomal fraction. *Biochemical and Biophysical Research Communications* **150**, 899–903.

Eckhart, W., Hutchinson, M.A. & Hunter, T. (1979). An activity phosphorylating tyrosine in polyoma T antigen immunoprecipitates. *Cell* **18**, 925–33.

Gallagher, S., Short, T.W., Ray, P.M., Pratt, L.H. & Briggs, W.R. (1988). Light-mediated changes in two proteins found associated with plasma membrane fractions from pea stem sections. *Proceedings of the National Academy of Sciences USA* **85**, 8003–7.

Garrison, J.C. (1983). Measurement of hormone-stimulated protein phosphorylation in intact cells. *Methods in Enzymology* **99**, 20–36.

Grimm, R., Gast, D. & Rudiger, W. (1989). Characterization of a protein-kinase activity associated with phytochrome from etiolated oat (*Avena sativa* L.) seedlings. *Planta* **178**, 199–206.

Guitton, C., Dorne, A.M. & Mache, R. (1984). *In organello* and *in vitro* phosphorylation of chloroplast ribosomal proteins. *Biochemical and Biophysical Research* **121**, 297–303.

Guitton, C. & Mache, R. (1987). Phosphorylation *in vivo* of the large subunit of a ribulose-1,5 bisphosphate carboxylase and glyceraldehyde-3-phosphate dehydrogenase. *European Journal of Biochemistry* **106**, 249–54.

Hardie, D.G. (1990). Roles of protein kinases and phosphatases in signal transduction. In *Hormone Perception and Signal Transduction in Animals and Plants* (ed. J. Roberts, C. Kirk & M. Venis), pp. 241–55. Cambridge: Company of Biologists.

Harmon, A.C., Putnam-Evans, C. & Cormier, M.J. (1987). A calcium-dependent but calmodulin-independent protein kinase from soybean. *Plant Physiology* **83**, 830–7.

Harper, J.F., Sussman, M.R., Schaller, G.E., Putnam-Evans, C., Charbonneau, H. & Harmon, A.C. (1991). A calcium-dependent protein kinase with a regulatory domain similar to calmodulin. *Science* **252**, 951–4.

Hetherington, A.M., Battey, N.H. & Millner, P.A. (1990). Protein kinase. In *Methods in Plant Biochemistry*, vol. 3 (ed. P.J. Lea), pp. 371–83. London: Academic Press.

Hetherington, A. & Trewavas, A.J. (1984). Regulation of a membrane-bound protein kinase in pea shoots by calcium ions. *Planta* **161**, 409–18.

Huber, S.C. & Huber, J.L.A. (1990). Regulation of spinach leaf sucrose-phosphate synthase by multisite phosphorylation. In *Current Topics in Plant Biochemistry and Physiology*, vol. 9 (ed. D.D. Randall & D.G. Blevins), pp. 329–43. Columbia: Missouri: Interdisciplinary Plant Group, University of Missouri.

Keates, R.A.B. & Trewavas, A.J. (1974). Protein kinase activity associated with isolated ribosomes from peas and *Lemna*. *Plant Physiology* **54**, 95–9.

Kim, I.-S., Bai, U. & Song, P.-S. (1989). A purified 124-kD oat phytochrome does not possess a protein kinase activity. *Photochemistry and Photobiology* **49**, 319–23.

Laing, W.A. & Christeller, J.T. (1984). Chloroplast phosphoproteins: distribution of phosphoproteins within spinach chloroplasts. *Plant Science Letters* **36**, 99–104.

Lin, Z.F., Lucero, H.A. & Racker, E. (1982). Protein kinases from spinach chloroplasts. I. Purification and identification of two distinct protein kinases. *Journal of Biological Chemistry* **257**, 12153–6.

McMichael, R.W. Jr & Lagarias, J.C. (1990). Polycation-stimulated phytochrome phosphorylation: Is phytochrome a protein kinase? In *Current Topics in Plant Biochemistry and Physiology*, vol. 9 (ed. D.D. Randall & D.G. Blevins), pp. 259–70. Columbia, Missouri: Interdisciplinary Plant Group, University of Missouri.

Melanson, D. & Trewavas, A.J. (1982). Changes in tissue protein pattern associated with the induction of DNA synthesis by auxin. *Plant Cell and Environment* **5**, 53–64.

Park, M.-H. & Chae, Q. (1989). Intracellular protein phosphorylation in oat (*Avena sativa* L.) protoplasts by phytochrome action (1) measurement of action spectra for the protein phosphorylation. *Biochemical and Biophysical Research Communications* **162**, 9–14.

Polya, G.M. & Chandra, S. (1990). Ca^{2+}-dependent protein phosphorylation in plants: Regulation, protein substrate specificity and product dephosphorylation. In *Current Topics in Plant Biochemistry and Physiology*, vol. 9 (ed. D.D. Randall & D.G. Blevins), pp. 164–80. Columbia, Missouri: Interdisciplinary Plant Group, University of Missouri.

Ralph, R.K., McCombs, P.J.A., Tener, G. & Wojcik, S.J. (1972). Evidence for modification of protein phosphorylation by cytokinins. *Biochemical Journal* **130**, 901–11.

Ranjeva, R. & Boudet, A.M. (1987). Phosphorylation of proteins in plants: Regulatory effects and potential involvement in stimulus-response coupling. *Annual Review of Plant Physiology* **38**, 73–93.

Roeske, C.A. & Chollet, R. (1989). Role of metabolites in the reversible light-activation of pyruvate, orthophosphate dikinase in *Zea mays* mesophyll cells *in vivo*. *Plant Physiology* **90**, 330–7.

Shacklock, P.S., Read, N.D. & Trewavas, A.J. (1992). Cytosolic free calcium mediates red light-induced photomorphogenesis. *Nature (London)* **358**, 153–5.

Short, T.W. & Briggs, W.R. (1990). Characterization of a rapid, blue light-mediated change in detectable phosphorylation of a plasma membrane protein from etiolated pea (*Pisum sativum* L.) seedlings. *Plant Physiology* **92**, 179–85.

Soll, J. (1985). Phosphoproteins and protein-kinase activity in isolated envelopes of pea (*Pisum sativum* L.) chloroplasts. *Planta* **166**, 394–400.

Soll, J. & Buchanan, B.B. (1983). Phosphorylation of chloroplast ribulose bisphosphate carboxylase/oxygenase small subunit by an envelope-bound protein kinase *in situ*. *Journal of Biological Chemistry* **258**, 6686–9.

Stadtman, E.R. & Chock, P.B. (1977). Superiority of interconvertible enzyme cascades in metabolic regulation: Analysis of monocyclic systems. *Proceedings of the National Academy of Sciences USA* **74**, 2761–5.

Sutherland, E.W. & Wosilait, W.D. (1955). Inactivation and activation of liver phosphorylase. *Nature (London)* **175**, 169–70.

Trewavas, A. (1976). Post-translational modification of proteins by phosphorylation. *Annual Review of Plant Physiology* **27**, 349–74.

Trewavas, A. (1979). Nuclear phosphoproteins in germinating cereal embryos and their relationship to the control of mRNA synthesis and the onset of cell division. In *Recent Advances in the Biochemistry of Cereals* (ed. D.L. Laidman & R.G. Wyn Jones), pp. 175–208. New York: Elsevier.

Trewavas, A. & Blowers, D.P. (1990). Protein kinase in the plant plasma membrane. In *Current Topics in Plant Biochemistry and Physiology*, vol. 9 (ed. D.D. Randall & D.G. Blevins), pp. 153–63. Columbia, Missouri: Interdisciplinary Plant Group, University of Missouri.

Warhepa, K.M.F., Hamm, H.E., Rasenick, M.M. & Kaufman, L.S. (1991). A blue-light-activated GTP-binding protein in the plasma membranes of etiolated peas. *Proceedings of the National Academy of Sciences USA* **88**, 8925–9.

Wielgat, B. & Kleczkowski, K. (1981). Gibberellic acid enhanced phosphorylation of pea chromatin proteins. *Plant Science Letters* **21**, 381–8.

Wong, Y.-S., Cheng, H.C., Walsh, D.A. & Lagarias, J.C. (1986). Phosphorylation of *Avena* phytochrome *in vitro* as a probe of light-induced conformational changes. *Journal of Biological Chemistry* **261**, 12089–97.

A.K. MATTOO, T.D. ELICH,
M.L. GHIRARDI, F.E. CALLAHAN
and M. EDELMAN

Post-translational modification of chloroplast proteins and the regulation of protein turnover

Our efforts are centred around the regulatory mechanisms controlling the assembly, function and turnover of the photosystem II (PSII) reaction centre. The PSII reaction centre consists of D1, D2, the α and β subunits of cytochrome b_{559} and the psbI gene product. Of these proteins, the metabolism of the D1 protein is relatively well understood. Because D1 is synthesised and rapidly degraded in the light, it is well suited for *in vivo* pulse–chase analysis. The D1 protein undergoes at least five post-translational modifications during its life cycle: C-terminal processing, removal of the initiating methionine residue, N-acetylation of the resulting N-terminal threonine residue, covalent palmitoylation, and O-phosphorylation of the N-terminal threonine of the mature protein. Processing of D1 occurs on stroma-exposed membranes while palmitoylation and phosphorylation occur in spatially distinct grana membranes. The palmitoylation is light-stimulated, inhibited by the herbicides atrazine and DCMU, and is apparently transient in nature. This modification might be important in the assembly of the protein in the PSII reaction centre and/or in its translocation from stroma membranes to grana. D1 phosphorylation increases with increasing light intensity, is inhibited by atrazine and DCMU and is redox regulated *in vitro*. *In vivo*, phosphorylated D1 undergoes light-dependent turnover as a function of light intensity; this process occurs under blue, green, and red illuminations and is inhibited by sodium fluoride but not by the PSII herbicides. We are investigating the role of phosphorylation–dephosphorylation in the function and dynamics of PSII.

Society for Experimental Biology Seminar Series 53: *Post-translational modifications in plants*, ed.
N.H. Battey, H.G. Dickinson & A.M. Hetherington. © Cambridge University Press 1993, pp. 65–78.

Introduction

Protein turnover is a common feature of the active metabolism taking place in the photosynthetic organelle, the chloroplast. It is now established that the chloroplast houses a unique genome that encodes a large portion of the chloroplast proteins and in addition imports discrete gene products encoded in the nucleus. These active processes entail that the chloroplast must have developed intricate mechanisms for turnover of the proteins encoded within the organelle and of those imported from other compartments of the plant cell. However, the signals that target chloroplast proteins for turnover and the mechanisms involved are not as yet well understood. We are studying the assembly, function and turnover of proteins associated with the photosystem II (PSII) reaction centre and also the signals involved in their dynamic flux within the chloroplast membranes.

The PSII consists of a heterodimer of homologous proteins, D1 and D2, α and β subunits of the cytochrome b_{559}, and the psbI gene product (for reviews see Rutherford, 1989; Mattoo, Marder & Edelman, 1989). The D1/D2 heterodimer is thought to contain the requisite electron carriers and ligands necessary for the electron transport through the PSII reaction centre. The functions of cytochrome b_{559} and the psbI gene product are unknown.

The metabolic life history of the D1 protein has been elucidated. The protein is synthesised as a precursor of 33.5 – 34 kDa (Reisfeld, Mattoo & Edelman, 1982) on unappressed, stromal lamellae, and is processed to the mature size on these membranes (Mattoo & Edelman, 1987). Following processing, the D1 protein translocates to the appressed, grana regions of thylakoids where it functions as a reaction centre component and also undergoes rapid, light-dependent turnover at rates that are several times higher than those of D2 or any other PSII protein (Mattoo *et al.*, 1984, 1989). This implies that the PSII reaction centres are constantly being assembled and disassembled in the thylakoids (Ghirardi *et al.*, 1990). This property of D1 makes it possible to follow its metabolism *in vivo* via pulse–chase experimental analysis. The D1 protein undergoes at least five post-translational modifications during its life cycle: C-terminal processing, removal of the initiating methionine residue, N-acetylation of the resulting N-terminal threonine residue (Michel *et al.*, 1988), covalent palmitoylation (Mattoo & Edelman, 1987), and O-phosphorylation of the N-terminal threonine of the mature protein (Michel *et al.*, 1988). We describe here some of our experiments dealing with the palmitoylation and phosphorylation of chloroplast proteins and the possible roles these post-translational modifications might confer on at least one of them, the D1 protein.

Post-translational acylation of chloroplast proteins

Spirodela oligorrhiza plants rapidly incorporate [³H]palmitate or [³H]myristate into acid-precipitable proteins, palmitate being a preferred substrate over myristate (Fig. 1). The slower incorporation of myristate into proteins is probably due to its elongation to palmitate, which in turn is incorporated into specific proteins. SDS–PAGE analysis of ³H-palmitate-labelled chloroplast proteins revealed acylation of two membrane proteins and two soluble proteins. These acylated proteins were identified as the D1 protein (Mattoo *et al.*, 1988), the light-harvesting Chl *a/b* apoprotein (Mattoo & Edelman, 1987), the large subunit of ribulose-1,5-bisphosphate carboxylase–oxygenase (Mattoo & Edelman, 1987), and the acyl carrier protein (Callahan *et al.*, 1989). Light stimulated palmitoylation of all four chloroplast proteins (Mattoo & Edelman, 1987).

Because of the low abundance of the acylated form of the proteins, a

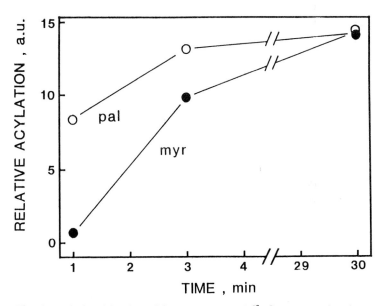

Fig. 1. Relative kinetics of incorporation of [³H]myristate (myr) and [³H]palmitate (pal) into total acid-insoluble *Spirodela* thylakoid proteins.

method was developed to identify the fatty acid ligand on the proteins. This method involved electrotransfer of the acylated proteins to nitrocellulose paper, direct hydrolysis of the immobilised proteins, derivatisation of the fatty acids with *p*-nitrophenacyl bromide, and final resolution by reversed-phase high-performance liquid chromatography (HPLC) (Callahan *et al.*, 1989). The use of a radioactive flow detector coupled to reversed-phase HPLC allowed qualitative analysis of the acyl moiety under conditions where the absolute amount of the fatty acid was too low for detection by HPLC alone. The results suggested that palmitate is a major fatty acid ligand of the four chloroplast proteins. Incubation of the SDS–polyacrylamide gels with 1 M NH_2OH at pH 6.6 indicated that only the [^{3}H]palmitate-labelled acyl carrier protein was sensitive to this treatment, suggesting that a thiol ester linkage was involved in this case and that the remaining three proteins were linked to palmitoyl group through a stronger, ether or amide bond (Mattoo & Edelman, 1987).

The actual site of palmitoylation on the D1 protein or on the other palmitoylated chloroplast proteins has not been determined. However, *in situ* trypsinisation of ^{3}H-acylated thylakoids revealed the association of the [^{3}H]-label with the two characteristic membrane-associated fragments, T22 and T20 (Marder *et al.*, 1986), of the D1 protein. These results map the acylation site to the amino-terminal two thirds of the D1 protein.

Physiological aspects of D1 palmitoylation

The herbicides atrazine and diuron (DCMU) inhibit PSII electron transport by displacing the quinone on the D1 protein (Trebst, 1986). In turn, they also inhibit the light-dependent degradation of D1 (Mattoo *et al.*, 1984). In preliminary experiments, we have checked whether these herbicides affect palmitoylation of D1. *Spirodela* plants were incubated in the absence and presence of 10 µM atrazine or DCMU and then labelled with [^{3}H]palmitate for 3 min. Membrane proteins were then subjected to SDS–PAGE/fluorography. Fig. 2 shows that both of the PSII herbicides inhibited palmitoylation of the D1 protein (32 kDa) but not that of the LHCP. These results suggest that active PSII electron transport is a requirement for palmitoylation of the D1 protein but not for LHCP. Alternatively, a conformational change in the D1 protein upon binding DCMU or atrazine (Mattoo *et al.*, 1981) might interfere with the palmitoylation reaction.

The involvement of acylation in the spatiotemporal dynamics of the D1 protein was investigated by radiolabelling *Spirodela* plants for a short time with [^{3}H]palmitate and then fractionating the thylakoids into grana

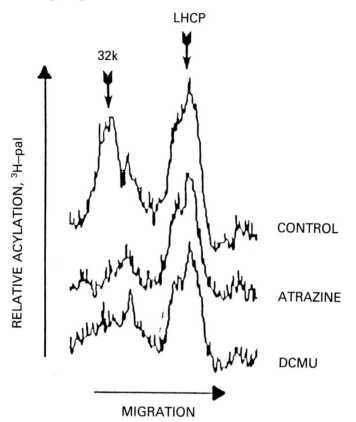

Fig. 2. Atrazine or DCMU inhibit palmitoylation of the D1 protein (32k) but not of LHCP.

and stromal lamellae. The experiment was designed to look at the form of the D1 protein, and the stage in its life history, with which palmitate associates. Palmitoylation was found to be associated solely with grana-localised mature D1 (Mattoo & Edelman, 1987). Ephemeral existence of the palmitate-labelled D1 in the stromal lamellae could not be ruled out in these experiments. Nevertheless, it is tempting to speculate that palmitoylation of D1 might help in the translocation of the newly processed D1 from the stromal lamellae to the grana. This modification could also be important in the assembly of the protein in the PSII reaction centre. Further experiments are warranted to define precisely the physiological role of palmitoylation in D1 dynamics as well as in the dynamics of the other acylated chloroplast proteins.

Post-translational phosphorylation of thylakoid proteins

Both D1 and D2 are known to be reversibly phosphorylated *in vitro* on their acetylated N-terminal threonine residues (Michel *et al.*, 1988). In addition, four other PSII-associated proteins are also reversibly phosphorylated on or near their N-terminus under similar conditions: CP43, the psbH gene product, and type I and II LHCII polypeptides (see Bennett, 1991, for review). Phosphorylation of the LHCII polypeptides has been studied in greater detail since this modification has been proposed to mediate changes in energy distribution to compensate for imbalances between PSII and photosystem I (PSI) (i.e. state transitions). Under conditions that favour PSII excitation, reduction of the plastoquinone pool leads to activation of the thylakoid protein kinase(s). Phosphorylation of so called 'mobile LHCII' enables dissociation of this antennae sub-population from granal PSII centres, leading to a decrease in the optical cross section of PSII. It is thought that the dissociated, mobile LHCII then migrates to the unstacked regions, where it may transfer energy to PSI.

The role of phosphorylation of the PSII core polypeptides is not as well established. A number of effects on PSII function have been reported to occur upon phosphorylation of thylakoid proteins *in vitro*, including: (i) an increase (Shochat *et al.*, 1982; Vermaas, Steinback & Arntzen, 1984), decrease (Habash & Baker, 1990) or no change (Hodges, Packham & Barber, 1985) in the affinity for PSII herbicides; (ii) an increase (Jursinic & Kyle, 1983) or decrease (Hodges, Boussac & Briantais, 1987) in the stability of Q_B^-; (iii) a decrease in light-saturated photosynthetic electron transport (Horton & Lee, 1984; Hodges *et al.*, 1985; Packham, 1987; Habash & Baker, 1990; Giardi *et al.*, 1991); (iv) protection against photoinhibition (Horton & Lee, 1985); and (v) a decrease in the connectivity of PSII (Kyle, Haworth & Arntzen, 1982). However, with the exception of Giardi *et al.* (1991), who reported a correlation between rates of electron transport and the relative extent of PSII core phosphorylation, in none of these cases was the extent of thylakoid protein phosphorylation established; nor were the observed effects correlated with phosphorylation of any specific polypeptide(s). Furthermore, all of these studies were performed on isolated thylakoids or osmotically shocked chloroplasts and thus their physiological relevance remains to be determined. In this regard, no studies to date have rigorously identified and characterised an *in vivo* phosphorylation of a PSII-associated protein. Therefore, it is not even clear that the same proteins are phosphorylated on the same sites *in vivo* as *in vitro*. Interestingly, in several studies where *in vivo* and *in vitro* phosphorylated thylakoids could be compared, qualitative and quantitative differences in the ^{32}P-radiolabelled protein patterns were apparent (Schuster *et al.*, 1986; Bhalla & Bennett, 1987; Bennett, Shaw & Michel, 1988; Elich, Edelman & Mattoo, 1992).

In vitro and *in vivo* D1 phosphorylation is identical

Recently, an *in vivo*-generated electrophoretic variant of D1, designated 32*, was resolved in *Spirodela oligorrhiza* as well as in four other higher plants (Callahan *et al.*, 1990). This form was generated specifically in reaction centres localised in the grana, and its appearance was correlated with the onset of D1 degradation (Fig. 3). Formation of 32* (D1*) was light-dependent and inhibited by DCMU (Callahan *et al.*, 1990), a herbicide known to inhibit the light-dependent degradation of D1. These results led to the hypothesis that the modification giving rise to 32* may be a signal targeting the protein for degradation.

Studies conducted to check the possible involvement of palmitoylation

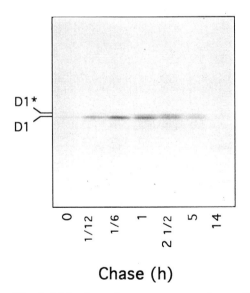

Grana Lamellae

Chase (h)

Fig. 3. Light-dependent appearance of a modified form of the D1 protein (D1*) in grana thylakoids as a function of time.

(see above) or irreversible cross-linking within the protein (Adir, Shochat & Ohad, 1990) in the modification that generated 32* were negative. Next we tested phosphorylation because charge alteration can confer decreased mobility of some proteins on SDS–PAGE.

The first indications that phosphorylation of D1 results in the formation of 32* came from *in vitro* studies. In these experiments, when isolated *Spirodela* thylakoids were incubated with [γ-^{32}P]ATP in the presence of light, quantitative conversion of D1 to a slower-migrating protein band co-electrophoresing with 32* generated *in vivo* was observed. The ^{32}P-label was associated solely with the less mobile D1 protein band (Fig. 4). Light was required for both processes; however, reducing power in the form of ferredoxin and NADPH could substitute for the light requirement (Fig. 4; compare lanes 8 and 9). Both the *in vitro* phosphorylation of D1 and the mobility shift of D1 were inhibited by the PSII herbicides DCMU and atrazine, and by an ATP analogue, FSBA, which is known to inhibit protein kinases.

The next question was to verify that the D1 phosphorylated *in vitro* was identical to the 32* generated *in vivo*. Following adaptation for 24 hours to a medium lacking phosphate, *Spirodela* plants were pulsed with carrier-free ^{32}P for 3 h and thylakoid protein patterns were compared with thylakoids phosphorylated *in vitro* or with thylakoids isolated from intact plants radiolabelled with [^{35}S]methionine (Fig. 5). Among the ^{35}S-labelled proteins, the D1 protein (32) and the 26-kDa LHCP were prominent (lane 1). In contrast, many more proteins were found to be labelled with ^{32}P, and qualitative and quantitative differences were evident between the *in vivo* and *in vitro* phosphorylated protein patterns (compare lane 2 with lane 3). However, there were co-electrophoresing bands in both cases, including D1 and putative D2 and LHCII proteins. DCMU was a more effective inhibitor of phosphorylation *in vitro* than *in vivo*. Further, with the exception of the 9 kDa phosphoprotein, DCMU inhibited the phosphorylation *in vivo* of the proteins that co-electrophoresed with those phosphorylated *in vitro* (Elich *et al.*, 1992).

The indication that *in vitro* and *in vivo* phosphorylation of D1 might be similar was subsequently analysed. The following common features between *in vitro* phosphorylated D1 and the putative *in vivo* phosphorylated D1 band rigorously demonstrated that they are identical:

1. Phosphoamino acid analysis of both the ^{32}P-labelled proteins revealed the presence of only phosphothreonine.

2. Two tryptic phosphopeptides, with identical mobilities and ratios of the radiolabel, were resolved from both phosphoproteins by TLE/TLC mapping experiments.

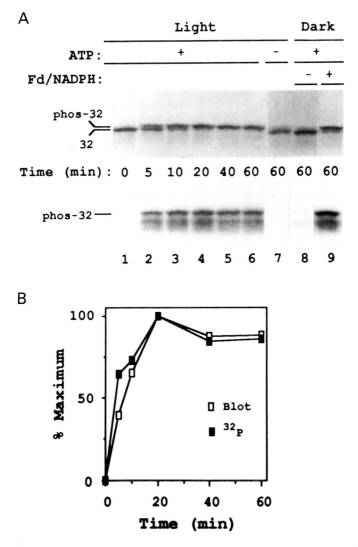

Fig. 4. Light-dependent phosphorylation of the D1 protein in isolated thylakoids. Isolated thylakoids were incubated in the light or darkness, in the presence or absence of [γ-^{32}P]ATP and Fd/NADPH as indicated. At the indicated times, samples were taken and analysed by SDS–PAGE, immunoblotting and autoradiography. (A) Immunoblot and its autoradiogram, with the positions of the D1 parent protein (32) and its phosphorylated form (phos-32) marked. (B) Densitometric quantification of the bands labelled phos-32 in the immunoblot (open square) and autoradiogram (solid square). Adapted from Elich *et al.* (1992).

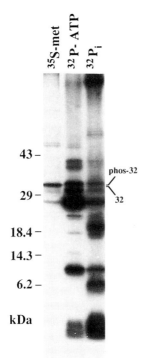

Fig. 5. Comparison of thylakoid proteins radiolabelled *in vivo* with [^{35}S]methionine or [^{32}P]orthophosphate, or *in vitro* with [γ-^{32}P]ATP. Thylakoids were isolated from *Spirodela* plants radiolabelled *in vivo* by incubation with 50 μCi ml^{-1} [^{35}S]methionine for 30 min under 90 μE m^{-2} s^{-1} white light (lane 1), or by incubation with 0.5 mCi ml^{-1} carrier-free [^{32}P]orthophosphate for 3 h under 50 μE m^{-2} s^{-1} white light (lane 3). Alternatively, isolated unlabelled thylakoids were phosphorylated *in vitro* with [γ-^{32}P]ATP (lane 2) as described in Fig. 1. The resulting thylakoid preparations were resolved on 10–20% linear gradient SDS gels, followed by autoradiography. The positions of the non-phosphorylated (32) and phosphorylated (phos-32) forms of the D1 protein are indicated. Adapted from Elich *et al.* (1992).

3. The phosphorylation site(s) were localised in both cases to within 1 kDa of the N-terminus by partial proteolytic mapping experiments and *in situ* trypsinisation of ^{32}P-labelled thylakoids.

Phosphorylation of D1 *in vitro* is known to be mediated by a redox-regulated protein kinase. That this is also the case *in vivo* is suggested by the similar dependence on light intensity, and inhibition by DCMU.

Does phosphorylation target D1 for degradation?

The ability to resolve phosphorylated D1 from non-phosphorylated form allowed us to examine the specific metabolism of the two forms. Under steady-state conditions, both forms exhibit similar light-dependent turnover (Fig. 6). However, in the presence of DCMU, only turnover of the phosphorylated DS1 was observed. This light-dependent, linear-electron-transport-independent turnover was proportional to light intensity, occurred throughout the visible spectrum, and was inhibited by sodium fluoride (A.K. Mattoo *et al.*, in preparation). We are presently investigating the role of light-dependent phosphorylation–dephosphorylation in the function and dynamics of PSII.

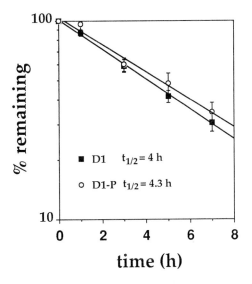

Fig. 6. Pulse–chase analysis of the steady-state light-dependent turnover of granal-localised D1 and phosphorylated D1 (D1-P). Illuminated (90 $\mu E\ m^{-2}\ s^{-1}$ white light) *Spirodela* plants were labelled for 30 min with 50 $\mu Ci\ ml^{-1}$ [^{35}S]methionine and chased for 3 h with 1 mM unlabelled methionine to ensure that all radiolabelled D1 was in the grana and that the radiolabel was distributed between D1 and its phosphorylated form. The plants were then further incubated for the times indicated (after the initial 3 h chase) after which they were frozen until thylakoid isolations were performed. Isolated thylakoids were resolved by SDS–PAGE and analysed by autoradiography. The average relative amounts of D1 (solid squares) and phosphorylated D1 (D1-P) (open circles) from six separate experiments were quantified by densitometry and plotted as a function of time (error bars depict ±1 SEM).

Acknowledgments

The results mentioned in this paper were supported in part by a BARD grant to M.E. and A.K.M. and a USDA/CRGO grant to A.K.M.

References

Adir, N., Shochat, S. & Ohad, I. (1990). Light-dependent D1 protein synthesis and translocation is regulated by reaction center II. *Journal of Biological Chemistry* **265**, 12563–8.

Bennett, J. (1991). Protein phosphorylation in green plant chloroplasts. *Annual Review of Plant Physiology and Plant Molecular Biology* **42**, 281–311.

Bennett, J., Shaw, E.K. & Michel, H. (1988). Cytochrome b_6f complex is required for phosphorylation of light-harvesting chlorophyll a/b complex II in chloroplast photosynthetic membranes. *European Journal of Biochemistry* **171**, 95–100.

Bhalla, P. & Bennett, J. (1987). Chloroplast phosphoproteins: phosphorylation of a 12-kDa stromal protein by the redox-controlled kinase of thylakoid membranes. *Archives of Biochemistry and Biophysics* **252**, 97–104.

Callahan, F.E., Ghirardi, M.L., Sopory, S.K., Mehta, A.M., Edelman, M. & Mattoo, A.K. (1990). A novel metabolic form of the 32kDa-D1 protein in the granal-localized reaction center of photosystem II. *Journal of Biological Chemistry* **265**, 15357–60.

Callahan, F.E., Norman, H.A., Srinath, T., St. John, J.B., Dhar, R. & Mattoo, A.K. (1989). Identification of covalently bound fatty acids on acylated proteins immobilized on nitrocellulose paper. *Analytical Biochemistry* **183**, 220–4.

Elich, T.D., Edelman, M. & Mattoo, A.K. (1992). Identification, characterization, and resolution of the *in vivo* phosphorylated form of the photosystem II reaction center protein. *Journal of Biological Chemistry* **267**, 3523–9.

Ghirardi, M.L., Callahan, F.E., Sopory, S.K., Elich, T.D., Edelman, M. & Mattoo, A.K. (1990). Cycling of the Photosystem II reaction center core between grana and stroma lamellae. In *Current Research in Photosynthesis*, vol. 1 (ed. M. Baltechefsky), pp. 733–8. The Netherlands: Kluwer Academic Publishers.

Giardi, M.T., Rigoni, F., Barbato, R. & Giacometti, G.M. (1991). Relationship between heterogeneity of the PSII core complex from grana particles and phosphorylation. *Biochemical and Biophysical Research Communications* **176**, 1298–305.

Habash, D.Z. & Baker, N.R. (1990). Demonstration of two sites of inhibition of electron transport by protein phosphorylation in wheat thylakoids. *Journal of Experimental Botany* **41**, 761–7.

Hodges, M., Boussac, A. & Briantais, J.-M. (1987). Thylakoid membrane phosphorylation modifies the equilibrium between Photosystem II quinone electron acceptors. *Biochimica et Biophysica Acta* **894**, 138–45.

Hodges, M., Packham, N.K. & Barber, J. (1985). Modification of photosystem II activity by protein phosphorylation. *FEBS Letters* **181**, 83–7.

Horton, P. & Lee, P. (1984). Phosphorylation of chloroplast thylakoids decreases the maximum capacity of photosystem II electron transfer. *Biochimica et Biophysica Acta* **767**, 563–7.

Horton, P. & Lee, P. (1985). Phosphorylation of chloroplast membrane proteins partially protects against photoinhibition. *Planta* **165**, 37–42.

Jursinic, P.A. & Kyle, D.J. (1983). Changes in the redox state of the secondary acceptor of photosystem II associated with light-induced thylakoid protein phosphorylation. *Biochimica et Biophysica Acta* **723**, 37–44.

Kyle, D.J., Haworth, P. & Arntzen, C.J. (1982). Thylakoid membrane protein phosphorylation leads to a decrease in connectivity between photosystem II reaction centers. *Biochimica et Biophysica Acta* **680**, 336–42.

Marder, J.B., Chapman, D.J., Telfer, A., Nixon, P.J. & Barber, J. (1987). Identification of psbA and psbD gene products, D1 and D2, as proteins of a putative reaction centre of photosystem II. *Plant Molecular Biology* **9**, 325–33.

Mattoo, A.K. & Edelman, M. (1987). Intramembrane translocation and posttranslational palmitoylation of the chloroplast 32-kDa herbicide binding protein. *Proceedings of the National Academy of Sciences USA* **84**, 1497–1501.

Mattoo, A.K., Hoffman-Falk, H., Marder, J.B. & Edelman, M. (1984). Regulation of protein metabolism: coupling of photosynthetic electron transport to in vivo degradation of the rapidly metabolized 32-kilodalton protein of the chloroplast membrane. *Proceedings of the National Academy of Sciences USA* **81**, 1380–4.

Mattoo, A.K., Marder, J.B. & Edelman, M. (1989). Dynamics of the photosystem II reaction center. *Cell* **56**, 241–6.

Mattoo, A.K., Norman, H.A., Callahan, F.E., St. John, J.B. & Edelman, M. (1988). Plant protein acylation: Identification of the modified proteins and analysis of the bound fatty acid ligand. In *Post-translational Modification of Proteins by Lipids: A Laboratory Manual* (ed. U. Brodbeck & C. Bordier), pp. 82–7. Heidelberg: Springer-Verlag.

Mattoo, A.K., Pick, U., Hoffman-Falk, H. & Edelman, M. (1981). The rapidly metabolized 32,000-dalton polypeptide of the chloroplast is the 'proteinaceous shield' regulating photosystem II electron transport and mediating diuron herbicide sensitivity. *Proceedings of the National Academy of Sciences USA* **78**, 1571–6.

Michel, H., Hunt, D.F., Shabanowitz, J. & Bennett, J. (1988). Tandom mass spectrometry reveals that three photosystem II proteins of spinach chloroplasts contain N-acetyl-O-phosphothreonine at their N-terminii. *Journal of Biological Chemistry* **263**, 1123–30.

Packham, N.K. (1987). Phosphorylation of the 9 kDa photosystem II-associated protein and the inhibition of photosynthetic electron transport. *Biochimica et Biophysica Acta* **893**, 259–66.

Reisfeld, A., Mattoo, A.K. & Edelman, M. (1982). Processing of a chloroplast-translated membrane protein *in vivo*. *European Journal of Biochemistry* **124**, 125–9.

Rutherford, A.W. (1989). Photosystem II, the water-splitting enzyme. *Trends in Biochemical Sciences* **14**, 227–32.

Schuster, G., Dewit, M., Staehelin, A. & Ohad, I. (1986). Transient inactivation of the thylakoid photosystem II light-harvesting protein kinase system and concomitant changes in intramembrane particle size during photoinhibition of *Chlamydomonas reinhardtii*. *Journal of Cell Biology* **103**, 71–80.

Shochat, S., Owens, G.C., Hubert, P. & Ohad, I. (1982). The dichlorophenyldimethylurea-binding site in thylakoids of *Chlamydomonas reinhardii*. Role of photosystem II reaction center and phosphorylation of the 32–35 kilodalton polypeptide in the formation of the high-affinity binding site. *Biochimica et Biophysica Acta* **681**, 21–31.

Trebst, A. (1986). The topology of the plastoquinone and herbicide binding peptides of photosystem II in thylakoid membrane. *Zeitschrift für Naturforschung* **41**c, 240–5.

Vermaas, W.F.J., Steinback, K.E. & Arntzen, C.J. (1984). Characterization of chloroplast thylakoid polypeptides in the 32-kDa region: polypeptide extraction and protein phosphorylation affect binding of photosystem II-directed herbicides. *Archives of Biochemistry and Biophysics* **231**, 226–32.

J. SOLL

Purification of a small phosphoprotein from chloroplasts and characterisation of its phosphoryl group

Introduction

The phosphorylation pattern of chloroplast proteins changes dramatically upon a switch in external conditions, e.g. a light–dark transition *in vivo* (Bennett, 1991) or a change in nucleoside triphosphate concentration *in vitro* (Ranjeva & Boudet, 1987). In the presence of very low concentrations of ATP or GTP (5–30 nM) and short labelling kinetics, only one protein of low molecular mass (about 19 kDa) is detected in soluble chloroplast proteins (Soll & Bennett, 1988; Soll, Berger & Bennett, 1989). This chapter describes the purification and analysis of this novel phosphoprotein.

Characterisation of the 19 kDa phosphoprotein

The 19 kDa protein was first detected as a phosphoprotein in highly purified intact chloroplasts from pea and spinach (Soll & Bennett, 1988). Upon subfractionation of chloroplasts on sucrose gradients, the phosphorylated protein could be detected in the total soluble extract of chloroplasts and the envelope membrane fraction. The total soluble extract contains stromal proteins as well as proteins from the intermembrane space, which are mostly liberated during lysis of the plastids. The hydrophilic nature of the 19 kDa protein was verified by Triton X-114/water phase partitioning: the protein was exclusively found in the water phase. Mild sonication of mixed envelope membranes from chloroplasts also resulted in an almost complete release of the protein from the membrane. Differential labelling of the 19 kDa protein in intact and lysed chloroplasts of this soluble or peripheral membrane protein suggests its localisation outside the stromal space, perhaps in the inter-envelope lumen or the outer envelope membrane (Soll, Steidl & Schröder, 1991).

The 19 kDa protein was purified from soluble chloroplast proteins by

Society for Experimental Biology Seminar Series 53: *Post-translational modifications in plants*, ed. N.H. Battey, H.G. Dickinson & A.M. Hetherington. © Cambridge University Press 1993, pp. 79–90.

Table 1. *Purification scheme of the 19 kDa protein from chloroplast soluble proteins*

The protein was applied to an ATP–agarose column under high salt conditions (Welch & Feramisco, 1985). The column was washed excessively with high-salt and low-salt buffer and finally eluted with 1 mM GTP in low-salt buffer. The protein fraction eluted with GTP was dialysed and applied to DEAE–cellulose, from which it was recovered in the 125 mM NaCl eluate (Soll *et al.*, 1991).

Step	Volume (ml)	Protein (mg)	cpm/mg	Recovery (%)	Purification (-fold)
Soluble extract	20	326	18 000	100	1
ATP–agarose GTP–eluate	30	0.210	n.d.[a]	—	—
DEAE–cellulose 125 mM NaCl	6	0.01	80×10^6	13.7	4470

[a]Owing to unlabelled GTP present in this fraction, it was not possible to determine the specific activity and the purification factor.

affinity chromatography on ATP-agarose (Welch & Feramisco, 1985). Most of the protein could be eluted from the matrix by 1 mM GTP. After dialysis this fraction was further purified by anion exchange chromatography on DEAE–cellulose, from which it eluted at 125 mM NaCl (Table 1) (Soll *et al.*, 1991). The purification protocol resulted in an almost 4500–fold enrichment and in an apparent homogeneity of the protein as judged by SDS–PAGE and silver staining.

Characterisation of the purified protein reveals a number of uncommon features. The 19 kDa protein is probably labelled by [γ-^{32}P]NTP in an autophosphorylation event (see below). Phosphorylation occurs in a trichloroacetic acid (TCA)- or acetone-precipitable form. The 19 kDa protein shows a very high specificity for ATP and GTP as demonstrated by the extremely low K_m values of 8 nM and 5 nM for ATP and GTP, respectively (Table 2). In the presence of [γ-^{32}P]ATP, inclusion of unlabelled GTP resulted in less labelling of the protein and vice versa. Unlabelled UTP, however, did not compete with the label in the 19 kDa protein in the presence of [γ-^{32}P]ATP, demonstrating its nucleotide-specificity. The pI of the purified protein is 5.2.

Phosphorylation of the 19 kDa protein was only slightly influenced by temperature. Maximal phosphorylation was obtained at 25 °C; a residual activity of 80% and 70% was measured at 4 °C and −18 °C (50% glycerol), respectively (Fig. 1A). Incorporation of ^{32}P from [γ-^{32}P]ATP or GTP is

Table 2. *Characterization of the* $[\gamma\text{-}^{32}P]$ *NTP-dependent phosphorylation of the 19 kDa protein*

	$[\gamma\text{-}^{32}P]$ ATP	$[\gamma\text{-}^{32}P]$ GTP
K_m	8 nM	5 nM
V_{max} (uncorrected)^a (μg^{-1} min^{-1})	8.2 fmol	33 fmol
K_i ADP	40 nM	55 nM
K_i GDP	—	34 nM
Mg^{2+} (optimal concentration)	5 mM	1 mM
Mn^{2+} (optimal concentration)	2 mM	0.25 mM
$Ca^{2+\ b}$	no stimulation	
pH (Tricine) (optimum)	7.9	—

^{a}The number is probably greatly underestimated. As described in the text, the phosphate linkage is labile to alkaline and acidic pH as encountered during SDS–PAGE; for example, staining and destaining of the SDS–PAGE gel within 1.5–2 h resulted routinely in 70–80% loss of protein-bound ^{32}P. TCA precipitation and washing did not improve the recovery rates.
^{b}Ca^{2+} could meet the divalent cation requirements of the phosphorylation reaction in reaction-containing crude extract.

linear for only 10–30 s at 25 °C and 30–50 s at lower temperatures. The protein is, however, progressively deactivated by heat treatment: 5 min at 60 °C results in about 50% loss of activity, and boiling at 100 °C for 5 min completely abolishes the enzyme activity. The phosphoryl-group turnover is extremely rapid, as deduced from a pulse–chase experiment (Fig. 1B). If the protein was labelled in the presence of 8 nM $[\gamma\text{-}^{32}P]$ATP for 60 s and cold ATP (10 μM) was then added, 90% of the labelled phosphoryl groups in the protein turned over within 15 s (Fig. 1B).

Inhibition of the purified 19 kDa protein with 5'-*p*-fluorosulphonylbenzoyladenosine (FSBA) resulted in a time- and concentration-dependent inhibition of the phosphorylation reaction (Fig. 1C). FSBA covalently attaches to lysine residues and can also be used for affinity labelling. Inhibition of the 19 kDa protein by FSBA was strongly dependent on the presence of Mg^{2+} (Buhrow & Staros, 1985). Half-maximal inhibition in the presence of Mg^{2+} is obtained at 0.6 mM FSBA. Mg^{2+} seems clearly involved in a structural role, rather than simply as a cofactor or to balance P_i charges in ATP. P_1P_5 diadenosine pentaphosphate (Lienhard & Secemski, 1973) is also a strong inhibitor of the autophosphorylation reaction of the purified 19 kDa protein (Fig. 1D). Half-maximal inhibition is observed at around 30 μM effector.

Experiments using standard HCl hydrolysis and high-voltage electro-

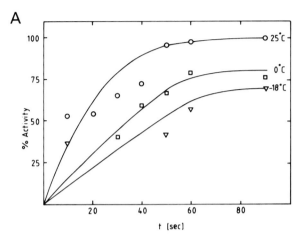

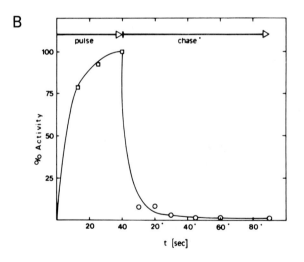

Fig. 1. Characterisation of the phosphorylation reaction of the purified 19 kDa protein. Reactions were done (if not stated otherwise) in 10 mM Tricine, pH 7.9, 5 mM MgCl$_2$, 30 nM [γ^{32}P]NTP for 60 s at room temperature. (A) Temperature dependence of the phosphorylation reaction. (B) The 19 kDa protein shows a very high phosphoryl group turnover. At the beginning of the chase period, 10 μM unlabelled NTP was added. (C) Fluorosulphonylbenzoyladenosine (FSBA) inhibits the

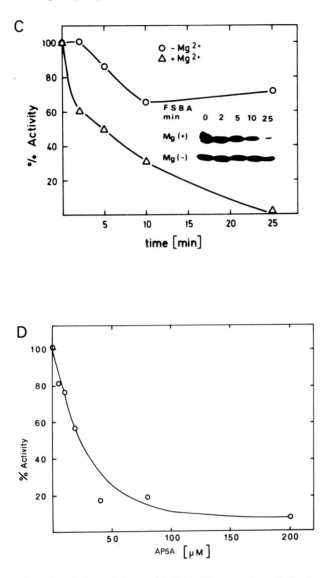

phosphorylation of the purified 19 kDa protein only in the presence of Mg^{2+}. The protein was pre-incubated with 2 mM FSBA in either the presence or the absence of 5 mM Mg^{2+} for the time points indicated. Subsequently the phosphorylation was carried out as above. The insert shows the autoradiograph that was quantified to give the graph. (D) Phosphorylation is inhibited by P_1P_5-diadenosine pentaphosphate (AP5A).

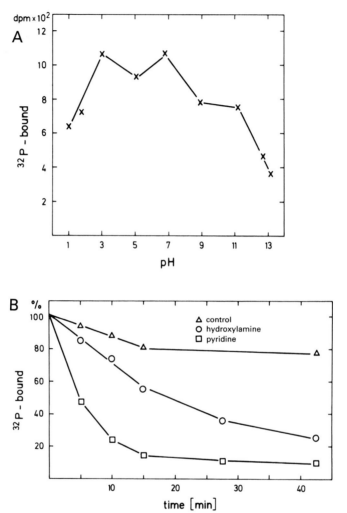

Fig. 2. Stability of the phosphate linkage in the 19 kDa protein under different conditions. Phosphorylated 19 kDa protein was spotted onto nitrocellulose filters and incubated in buffer solution of different pH (A) or washed in acetate buffer, pH 5.5, in the presence of 1 M hydroxylamine or pyridine, respectively (B).

phoresis (Hunter & Sefton, 1980) demonstrated that neither of the commonly phosphorylated amino acids such as serine, tyrosine or threonine became phosphorylated. The pH stability profile of the phosphate linkage demonstrated that the phosphate bond was labile to acidic conditions

Table 3. *Autophosphorylation activity of the purified 19 kDa protein is altered by amino acid-modifying agents*

At least five different substrate concentrations were used to determine this influence on the phosphorylation reaction.

Substrate	Concentration	Preincubation (min)	Inhibition (%)
DIDS[a]	50 μM	10	50
GSSG[b]	5 mM	5	70
DTT[c]	10 mM	30	*ca.* 0
NEM[d]	5 mM	30	80
ETSH[e]	20 mM	30	*ca.* 0
Pyridoxal phosphate	5 mM	20	75
Phenylglyoxal[f]	15 mM	20	*ca.* 0
Bis(sulphosuccinimydyl)suberate	0.55 mM	10	85

[a]4,4'-diisothiocyanostilbene-2,2'-disulphonic acid
[b]glutathione (oxidized)
[c]diothiothreitol
[d]N'-ethylmaleimide
[e]mercaptoethanol
[f]Purified enzyme fractions in Tricine buffer where extensively dialysed against phosphate buffer, pH 7.9, to avoid possible cross-reactions. P_i does not influence the enzyme activity.

(Fig. 2A), supporting our finding that none of the hydroxylated amino acids are phosphorylated. The label was, however, labile at alkaline pH, which excluded histidine as the phosphoryl group acceptor (DeLuca *et al.*, 1963; Post & Kume, 1973; Stelte & Witzel, 1986). Exposure of the phosphoprotein to hydroxylamine or pyridine buffered in acetate (Di Sabato & Jencks, 1961; Fujitaki & Smith 1984) demonstrated a concentration-dependent, base-catalysed enhancement of the hydrolysis rate (Fig. 2B). Pyridine (1 M) was much more effective than hydroxylamine at identical conditions. It is known from the literature (Fujitaki & Smith, 1984) that aspartyl- and glutamylphosphate are susceptible to hydroxylamine treatment in acetate buffer (pH 5.5). It was, however, clearly shown that neither aspartyl- nor glutamylphosphate was pyridine-labile (Di Sabato & Jencks, 1961). Using the model compound acetylphosphate, we verified the results and found also that acetylphosphate is not susceptible to pyridine treatment but only hydroxylamine-labile. These data tend to exclude aspartyl or glutamyl residues as phosphate acceptors. The phosphate linkage was not attacked by iodine treatment, indicating that cysteine residues are not phosphorylated. These data might indicate that lysine serves as a phosphoryl group acceptor.

The results described in Table 3 point also to an involvement of NH_2-groups, either as ^{32}P acceptors or as residues vital for enzymatic activity. Substances such as 4,4-diisothiocyanatostilbene-2,2'-disulphonic acid, pyridoxal phosphate or bis-(sulphosuccinimydyl)suberate, which are specific for free $-NH_2$ groups such as are found in lysines, severely inhibited the phosphorylation of the 19 kDa protein. The arginine-modifying agent phenylglyoxal was ineffective, indicating that arginines are not involved in the phosphorylation (Vanoni et al., 1987). At the moment it is unclear whether the treatments described above block a putative phosphorylation site or a site involved in catalytic activity. Reagents that oxidise $-SH$ groups, such as N-ethylmaleimide, glutathione (oxidised form) or 4,4'-diisothiocyanatostilbene-2,2'-disulphonic acid, resulted in pronounced inhibition (Table 3). Pre-incubation with mercaptoethanol or dithiothreitol, however, did not stimulate the enzymatic activity. Larger quantities of phosphorylated protein must be isolated and subjected to protein sequencing to identify unequivocally the phosphorylated amino acid residue. Experiments are under way to resolve this problem. It is, however, evident from the data that phosphorylation occurs at a very unusual site, which is probably covalently attached to the protein. Strong binding of $[\gamma^{-32}P]GTP$ to the protein, which would mimic a covalent modification by P_i transfer from the γ-position of NTP, was excluded by the following observations. (i) Labelled protein could be precipitated by TCA or acetone; tightly bound GTP would be separated by this treatment. (ii) Extraction of the labelled protein with acidified chloroform did not result in the recovery of label from the organic solvent phase or the water phase but was recovered together with the precipitated protein from the interface.

Selective inhibitor substrates for a number of different cellular ATPases were also without influence on the enzymatic activity (Soll et al., 1991); this observation supports the notion that the 19 kDa protein does not exhibit significant NTPase activity, or form a phosphorylated enzyme intermediate as found for ATPases (Sze et al., 1987; Serrano, 1988).

The 19 kDa protein is present in a high-molecular-mass form in vitro. The purified protein was subjected to non-denaturing gel electrophoresis (Smith & Ellis, 1979). One aliquot was electrophoresed after incubation with $[\gamma^{-32}P]ATP$; a second aliquot was incubated with $[\gamma^{-32}P]ATP$ in gelo after electrophoresis had been completed. In both cases ^{32}P-labelled bands were detected at an approximate molecular mass of 120–150 kDa (Fig. 3). This suggests that the 19 kDa protein is present as a high-molecular-mass complex in its purified form. These data also indicate that

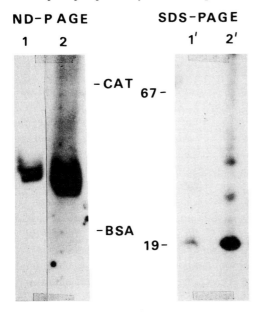

Fig. 3. The 19 kDa protein forms a high-molecular-mass oligomer. Purified 19 kDa protein was submitted to non-denaturing PAGE after phosphorylation (lane 1). Phosphorylation was carried out in gelo (lane 2). Radioactive bands localised by autoradiography were rerun on denaturing SDS–PAGE and again analysed by autoradiography.

^{32}P incorporation from [γ-^{32}P]ATP occurs by an autophosphorylation event. To characterise these labelled proteins further, the labelled bands were localised on the dried, non-denaturing PAGE by autoradiography, cut out, re-swollen in SDS sample buffer (Laemmli, 1970) and run on a denaturing SDS–PAGE to determine the monomeric size of these high-molecular-mass components. Upon this second electrophoresis, only the 19 kDa protein was labelled (Fig. 3).

We have used limited proteolysis employing endoprotease gluc C (Cleveland, 1983) to demonstrate that the 19 kDa phosphoprotein seen on SDS–PAGE is indeed identical to the 19 kDa phosphoprotein component of the 120–150 kDa protein from the non-denaturing gel. To do this, a gel slice from the non-denaturing gel containing a 210–230 kDa phosphoprotein was treated with endoprotease gluc C. In a second experiment, purified and labelled 19 kDa protein, which was run on SDS–PAGE only (cf. Fig. 3), was treated with endoprotease gluc C in the same manner. The degradation pattern of labelled proteins was identical, strongly indicating that the 19 kDa protein on SDS–PAGE is a monomer of the 120–150 kDa protein labelled in the non-denaturing gel.

Conclusions

Low K_m values, as found in our study, are not common for protein kinases, which usually have a K_m 100–1000-fold higher (Edelman, Blumenthal & Krebs, 1987). Our attempts to demonstrate phosphoryl group transfer from the phosphorylated 19 kDa protein after removal of unreacted ATP to soluble envelope or stromal proteins gave no indication for such a transfer reaction (J. Soll, unpublished). Proteins that exhibit very low K_m for their nucleoside triphosphate substrate belong to the G-protein family (Kaziro et al., 1991).

Differential labelling in intact and lysed chloroplasts (Soll et al., 1991) of this soluble or peripheral 19 kDa membrane protein suggests its localisation outside the stromal space, perhaps in the inter-envelope lumen or the outer envelope membrane. A location in the inter-envelope lumen would make the protein accessible to low-molecular-mass effectors, which can enter this space from the cytoplasm through large pore proteins present in the outer envelope (Flügge & Benz, 1984).

The suggestion that phosphorylation is influenced by low-molecular-mass metabolites or second messengers is discussed by Soll et al. (1991). Perhaps even more interestingly, in situ phosphorylation is stimulated by Ca^{2+} but this is not the case for the purified enzyme (Table 1). This observation might indicate that in vivo the 19 kDa protein is involved in a signal transduction chain operative between chloroplasts and cytosol, and is triggered by a second messenger such as Ca^{2+}. Clearly more work is needed to elucidate the interesting function of this protein in chloroplast metabolism.

Acknowledgements

The excellent technical assistance of A. Steidl and V. Berger is gratefully acknowledged. I thank Dr John Bennett for his continuous support and comments during the course of this work.

References

Bennett, J. (1991). Protein phosphorylation in green plant chloroplasts. *Annual Review of Plant Physiology and Plant Molecular Biology* **42**, 281–311.

Buhrow, S.A. & Staros, J.V. (1985). 5′-p-fluorosulfonylbenzoyl adenosine as a probe of ATP-binding sites in hormone receptor-associated kinases. *Methods in Enzymology* **109**, 817–26.

Cleveland, P.W. (1983). Peptide mapping in one dimension by limited proteolysis of sodium-dodecylsulfate solubilized proteins. *Methods in Enzymology* **96**, 222–9.

DeLuca, M., Ener, K.E., Hultquist, D.E., Kreil, G., Peter, J.B., Moyer, R.W. & Boyer, P.D. (1963). The isolation and identification of phosphohistidine from mitochondrial protein. *Biochemische Zeitschrift* **338**, 512–25.

Di Sabato, G. & Jencks, W.P. (1961). Mechanism and catalysis of reactions of acyl phosphates. I. Nucleophilic reactions. *Journal of the American Chemical Society* **83**, 4393–400.

Edelman, A.M., Blumenthal, O.K. & Krebs, E.G. (1987). Protein serine/threonine kinases. *Annual Review of Biochemistry* **56**, 567–613.

Flügge, U.I. & Benz, R. (1984). Pore forming activity in the outer membrane of the chloroplast envelope. *FEBS Letters* **169**, 85–9.

Fujitaki, J.M. & Smith, R.A. (1984). Techniques in the detection and characterization of phosphoramidate-containing proteins. *Methods in Enzymology* **107**, 23–36.

Hunter, T. & Sefton, B.M. (1980). Transforming gene product of Rous sarcoma virus phosphorylates tyrosine. *Proceedings of the National Academy of Sciences USA* **77**, 1311–15.

Kaziro, Y., Itoh, H., Kozasa, T., Nakafuku, M. & Satoh, T. (1991). Structure and function of signal-transducing GTP-binding proteins. *Annual Review of Biochemistry* **60**, 349–400.

Laemmli, U.K. (1970). Cleavage of structural proteins during the assembly of the head of bacteriophage T4. *Nature (London)* **227**, 680–5.

Lienhard, G.E. & Secemski, I.I. (1973). P¹, P⁵-Di(adenosine-5′) pentaphosphate, a potent multisubstrate inhibitor of adenylate kinase. *Journal of Biological Chemistry* **248**, 1121–3.

Post, R.L. & Kume, S. (1973). Evidence for an aspartyl phosphate residue at the active site of sodium and potassium ion transport adenosine triphosphatase. *Journal of Biological Chemistry* **248**, 6993–7000.

Ranjeva, R. & Boudet, A.M. (1987). Phosphorylation of proteins in plants: regulatory effects and potential involvement in stimulus response coupling. *Annual Review of Plant Physiology* **38**, 73–93.

Serrano, R. (1988). Structure and function of proton translocating ATPase in plasma membranes of plants and fungi. *Biochimica et Biophysica Acta* **947**, 1–28.

Smith, S.M. & Ellis, R.J. (1979). Processing of the small subunit of ribulose bisphosphate carboxylase and its assembly into whole enzyme are stromal events. *Nature (London)* **278**, 662–4.

Soll, J. & Bennett, J. (1988). Localization of a 64-kDa phosphoprotein in lumen between the outer and inner envelopes of pea chloroplasts. *European Journal of Biochemistry* **175**, 301–7.

Soll, J., Berger, V. & Bennett, J. (1989). Adenylate effects on protein phosphorylation in the interenvelope lumen of pea chloroplast. *Planta* **177**, 393–400.

Soll, J., Steidl, A. & Schröder, I. (1991). Dynamic phosphorylation of a small chloroplast protein exhibiting so far undescribed labelling properties. In *NATO ASI series*, vol. H 56, *Cellular regulation by protein phosphorylation* (ed. L.M.G. Heilmeyer, Jr), pp. 73–9. Berlin, Heidelberg: Springer.

Stelte, B. & Witzel, H. (1986). Formation of an aspartyl phosphate intermediate in the reactions of nucleoside phosphotransferase from carrots. *European Journal of Biochemistry* **155**, 121–4.

Sze, H., Randall, S.K., Kaestner, K.H. & Lai, S. (1987). Vacuolar H$^+$-ATPase from oat roots. In *Plant Membranes: Structure, Function, Biogenesis* (ed. C. Leaver & H. Sze), pp. 195–207. New York: Alan R. Liss.

Vanoni, M.A., Simonetta, M.P., Curti, B., Negri, A. & Ronchi, S. (1987). Phenylglyoxal modification of arginines in mammalian D-amino-acid oxidase. *European Journal of Biochemistry* **167**, 261–7.

Welch, W.J. & Feramisco, J.R. (1985). Rapid purification of mammalian 70,000-Dalton stress protein: Affinity of the proteins for nucleotides. *Molecular and Cellular Biology* **5**, 1229–37.

I.R. WHITE, I. ZAMRI, A. WISE
and P.A. MILLNER

Use of synthetic peptides to study G proteins and protein kinases within plant cells

Introduction

The recent advances in synthetic peptide technology have presented new opportunities to study plant signal transduction. In particular, the use of synthetic peptides has the advantage that relatively large quantities (10–100 mg) of pure site-specific reagents can be prepared with relative ease and can be utilised for a number of purposes. The most evident use is perhaps as synthetic antigens, where it is impossible or impracticable to isolate sufficient pure protein to permit immunisation. Such anti-peptide antibodies have proved extremely useful in identification of the plant homologues of their animal G protein (mainly Gα subunit) counterparts. However, there are other areas where synthetic peptides can be usefully employed. Important information concerning the substrate-specificity of protein kinases can be gained from observing the relative efficiency with which members of a series of peptides, each differing slightly in structure, can act as phosphate acceptors for the target kinase. Additionally, synthetic peptides can be used as direct mimics of protein structure in order to probe protein : protein interaction. For example, in our work we have utilised a peptide of moderate length (15 residues), whose sequence corresponds to a domain within the Gα subunit which is known to specify G-protein–receptor interaction, as an affinity ligand for receptor isolation. In other work, peptides that represent sections of the *petD* gene product have been used to delineate the domains of the petD protein, which may be important in LHC II kinase regulation. In this chapter, some of the above uses of synthetic peptides are described. In addition, consideration is given to the appropriate controls that must be performed and pitfalls that should be avoided if meaningful data are to be generated.

Society for Experimental Biology Seminar Series 53: *Post-translational modifications in plants*, ed. N.H. Battey, H.G. Dickinson & A.M. Hetherington. © Cambridge University Press 1993, pp. 91–108.

Preparation of synthetic peptides

Solid-phase peptide synthesis, initially introduced by Merrifield (1963) and which employed the acid-labile protecting group t-butoxycarbonyl (t-Boc), remained the mainstay of peptide synthetic chemistry for a number of years. However, more recently, milder and more rapid peptide synthesis protocols have emerged with the advent of the fluorenylmethoxycarbonyl (FMOC) protecting group and activation via pentafluorophenyl (OPFP) esters (Atherton & Sheppard, 1989). These developments, along with automation of the repetitive deprotection, coupling and washing cycles, have brought peptide synthesis within the remit of the non-specialist chemist and have enabled the use of synthetic peptides to become routine. However, even with these advances, peptide synthesis remains moderately time-consuming and relatively expensive; there are a number of points that can usefully be made concerning the design of peptides in order to maximise their usefulness. The most obvious consideration is perhaps the actual sequence chosen for synthesis; this will be considered in further detail within subsequent sections. A further consideration is the addition of extra residues, in addition to the native sequence. For example, if the sequence of interest lacks cysteine it is often useful to add this residue at the N- or C-terminus. This provides a reactive group (Glazer, De Lange & Sigman, 1982) for derivatisation of the peptide or to permit its coupling to carrier proteins or to chromatographic matrices (Nilsson & Mosbach, 1987). As peptides are synthesised in a C-terminal to N-terminal direction, when cysteine is to be included at the N-terminus it is also possible to remove part of the growing peptide after the penultimate coupling step and prepare peptide lacking the additional cysteine.

Anti-peptide antibodies

The most evident reason for preparation of antibodies directed against synthetic peptides is where the antigen cannot be isolated in sufficient quantity. However, a further advantage of such antibodies is that they are directed against a limited subset of epitopes presented by the synthetic peptide. In cases where a number of similar proteins, which may be troublesome to resolve completely, are present, for example G-protein subunits (Sternweiss & Pang, 1990) or where highly immunogenic contaminants may be present (Wistow *et al.*, 1986), this may be a distinct advantage.

The choice of peptide to be synthesised can be guided by various considerations, including its amino acid sequence, and the hydrophilicity, surface probability and segmental flexibility of the sequence within the protein. In taking such factors into account the use of secondary structural modelling software, such as the WISCONSIN suite of programmes (Devereux, Haeberli & Smithies, 1984) can be very useful in presenting parameters in graphical fashion. The WISCONSIN software suite also permits calculation of an antigenic index, which is effectively a composite of the above parameters.

An additional consideration is whether the antibody is to be directed against highly conserved regions of sequence, thus generating a pan-specific antibody, or against less conserved sequence, thus allowing identification of specific protein subclasses. Both the G protein (Fig. 1) (Mumby *et al.*, 1986; Goldsmith *et al.*, 1987) and protein kinase families of proteins (Hanks, Quinn & Hunter, 1988; Hagendorn, Tettelbach & Panella, 1990) possess domains that have been used in this fashion. Once synthesised, the peptide is usually conjugated to a carrier prior to immunisation of the experimental animal.

It is impossible here to cover the above points in greater depth; a fuller discussion of them can be found elsewhere (Doolittle, 1986; van Regenmortel *et al.*, 1988). However, in our experience to date, a number of operative 'rules' have emerged with respect to preparation of antisera for the identification of plant G proteins. First, although relatively small peptides (fewer than 10 residues) will act as antigens, slightly larger peptides (15–25 residues) have provided antisera of higher titre and with apparently greater specificity. This is perhaps unsurprising since a more substantially defined region of protein sequence was presented as the antigen. Second, although a wide variety of carriers (e.g. BSA, keyhole limpet haemocyanin) and coupling reagents (e.g. glutaraldehyde (Doolittle, 1986)) are available, we have found the strategy of Lachmann *et al.* (1986) to provide the cleanest antibodies. Briefly, this involved the use of BCG-hypersensitised animals and coupling of the peptide antigen to the carrier (PPD) in a defined orientation using the heterobifunctional cross-linker sSMCC. The principal advantage is that PPD has negligible anti-genicity, thus ensuring that antibodies were mainly directed against the target peptide. Finally, we have found that peptides which represented N- or C-terminal domains of proteins were usually better antigens than those corresponding to internal domains within the primary amino acid sequence.

```
GBA1_ARATH    MGLLC-------------------------------------------------------S
GBAS_BOVIN    MG--CLGNSKTEDQRNEEKAQREANKK---------------------------------
GBI1_BOVIN    M--------GCTLSAEDKAAVERSKM----------------------------------
GBA2_YEAST    MGL-CASSEKNGSTPDTQTASAGSDNVGKAKVPPKQEPQKTVRTVNTANQQEKQQQRQQQ
GBI2_RAT      M--------GCTVSAEDKAAAERSKM----------------------------------
GBT1_BOVIN    M--------GAGASAEEK----HSRE----------------------------------
GB01_BOVIN    M--------GCTLSAEERAALERSKA----------------------------------
GB11_MOUSE    MT---LESMMACCLSDEVKESKRINAE---------------------------------
GBQ_MOUSE     MT---LESIMACCLSEEAKEARRINDE---------------------------------
GBAZ_RAT      M--------GCRQSSEEKEAARRSRR----------------------------------
              *


GBA1_ARATH    RSRHHTEDTDEN-----------------TQAAEIERRIEQEAK-------------A
GBAS_BOVIN    ---------------------------------IEKQLQKDKQV--------------
GBI1_BOVIN    ---------------------------------IDRNLREDGEK--------------
GBA2_YEAST    PSPHNVKDRKEQNGSINNAISPTATANTSGSQQINIDSALRDRSSNVAAQPSLSDASSGS
GBI2_RAT      ---------------------------------IDKNLREDGEK--------------
GBT1_BOVIN    ---------------------------------LEKKLKEDAEK--------------
GB01_BOVIN    ---------------------------------IEKNLKEDGIS--------------
GB11_MOUSE    ---------------------------------IEKQLRRDKRD--------------
GBQ_MOUSE     ---------------------------------IERHVRRDKRD--------------
GBAZ_RAT      ---------------------------------IDRHLRSESQR--------------
                                                   ..  .


                                   A
GBA1_ARATH    EKHIRKI⌈LLLGAGESGKST⌉IFKQIKLLFQTGFDEGELKSYVPVIHANVY-QTIKLLHDGT
GBAS_BOVIN    YRATHRILLLGAGESGKSTIVKQMRILHVNGFNGEGGEEDPQAARSNSDGEKATKVQDIK
GBI1_BOVIN    AAREVKILLLGAGESGKSTIVKQMKIIHEAGYSEEECKQYKAV---------------VY
GBA2_YEAST    NDKELKVLLLGAGESGKSTVLQQLKILHQNGF●EQEIKEYIPL---------------IY
GBI2_RAT      AAREVKILLLGAGESGKSTIVKQMKIIHEDGYSEEECRQYRAV---------------VY
GBT1_BOVIN    DARTVKILLLGAGESGKSTIVKQMKIIHQDGYSLEECLEFIAI---------------IY
GB01_BOVIN    AAKDVKILLLGAGESGKSTIVKQMKIIHEDGFSGEDVKQYKPV---------------VY
GB11_MOUSE    ARRELKL⌊LLL⌈GTGE⌋SGKS⌋TFIKQMRIIHGAGYSEEDKRGFTKL---------------VY
GBQ_MOUSE     ARRELKILLLGTGESGKSTFIKQMRIIHGSGYSDEDKRGFTKL---------------VY
GBAZ_RAT      QRREIKL⌊LLL⌋GTSN⌊SGKS⌋TIVKQMKIIHSGGFNLEACKEYKPL---------------II
              ..****...*****.  .*....  *.


GBA1_ARATH    KEFAQNETDSAKYMLSSESIAIGEKLSEIGGRL---------DYP-RLTKDIAEGIETL
GBAS_BOVIN    NNLKEAIETIVAAM-SNLVPPVELANPENQFRVDYILSVMNVP--DFDFPPEFYEHAKAL
GBI1_BOVIN    SNTIQSIIAIIRAM-GRLKIDFGDSARADDARQLFVLAGAAEE---GFMTAELAGVIKRL
GBA2_YEAST    QNLLEIGRNLIQAR-TRFNVNLEPECELTQQDLSRTMSYEMPNNYTGQFPEDIAGVISTL
GBI2_RAT      SNTIQSIMAIVKAM-GNLQIDFADPQRADDARQLFALSCAAEE--QGMLPEDLSGVIKRL
GBT1_BOVIN    GNTLQSILAIVRAM-TTLNIQYGDSARQDDARKLMHMADTIEE---GTMPKEMSDIIQRL
GB01_BOVIN    SNTIQSLAAIVRAM-DTLGIEYGDKERKADAKMVCDVVSRMED--TEPFSPELLSAMMRL
GB11_MOUSE    QNIFTAMQAMVRAM-ETLKILYKYEQNKANALLIREV--DVEK--VTTFEHQYVNAIKTL
GBQ_MOUSE     QNIFTAMQAMIRAM-DTLKIPYKYEHNKAHAQLVREV--DVEK--VSAFENPYVDAIKSL
GBAZ_RAT      YNAIDSLTRIIRAL-AALKIDFHNPDRAYDAVQLFALTGPAES--KGEITPELLGVMRRL
              .                       . *  .*.     .   ** *..*.* *  *.*. .


GBA1_ARATH    WKDPAIQETC---ARGNELQVPDCTKYLMENLKRLSDINYIPTKEDVLYARVRTTGVVEI
GBAS_BOVIN    WEDEGVRA-CYE--RSNEYQLIDCAQYFLDKIDVIKQDDYVPSDQDLLRCRVLTSGIFET
GBI1_BOVIN    WKDSGVQA-CFN--RSREYQLNDSAAYYLNDLDRIAQPNYIPTQQDVLRTRVKTTGIVET
GBA2_YEAST    WALPSTQD-LVNGPNASKFYLMDSTPYFMENFTRITSPNYRPTQQDILRSRQMTSGIFDT
GBI2_RAT      WADHGVQA-CFG--RSREYQLNDSAAYYLNDLERIAQSDYIPTQQDVLRTRVKTTGIVET
GBT1_BOVIN    WKDSGIQA-CFD--RASEYQLNDSAGYYLSDLERLVTPGYVPTEQDVLRSRVKTTGIIET
GB01_BOVIN    WGDSGIQE-CFN--RSREYQLNDSAKYYLDSLDRIGAADYQPTEQDILRTRVKTTGIVET
GB11_MOUSE    WSDPGVQA-CYD--RRREFQLSDSAKYYLTDVDRIATVGYLPTQQDVLRVRVPTTGIIEY
GBQ_MOUSE     WNDPGIQE-CYD--RRREYQLSDSTKYYLNDLDRVADPSYLPTQQDVLRVRVPTTGIIEY
GBAZ_RAT      WADPGAQA-CFG--RSSEYHLEDNAAYYLNDLERIAAPDYIPTVEDILRSRDMTTGIVEN
              *   . .         .  *  .*   .     .  ** *..*.* *  *.*. .
```

```
GBA1_ARATH   QFSPVGENKKSGEVYRLFDVGGQRNERRKWIHLFEGVTAVIFCAAISEYDQTLFEDEQKN
GBAS_BOVIN   KFQ-V-DKVN----FHMFDVGGQRDERRKWIQCFNDVTAIIFVVASSSYNMVIREDNQTN
GBI1_BOVIN   HFT-F-KDLH----FKMFDVGGQRSERKKWIHCFEGVTAIIFCVALSDYDLVLAEDEEMN
GBA2_YEAST   VID-MGSDIK----MHIYDVGGQRSERKKWIHCFDNVTLVIFCVSLSEYDQTLMEDKNQN
GBI2_RAT     HFT-F-KDLH----FKMFDVGGQRSERKKWIHCFEGVTAIIFCVALSAYDLVLAEDEEMN
GBT1_BOVIN   QFS-F-KDLN----FRMFDVGGQRSERKKWIHCFEGVTCIIFIAALSAYDMVLVEDDEVN
GBO1_BOVIN   HFT-F-KNLH----FRLFDVGGQRSERKKWIHCFEDVTAIIFCVALSGYDQVLHEDETTN
GBB11_MOUSE  PFD-L-ENII----FRMVDVGGQRSERRKWIHCFENVTSIMFLVALSEYDQVLVESDNEN
GBQ_MOUSE    PFD-L-QSVI----FRMVDVGGQRSERRKWIHCFENVTSIMFLVALSEYDQVLVESDNEN
GBAZ_RAT     KFT-F-KELT----FKMVDVGGQRSERKKWIHCFEGVTAIIFCVELSGYDLKLYEDNQTS
              .           . ****** **.***.*. ** ..* * *. .*   .
```

```
GBA1_ARATH   RMMETKELFDWVLKQPCFEKTSFMLFLNKFDIFEKKVLDVPLNVCEWFRDYQ--------
GBAS_BOVIN   RLQEALNLFKSIWNNRWLRTISVILFLNKQDLLAEKVLAGKSKIEDYFPEFARYTTPEDA
GBI1_BOVIN   RMHESMKLFDSICNNKWFTDTSIILFLNKKDLFEEKIK--KSPLTICYPEYA--------
GBA2_YEAST   RFQESLVLFDNIVNSRWFARTSVVLFLNKIDLFAEKLR--KVPMENYFPDYT--------
GBI2_RAT     RMHESMKLFDSICNNKWFTDTSIILFLNKKDLFEEKIT--QSPLTICFPEYT--------
GBT1_BOVIN   RMHESLHLFNSICNHRYFATTSIVLFLNKKDVFSEKIK--KAHLSICFPDYN--------
GBO1_BOVIN   RMHESLMLFDSICNNKFFIDISIILFLNKKDLFGEKIK--KSPLTICFPEYT--------
GBB11_MOUSE  RMEESKALFRTIITYPWFQNSSVILFLNKKDLLEDKIL--HSHLVDYFPEFD--------
GBQ_MOUSE    RMEESKALFRTIITYPWFQNSSVILFLNKKDLLEEKIM--YSHLVDYFPEYD--------
GBAZ_RAT     RMAESLRLFDSICNNNWFINTSLILFLNKKDILSEKIR--RIPLSVCFPEYK--------
              * *. **    .     * .****** *.. *.
```

```
GBA1_ARATH   -PVSSGKQEIEHAYEFVKKKFEELYYQNTAPDRVDRVFKIYRTTALDQKLVKKTFKLVDE
GBAS_BOVIN   TPEPGEDPRVTRAKYFIRDEFLRI--STASGDGRHYCYPHF-TCAVDTENIRRVFNDCRD
GBI1_BOVIN   ----GSNTYEEAA-AYIQCQFEDL--NK-RKDTKEI-YTHF-TCATDTKNVQFVFDAVTD
GBA2_YEAST   ----GGSDINKAAK-YILWRFVQL--NRANLSI----YPHV-TQATDTSNIRLVFAAIKE
GBI2_RAT     ----GANKYDEAA-SYIQSKFEDL--NK-RKDTKEI-YTHF-TCATDTKNVQFVFDAVTD
GBT1_BOVIN   ----GPNTYEDAG-NYIKVQFLEL--NM-RRDVKEI-YSHM-TCATDTQNVKFVFDAVTD
GBO1_BOVIN   ----GSNTYEDAA-AYIQAQFES---KN-RSPNKEI-YCHM-TCATDTNNIQVVFDAVTD
GBB11_MOUSE  ----GPQRDAQAAREFILKMFVDL--NP-DSD-KII-YSHF-TCATDTENIRFVFAAVKD
GBQ_MOUSE    ----GPQRDAQAAREFILKMFVDL--NP-DSD-KII-YSHF-TCATDTENIRFVFAAVKD
GBAZ_RAT     ----GQNTYEEAA-VYIQRQFEDL--NR-NKETKEI-YSHF-TCATDTSNIQFVFDAVTD
                         *                .   * * *  .. *   .
```

```
GBA1_ARATH   TLRRRNLLEAGLL-
GBAS_BOVIN   IIQRMHLRQYELL-
GBI1_BOVIN   VIIKNNLKDCGLF-
GBA2_YEAST   TILENTLKDSGVLQ
GBI2_RAT     VIIKNNLKDCGLF-
GBT1_BOVIN   IIIKENLKDCGLF-
GBO1_BOVIN   IIIANNLRGCGLY-
GBB11_MOUSE  TILQLNLKEYNLV-
GBQ_MOUSE    TILQLNLKEYNLV-
GBAZ_RAT     VIIQNNLKYIGLC-
              .    *    .
```

Fig. 1. Alignment of selected Gα subunits. Highly conserved regions, where residues are identical, are boxed. A, 'α_COMMON' domain involved in binding the GTP γ-phosphate. The identities of the aligned sequences are as follows: GBA1_ARATH, *Arabidopsis thaliana* GPA1; GBAS_BOVIN, bovine Gαs; GBI1_BOVIN, bovine Gαi1; GBA2_YEAST, *Saccharomyces cerevisiae* Gα; GBI2_RAT, rat Gαi2; GBT1_BOVIN, bovine Gαt1; GBO1_BOVIN, bovine Gαo1; GBB11_MOUSE, mouse Gα11; GBQ_MOUSE, mouse Gαq; GBAZ_RAT, rat Gαz. Asterisks indicate amino acid identity; full points, amino acid similarity.

Immunological identification of plant G proteins

In many cases anti-peptide antisera have been utilised to identify proteins that have hitherto been known only from their derived (from DNA) amino acid sequences. In animal G proteins, antiserum raised against the highly conserved sequence –LLLGAGESGKST–, also termed the α_{COMMON} sequence and which is present within most Gα subunits, was used to identify Gα subunits in a number of tissues (Mumby *et al.*, 1986). Indeed, the Gα homologue encoded by the gene *GPA1* in *Arabidopsis thaliana* (Ma, Yanofsky & Meyerowitz, 1990) also possesses this motif. Peptides corresponding to other domains, e.g. the C-terminus of Giα,

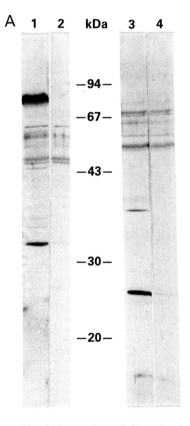

Fig. 2. Detection of Gα subunits in *Pisum sativum* microsomal membrane polypeptides separated by SDS–PAGE. Approximately 40 µg of protein, released by incubation of pea microsomal membranes in 1% (w/v) sodium cholate, were electrophoresed prior to transfer to Immobilon-

have also been utilised to prepare antisera specific to a particular G protein subtype (Goldsmith *et al.*, 1987). Applications of these antibodies to a variety of plant membrane preparations, including highly purified plasma membranes, have revealed a number of candidate Gα subunits (Blum *et al.*, 1988; Jacobs *et al.*, 1988). These have varied in apparent molecular mass from approximately 30 kDa to over 50 kDa, in line with the molecular masses observed for animal Gα subunits (Mumby *et al.*, 1986; Goldsmith *et al.*, 1987). Typical Western blot data using these antisera and solubilised *Pisum sativum* microsomal membranes are presented in Fig. 2; Table 1 summarises the principal polypeptides which in our experiments were shown to cross-react specifically with these

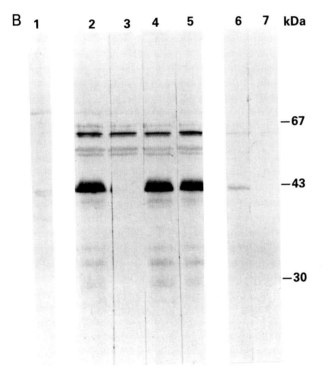

P membrane and immunolabelling. (A) Lanes 1 and 2, Gα$_{COMMON}$ antiserum at 1/100 dilution; lanes 3 and 4, Gαi antiserum at 1/120 dilution. The antibodies were incubated at 37 °C for 45 min in the absence (1,3) or presence (2,4) of the appropriate synthetic peptide before use. (B) Lane 1, preimmune IgG fraction at 1/200 dilution; lanes 2–5, Gα$_{ARA2}$ IgG fraction preincubated with the peptides indicated (2: no peptide; 3: Gα$_{ARA2}$; 4: pNDK; 5: Gα$_{ARA1}$); lanes 6 and 7, affinity-purified Gα$_{ARA2}$ IgG fraction (−peptide 6; +peptide 7).

Table 1. *Summary of polypeptides within plasma membranes derived from*
P. sativum *and* A. thaliana *that are specifically recognised by anti-peptide
antibodies directed against G protein sequences*

$G\alpha_{ARA2}$ and $G\alpha_{ARA1}$ represent C-terminal and internal sequence from the
Arabidopsis $G\alpha$ homologue, GPA1 (Ma *et al.*, 1990). The figures given are
molecular masses of the proteins in kilodaltons (kDa); (?), preliminary identification;
—, no cross-reacting component; n.d., not determined.

Antibody	*P. sativum*	*A. thaliana*
$G\alpha_{COMMON}$	33	—
	43	
$G\alpha i$	37	37
$G\alpha_{ARA2}$	43	43
	37	
$G\alpha_{ARA1}$	—	43
$G\alpha o$	36 (?)	—
$G\beta$	—	n.d.

peptide-directed antibodies to G protein sequences. The latter point is an
important one, established by showing that cross-reactivity could be
abolished (Fig. 2A,B) by pre-incubation of the antibody with excess
peptide to which it was directed. In some studies, this control was not
carried out and led to the misidentification of polypeptides as G proteins.
This artefactual immune labelling presumably arises because of the
presence of IgG directed towards part of the synthetic epitope, the carrier, or the junction between the two. The converse observation, i.e. that
cross-reactivity was not abolished by unrelated peptides, was also shown
(Fig. 2B, lanes 2,4 and 5). Finally, affinity purification of the antibodies,
using immobilised synthetic peptides, usually led to a substantial increase
in the specificity of cross-reaction (Fig. 2B, lanes 6 and 7); for example,
the approximately 60 kDa polypeptide labelled by unfractionated IgG
was not seen with affinity-purified IgG. However, a decrease in the effective titre of the antibody was usually found after the affinity purification
procedure.

It should be noted that this type of experiment, although important,
principally serves to demonstrate the presence of plant polypeptides that
possess identical, or substantially similar, regions of amino acid sequence
to those found in animal $G\alpha$ subunits. In particular, since the $G\alpha_{COMMON}$
domain is normally involved in binding the γ-phosphate of GTP, other
GTP-binding (but not necessarily signal-transducing) G proteins may also
contain this domain. Indeed, proteins of less than 30 kDa which are most

likely members of the monomeric 'small' G-protein family have been identified by us (not shown) and by others (Blum *et al.*, 1988). Large (ca. 90 kDa) proteins bearing this sequence have also been observed and are probably analogous to the large G proteins found in animal cells (Udrisar & Rodbell, 1990). The function of the latter proteins is presently unknown.

Synthetic peptides as substrates

Clearly, synthetic peptides can be used as substrates to aid the isolation and characterisation of both proteases and protein kinases. However, only the latter group of enzymes are considered here. Various proteins have been utilised as substrates for plant protein kinases, e.g. caseins, for kinases requiring acidic substrates (Erdmann, Bocher & Wagner, 1985; Muszynska, Dobrowolska & Ber, 1983; Dobrowolska, Ber & Muszynska, 1986) and histones, for those needing basic substrates (Coughlan & Hind, 1986; Elliot & Kokke, 1987; Schafer *et al.*, 1985). Although these proteins were often very effective substrates, their use revealed little about the substrate specificity of the protein kinases studied. Earlier, in work with animal protein kinases, the use of defined series of homologous synthetic peptides, in which the amino acid composition around the phosphoacceptor residue was systematically varied, proved fruitful in delineating the phosphorylation site requirements of the casein-kinase (Pinna *et al.*, 1984; Meggio *et al.*, 1984; Marin *et al.*, 1986), cAMP-dependent (Pinna *et al.*, 1984) and protein kinase C (Woodgett, Gould & Hunter, 1986) families. This approach has also been utilised more recently in delineating the substrate requirement of the thylakoid LHC kinase (Michel & Bennett, 1989; White *et al.*, 1990). From these studies, the principal requirements for phosphorylation were shown to be the presence of basic residues located N-terminally to the phosphoacceptor threonine residues (Fig. 3). Substitution of the lysine residue located N-terminal to the threonines in peptide S, which is equivalent to the N-terminus of LHC II, with the neutral lysine analogue norleucine (i.e. peptide N) or with the acidic residue aspartate (i.e. peptide A) severely affected the ability of the latter peptides to act as substrates. Conversely, when an additional basic residue was inserted (i.e. peptide B) the resulting peptide behaved as a much more effective substrate than peptide S. The effectiveness of peptides A, N, S and B in inhibiting phosphorylation of LHC II itself was directly correlated with their ability to act as substrates (White *et al.*, 1990), showing that they were competing with LHC II as substrates for the LHC kinase. These studies indicated that the substrate requirements of LHC kinase were more like those of the

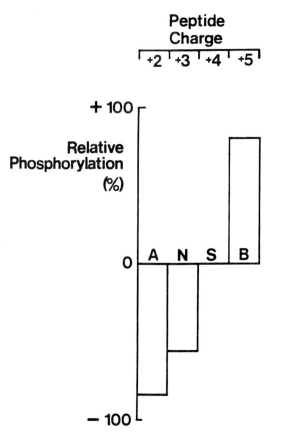

Fig. 3. Phosphorylation of synthetic peptide homologues of the LHC II N-terminus. Phosphorylation assays typically comprised a 50 μl total volume containing *P. sativum* thylakoids equivalent to 10 μg chlorophyll, 200 μM ATP, 3.7 kBq nmol^{-1} [γ-^{32}P]ATP and 100 μM of peptide A, N, S or B. Further assay details are given by White *et al.* (1990). The amino acid sequence of peptide S represents the native sequence; peptide B or peptides N and A are more basic or acidic peptides, respectively. Phosphorylation of peptides B, N and A is expressed relative to that of peptide S.

cAMP-dependent and PKC families of kinases than those of the casein kinases (see above). Other experiments with peptides of different lengths compared with peptide S indicated that the region C-terminal to the phosphoacceptor residues did not seem to be an important determinant of substrate effectiveness.

Peptides as site-directed reagents and affinity ligands

In addition to the roles indicated above, more recently synthetic peptides have been utilised in ways intended to mimic directly domains of signal-transducing proteins. Palm *et al.* (1989) found that synthetic peptides corresponding to first and second cytoplasmic loops of the turkey erythrocyte β-adrenergic receptor were able, at micromolar concentrations, to inhibit hormone-dependent adenylate cyclase activity. Conversely, a peptide corresponding to the C-terminal region of the third cytoplasmic loop of the same receptor directly stimulated (i.e. in the absence of hormone) both GTPγS binding and adenylate cyclase activity. Similar experiments with peptides corresponding to regions of the hamster β-adrenergic receptor (Cheung *et al.*, 1991) confirmed these data and indicated that the third cytoplasmic loop was important in governing recognition of the appropriate subclass of G protein by its respective receptor.

In our work (Zamri, White & Millner, 1991) this approach was applied to probe the interaction between LHC-kinase within tobacco thylakoids and the *petD* gene product. The LHC-kinase that phosphorylates LHC II (Bennett, 1983; Allen, 1992) is known to be activated in a redox-dependent manner. Early observations suggested that the kinase activity was regulated by the redox state of the electron carrier plastoquinone (Horton *et al.*, 1981; Millner *et al.*, 1982). However, more recent studies have implicated the cytochrome b_6f complex, which normally acts as a plastoquinol oxidase, in regulation of the LHC-kinase. Mutants of *Zea* and *Lemna* (Bennett, Shaw & Michel 1988; Coughlan, 1988) which possessed plastoquinone but lacked cytochrome b_6f were not able to phosphorylate LHC II. The same authors also showed that inhibitors of electron flow through the same complex inhibited phosphorylation. Finally, the co-isolation of cytochrome b_6f and LHC-kinase has also been demonstrated (Gal *et al.*, 1990).

In our experiments we were interested in the molecular mechanism of LHC-kinase activation, in particular the interaction of the kinase with the *petD* gene product. The latter protein is responsible for binding plastoquinol (Doyle *et al.*, 1989) and was predicted to possess both quinone reduction (Qi) and quinol oxidation (Qo) sites on the basis of sequence analysis and the use of mutants that were defective in cytochrome b_6f (or bc_1) complex operation (Hauska, Nitschke & Herrmann, 1988). Based on secondary structural predictions (Devereux *et al.*, 1984) using the tobacco petD sequence (Shinozaki *et al.*, 1986) and on the likely positions of residues important in the operation of the Qi and Qo sites, peptides were synthesised which were thought to form part of the Qi site (petD1 peptide), Qo site (petD2 and petD3 peptides) or neither site (petD4 peptide). The predicted folding pattern for the petD protein also indicated that petD1 and petD4 were on the same, probably stromal, side of the membrane. Fig. 4 shows the effects of these peptides on light-activated phosphorylation of LHC II. Of the peptides tested, petD1 and petD3 stimulated phosphorylation, by about 30% and 60% respectively, at 1 μM. By comparison, petD2 produced a two- to threefold stimulation of phosphorylation at the tenfold lower concentration of 0.1 μM. Finally, although at higher concentrations the stimulation by petD1, petD2 and petD3 was reduced, inhibition was only usually observed with petD4. In this case approximately 50% inhibition was produced by 100 μM petD4. Since petD4 was the only peptide of the petD series that possessed a potential phosphate acceptor residue, the possibility that petD4 could be acting as a competitive inhibitor of LHC-kinase activity was considered. Phosphorylation assays revealed that, under conditions where petD4 produced inhibition of LHC II phosphorylation, the petD4 peptide was itself phosphorylated.

In addition to their use to probe specific protein–protein interactions, synthetic peptides have also been employed as affinity reagents for the isolation and purification of signal transduction elements. Woodgett (1989) utilised synthetic peptides and phosphopeptides as affinity ligands for the chromatographic separation of glycogen synthase kinase-3 and casein kinase II from rabbit skeletal muscle; we have used immobilised peptide N (see Fig. 3) to resolve LHC-kinases from solubilised thylakoid membrane proteins (P.A. Millner *et al.*, unpublished data). The peptides were immobilised via amine or sulphydryl groups, respectively, using the appropriate activated Sepharose matrix. One difficulty in this approach, revealed by the latter attempts, was that peptide NJ also behaved as a non-specific anion exchanger, thus lowering its effectiveness as an affinity reagent. In addition, we have also used a fifteen-residue peptide corresponding in sequence to the C-terminus of the *Arabidopsis* Gα-homologue, GPA1 (Ma *et al.*, 1990) as affinity ligand to isolate a putative receptor protein which interacted with GPA1. The basis for this isolation strategy was that the C-terminal domain of Gα subunits has been shown by a number of studies to be of major importance in mediating receptor–G-protein interaction. From these studies emerged the observations that ADP-ribosylation of a cysteine residue close to the C-terminus of Gαi

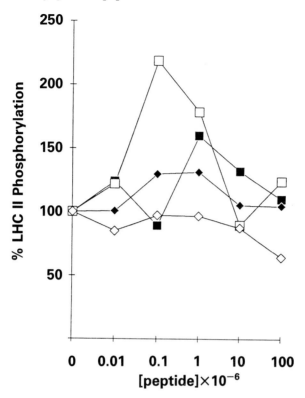

Fig. 4. Modulation of LHC-kinase activity in *Nicotiana tabacum* thylakoids by synthetic peptides which correspond to regions of protein petD. The sequence of the petD peptides is as follows: petD1, $_1$MGHNYYGEPAWPNDLLY$_{17}$; petD2, $_{38}$PSMIGEPADPFATPLE$_{53}$; petD3, $_{51}$PLEILPEWYFFPVF$_{64}$; petD4, $_{65}$QILRTVPNKLLGVLL$_{79}$. The assay conditions were essentially the same as for Fig. 3, except that the appropriate petD peptide indicated was present in the assay instead of peptides A, N, S or B. Following the incubation, the thylakoid membrane proteins were separated by SDS–PAGE and [^{32}P]-labelled proteins monitored by autoradiography and subsequent densitometry of the autoradiograph. Filled squares, petD1; open squares, petD2; filled diamonds, petD3; open diamonds, petD4.

blocked its ability to interact with its receptor (Ui, 1984; West *et al.*, 1985) as did mutations in this region of Gαs (Rall & Harris, 1987; Sullivan *et al.*, 1987).

Preparation of chimaeric Gα subunits comprising, for example, the N-terminal portion of Gαs and the C-terminal domain of Gαi, also indicated

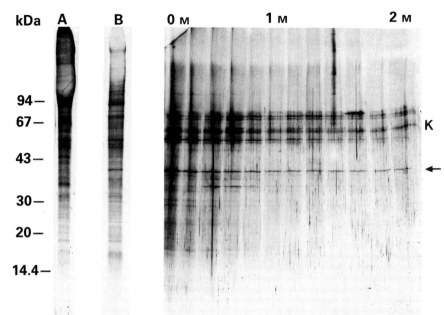

Fig. 5. Affinity chromatography of detergent-solubilised maize micro-somal membranes using immobilised peptide $G\alpha_{ARA2}$. Microsomal membranes from etiolated *Zea mays* seedlings were solubilised in 30 mM nonanoyl-N-methylglucamide prior to application to a 1 ml column containing the Ara2 peptide (DETLRRRNLLEAGLL). After washing to remove unbound material, bound polypeptides were eluted using a 0–2 M NaCl gradient. Samples of the membranes (A), detergent-solubilised polypeptides (B), and gradient fractions (0 M to 2 M) were separated by SDS–PAGE and the resulting gel stained with silver. The 37 kDa putative receptor is indicated. K, contaminating keratins.

that receptor–Gα coupling was dictated by this region (Kaziro *et al.*, 1991). In our experiments, detergent-solubilised maize microsomal membranes were applied to an affinity column containing immobilised GPA1 C-terminal peptide, followed by extensive washing and subsequent elution in a NaCl gradient. Under these conditions (Fig. 5) a 37 kDa polypeptide was observed to bind strongly to the column and to be retained on the column up to quite high NaCl concentrations. Protein microsequencing of this polypeptide revealed a short stretch of N-terminal amino acid sequence, which has subsequently been employed to design oligonucleotide primers for cloning and sequencing studies.

References

Allen, J.F. (1992). Protein phosphorylation in regulation of photosynthesis. *Biochimica et Biophysica Acta* **1098**, 275–335.

Atherton, E. & Sheppard, R.C. (1989). *Solid Phase Peptide Synthesis: A Practical Approach*. Oxford: IRL Press.

Bennett, J. (1983). Regulation of photosynthesis by reversible phosphorylation of the light harvesting chlorophyll a/b protein. *Biochemical Journal* **212**, 1–13.

Bennett, J., Shaw, E.K. & Michel, H. (1988). Cytochrome b_6f complex is required for phosphorylation of light-harvesting chlorophyll a/b complex II in chloroplast photosynthetic membranes. *European Journal of Biochemistry* **171**, 95–100.

Blum, W., Hinsch, K.-D., Schultz, G. & Weiler, E.W. (1988). Identification of GTP-binding proteins in the plasma membrane of higher plants. *Biochemical and Biophysical Research Communications* **156**, 954–8.

Cheung, A.H., Huang, R.-R.C., Graziano, M.P. & Strader, C.D. (1991). Specific activation of G_s by synthetic peptides corresponding to an intracellular loop of the β-adrenergic receptor. *FEBS Letters* **279**, 277–80.

Coughlan, S.J. (1988). Chloroplast thylakoid protein phosphorylation is influenced by mutations in the cytochrome bf complex. *Biochimica et Biophysica Acta* **933**, 413–22.

Coughlan, S.J. & Hind, G. (1986). Purification and characterization of a membrane-bound protein kinase from spinach thylakoids. *Journal of Biological Chemistry* **261**, 11378–85.

Devereux, J., Haeberli, P. & Smithies, O. (1984). A comprehensive set of sequence-analysis programs for the vax. *Nucleic Acids Research* **12**, 387–95.

Dobrowolska, G., Ber, E. & Muszynska, G. (1986). Separation and purification of maize casein kinases by affinity chromatography. *Journal of Chromatography* **376**, 421–7.

Doolitle, R.F. (1986). *Of URFS and ORFS – A Primer on How to Analyse Derived Amino Acid Sequences*. Oxford: University Science Books/Oxford University Press.

Doyle, M.P., Li, L.-B., Yu, L. & Yu, C.-A. (1989). Identification of a Mr= 17,000 Protein as the plastoquinone binding protein in the cytochrome b_6f complex from spinach chloroplasts. *Journal of Biological Chemistry* **264**, 1387–92.

Elliot, D.C. & Kokke, Y. (1987). Partial purification and properties of a protein kinase C type enzyme from plants. *Phytochemistry* **26**, 2929–35.

Erdmann, H., Bocher, M. & Wagner, K.G. (1985). The acidic peptide specific protein kinases from suspension-cultured tobacco cells: a comparison of the enzymes from whole cells and isolated nuclei. *Plant Science* **41**, 81–9.

Gal, A., Hauska, G., Herrmann, R. & Ohad, I. (1990). Interaction between light harvesting chlorophyll-a/b protein (LHC II) kinase and cytochrome b_6f complex. *Journal of Biological Chemistry* **265**, 19742–9.

Glazer, A.N., DeLange, R.J. & Sigman, D.S. (1982). Chemical modifications of proteins. In *Laboratory Techniques in Biochemistry and Molecular Biology* (ed. T.S. Work & E. Work), pp. 68–119. Amsterdam: Elsevier Biomedical Press.

Goldsmith, P., Gierschik, K., Milligan, G., Unson, C.G., Vinitsky, R., Malek, H.L. & Spiegel, A.M. (1987). Antibodies directed against synthetic peptides distinguish between GTP-binding proteins in neutrophil and brain. *Journal of Biological Chemistry* **262**, 14683–8.

Hagendorn, C.T., Tettelbach, W.H. & Panella, H.L. (1990). Development and characterization of polyclonal antibodies against a conserved sequence in the catalytic domain of protein kinases. *FEBS Letters* **264**, 59–62.

Hanks, S.K., Quinn, A.M. & Hunter, T. (1988). The protein kinase family: conserved features and deduced phylogeny of the catalytic domains. *Science* **241**, 42–52.

Hauska, G., Nitschke, W. & Herrmann, R.G. (1988). Amino acid identities in the three redox centre carrying polypeptides of cytochrome bc_1/b_6f complexes. *Journal of Bioenergetics and Biomembranes* **20**, 211–28.

Horton, P., Allen, J.F.A., Black, M.T. & Bennett, J. (1981). Regulation of phosphorylation of chloroplast membrane polypeptides by the redox state of plastoquinone. *FEBS Letters* **125**, 193–6.

Jacobs, M., Thelen, M.P., Farndale, R.W., Astle, M.C. & Rubery, P.H. (1988). Specific quanine nucleotide binding by membranes from *Cucurbita pepo* seedlings. *Biochemical and Biophysical Research Communications* **155**, 1478–84.

Kaziro, Y., Itoh, H., Kozassa, T., Nakafuku, M. & Satoh, T. (1991). Structure and function of signal transducing GTP-binding proteins. *Annual Review of Biochemistry* **60**, 349–400.

Lachmann, P.J., Strangeways, L., Vyarkarnam, A. & Evans, G. (1986). Raising antibodies by coupling peptides to PPD and immunizing BCG-sensitized animals. In *Synthetic Peptides as Antigens* (Ciba Foundation Symposium 119) (ed. R. Porter & J. Whelan), pp. 25–7. Chichester: John Wiley.

Ma, H., Yanofsky, M.Y. & Meyerowitz, E.M. (1990). Molecular cloning and characterisation of *GPA1*, a G protein subunit from *Arabidopsis thaliana*. *Proceedings of the National Academy of Sciences USA* **87**, 3821–5.

Marin, O., Meggio, F., Marchiori, F., Borin, G. & Pinna, L.A. (1986).

Site specificity of casein kinase-2 (TS) from rat liver cytosol. *European Journal of Biochemistry* **160**, 239–44.

Meggio, F., Marchiori, F., Borin, G., Chessa, G. & Pinna, L. (1984). Synthetic peptides including acidic clusters as substrates and inhibitors of rat liver casein kinase TS (Type-2). *Journal of Biological Chemistry* **259**, 14576–9.

Merrifield, R.B. (1963). Solid phase peptide synthesis. I. The synthesis of a tetrapeptide. *Journal of the American Chemical Society* **85**, 2149–54.

Michel, H. & Bennett, J. (1989). Use of synthetic peptides to study substrate specificity of a thylakoid protein kinase. *FEBS Letters* **254**, 165–70.

Millner, P.A., Widger, W.R., Abbott, M.S., Cramer, W.A. & Dilley, R.A. (1982). The effect of adenine nucleotides on inhibition of the thylakoid protein kinase by sulfhydryl directed reagents. *Journal of Biological Chemistry* **257**, 1736–42.

Mumby, S.M., Kahn, R.A., Manning, D.R. & Gilman, A.G. (1986). Antisera of designed specificity for subunits of guanine nucleotide-binding regulatory proteins. *Proceedings of the National Academy of Sciences USA* **83**, 265–9.

Muszynska, G., Dobrowolska, G. & Ber, E. (1983). Polypeptides from maize seedling with protein kinase functions. *Biochimica et Biophysica Acta* **757**, 315–23.

Nilsson, K. & Mosbach, K. (1987). Tresyl chloride activated supports for enzyme immobilization. *Methods in Enzymology* **135**, 65–78.

Palm, D., Munch, G., Dees, C. & Hekman, M. (1989). Mapping of β-adrenoceptor coupling domains to G_s-protein by site specific synthetic peptides. *FEBS Letters* **254**, 89–93.

Pinna, L.A., Meggio, F., Marchiori, F. & Borin, G. (1984). Opposite and mutually incompatible structural requirements of type-2 casein kinase and cAMP-dependent protein kinase as visualised with synthetic peptide substrates. *FEBS Letters* **171**, 211–14.

Rall, T. & Harris, B.A. (1987). Identification of the lesion in the stimulatory GTP-binding protein of the uncoupled S49 lyphoma. *FEBS Letters* **224**, 365–71.

van Regenmortel, M.H.V., Briand, J.P., Muller, S. & Plaue, S. (1988). *Synthetic Peptides as Antigens.* (*Laboratory Techniques in Biochemistry and Molecular Biology*, vol. 19.) Amsterdam: Elsevier.

Schafer, A., Bygrave, F., Matzenauer, S. & Marmé, D. (1985). Identification of a calcium and phospholipid-dependent protein kinase in plant tissue. *FEBS Letters* **187**, 25–8.

Shinozaki, K., Ohne, M., Tanaka, M., Wakasugi, T., Hayashida, N., Matsubayashi, T., Zaita, N., Chunwongse, J., Obokata, J., Yamaguchi-Shinozaki, K., Ohto, C., Torozawa, K., Meng, B.Y., Sugita, M., Deno, H., Kamogashira, T., Yamada, K., Kususda, J., Takaiwa,

F., Kato, A., Tohdoh, N., Shimada, H. & Sugiura, M. (1986). The complete nucleotide-sequence of the tobacco chloroplast genome – its gene organization and expression. *EMBO Journal* **5**, 2043–9.

Sternweiss, P.C. & Pang, I.H. (1990). Preparation of G proteins and their subunits. In *Receptor-Effector Coupling: A Practical Approach* (ed. E.C. Hulme), pp. 1–30. Oxford: IRL Press.

Sullivan, K.A., Miller, R.T., Masters, S.B., Beiderman, B. & Heidman, W. (1987). Identification of receptor contact site involved in receptor-G protein coupling. *Nature (London)* **330**, 758–60.

Udrisar, D. & Rodbell, M. (1990). Microsomal and cytosolic fractions of guinea pig hepatocytes contain 100-kilodalton GTP-binding proteins reactive with antisera against α-subunits of stimulatory and inhibitory heterotrimeric GTP-binding proteins. *Proceedings of the National Academy of Sciences USA* **87**, 6321–5.

Ui, M. (1984). Islet-activating protein, pertussis toxin – a probe for the functions of the inhibitory guanine-nucleotide regulatory component of adenylate cyclase. *Trends in Pharmacological Sciences* **5**, 277–9.

West, R.E., Moss, J., Vaughan, M., Liu, T. & Liu, T.-Y. (1985) Pertussis toxin-catalysed ADP-ribosylation of transducin – cysteine 347 is the ADP-ribose acceptor site. *Journal of Biological Chemistry* **260**, 4428–30.

White, I.R., O'Donnell, P.J., Keen, J.N., Findlay, J.B.C. & Millner, P.A. (1990). Investigation of the substrate specificity of thylakoid protein kinase using synthetic peptides. *FEBS Letters* **269**, 49–52.

Wistow, G.J., Katial, A., Cheryl, C. & Shinohara, T. (1986). Sequence analysis of retinal S-antigen. *FEBS Letters* **196**, 23–8.

Woodgett, J.R. (1989). Use of peptide substrates for affinity purification of protein serine kinases. *Analytical Biochemistry* **180**, 237–41.

Woodgett, J.R., Gould, K.L. & Hunter, T. (1986). Substrate specificity of protein kinase C. *European Journal of Biochemistry* **161**, 177–84.

Zamri, I., White, I.R. & Millner, P.A. (1991). Sensitivity of protein phosphorylation in tobacco thylakoids to protein kinase C inhibitors and synthetic peptides. *Biochemical Society Transactions* **20**, 8S.

G.F.E. SCHERER, A. FÜHR and M. SCHÜTTE

Activation of membrane-associated protein kinase by lipids, its substrates, and its function in signal transduction

Biochemical signal transduction is a term used for chains, or rather for networks, of reactions induced by chemical or physical signals. Chemicals as signals in plants may originate internally, for instance hormones or other compounds acting at low concentrations, or externally, for instance by cell–cell interactions (Scherer, 1990*a*). Typical physiological signals (besides hormones) in plants are gravity, light or touch. Usually, signal transduction reactions are triggered by a conformational change induced by binding of the signalling molecule or by other induced physical changes to a receptor structure, which are in turn transduced by conformational changes in proteins interacting with the receptor structure. A typical reaction chain in eukaryotic signal transduction leads from a membrane-bound receptor to interactions with G proteins, which may activate or inhibit second-messenger-generating enzymes (Gilman, 1984). The enzymes known to generate second messengers in animal cells are phospholipase C, phospholipase A_2 and phospholipase D, all of which generate lipid breakdown products as second messengers, and adenylate cyclase and guanylate cyclase, which generate cAMP and cGMP, respectively. Cytosolic Ca^{2+} ions also function as second messengers, the concentration of which is regulated by several processes, including the action of other second messengers. Hence, it is more correct to envisage signal transduction in plants as a network rather than as a linear chain of reactions.

Second messengers often activate protein kinases specifically, and these in turn can regulate enzymatic activities by regulatory phosphorylation, leading eventually to cellular responses (Ranjeva & Boudet, 1987). Reversibility of the responses is ensured by protein phosphatases but also by reversibility of other reactions in the signal transduction chain.

Society for Experimental Biology Seminar Series 53: *Post-translational modifications in plants*, ed. N.H. Battey, H.G. Dickinson & A.M. Hetherington. © Cambridge University Press 1993, pp. 109–121.

Many well-studied examples of signal transduction chains in animal cells are known, but for plants no general schemes exist. However, homologous and analogous examples for all categories of molecules or enzymes known to function in animal signal transduction have been described. Receptors are known for the hormones auxin (Klämbt, 1990; Napier & Venis, 1991), abscisic acid (Hornberg & Weiler, 1984) and gibberellic acid (Hooley, Beale & Smith, 1991); for red light (Bruce *et al.*, 1989) and for oligosaccharides (Cosio *et al.*, 1990), to mention only some. An α-subunit of a G protein completely homologous to animal proteins has been sequenced (Ma *et al.*, 1990). Cytosolic Ca^{2+} ions are recognised to have a second-messenger function (Hepler & Wayne, 1985), and a calcium-dependent protein kinase has been sequenced (Harper *et al.*, 1991) as have other protein kinases (Lawton *et al.*, 1989). The role of protein phosphorylation in the regulation of some cellular activities has been established (Ranjeva & Boudet, 1987). Once we accept as a general framework that the pieces of the puzzle are indicative of a more complete picture, even though it is borrowed from concepts in animal signal transduction, the argument can be turned around, and the conclusion drawn that any ligand-activated protein kinase from a plant source is likely to be a second-messenger-activated protein kinase. This also means that such an activator or ligand for a protein kinase is a potential second messenger for plants (Scherer, 1989). This argument is even more persuasive if the activator is generated by a known typical second-messenger-generating enzyme or functions in a reaction chain involving a receptor or a G protein or both.

Using these criteria, we hypothesised that the lysophospholipid/Ca^{2+}-activated protein kinase described by us is a second-messenger-activated protein kinase (Scherer, André & Martiny-Baron, 1990). The activator lysophosphatidylcholine is generated by an auxin-activated phospholipase A_2 by a reaction that involves (i) an auxin-binding protein acting as a receptor and a G-nucleotide-sensitive site, possibly a G protein, and (ii) lysophospholipids to activate a protein kinase. The similar phospholipid platelet-activating factor may serve as a substitute second messenger for plants (Scherer, Martiny-Baron & Stoffel, 1988; Scherer & Nickel, 1988; Nickel *et al.*, 1991) even though it is a genuine hormone with completely different functions in animals (Hanahan, 1986).

To verify the hypothesis that the lysophospholipid/Ca^{2+}-activated protein kinase has a function in plant signal transduction, we used physiological and biochemical approaches. In physiological experiments the molecules of the proposed signal transduction chain should exert characteristic effects if applied exogenously to plant cells; in biochemical experiments purification of these proteins should prove valuable in determining their further properties. Such biochemical knowledge, in turn, should help to define new physiological experiments.

In physiological experiments we applied platelet-activating factor as a substitute second messenger to cultured soybean cells, which responded specifically to this lipid with an increased proton excretion (Scherer & Nickel, 1988; Nickel *et al.*, 1991). Platelet-activating-factor-induced proton secretion is dependent on external calcium ions, which may either be needed as an unspecific charge balance or have a more specific role, perhaps influencing the cytosolic calcium level. Since potassium ions, which usually serve as a charge balance for proton efflux in plant cells, do not increase the proton efflux, a more specific role of calcium is indicated (Nickel *et al.*, 1991).

Another physiological approach was to apply auxin to cultured soybean cells pre-labelled with [^{32}P]phosphate and to monitor changes in protein phosphorylation (M. Schütte and G. Scherer, unpublished). After treatment, cells were quickly homogenised under conditions that preserved the phosphorylation status of proteins. Membrane-associated and soluble proteins were separated by centrifugation; both preparations exhibited totally different phosphorylation patterns after sodium dodecyl sulphate polyacrylamide gel electrophoresis (SDS–PAGE) and autoradiography. Notably, one membrane-associated protein with an apparent molecular mass of 65 kDa was consistently more strongly labelled even after a very brief hormone treatment of 5 minutes (Fig. 1). In isolated membranes from cultured soybean cells and from zucchini hypocotyls it was shown that several protein kinases were present, one of which had an apparent molecular mass of about 65 kDa. The presence of membrane-associated protein kinases of a very similar range of molecular mass in zucchini microsomes was shown by a renaturation assay in combination with phosphorylation in the gel after SDS–PAGE (Fig. 2). In this type of assay, proteins do not diffuse, so the renatured kinase can only phosphorylate proteins in its vicinity. This means that autophosphorylation is measured; this test seems to be a consistent property of all protein kinases (Blowers & Trewavas, 1987; Harmon, Putnam-Evans & Cormier, 1987; Soderling, 1990). Since autophosphorylation leads to kinase activation, the observed rapid and auxin-induced phosphorylation in soybean cells could be linked to activation of a 65 kDa membrane-associated protein kinase. Rapid phosphorylation has also been found in systems where elicitors were the chemical signals (Dietrich, Mayer & Hahlbrock, 1990; Farmer, Moloshok & Ryan, 1991; Felix *et al.*, 1991).

Signal-induced autophosphorylation resulting in an activated protein kinase, at present, must be regarded as a working model. In order to test this working model for our system we want to develop tools to detect the

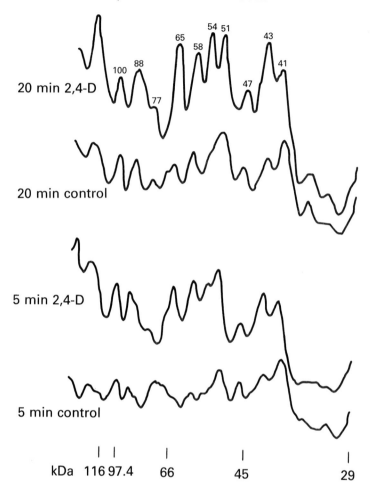

Fig. 1. Effect of the auxin 2,4-dichlorophenoxyacetic acid on membrane protein phosphorylation in cultured soybean cells. Cells were pre-labelled with [³²P]phosphate and then treated for either 5 or 20 min with the hormone (0·5 mM 2,4-D). Cells were killed, and soluble and microsomal membrane proteins separated by centrifugation and subjected to SDS–PAGE. Densitometric scans of the autoradiograms are shown. Several prominent phosphoproteins showed increased incorporation of label after hormone treatment, among which a 65 kDa phosphoprotein was most obvious.

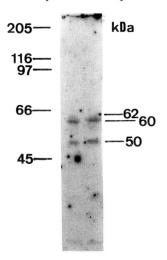

Fig. 2. Autophosphorylation of microsomal membrane proteins from zucchini hypocotyl as an indication for membrane-associated protein kinases. Microsomes were first separated according to their molecular mass by SDS–PAGE; the detergents were then washed out and the proteins allowed to renature Kinase activity was tested by the addition of [^{32}P]ATP; after additional washing, an autoradiogram from the gels was made. Five bands are visible, a doublet at 62/60 kDa, a 50 kDa, a 29 kDa, a 25 kDa and a 20 kDa phosphoprotein, all indicating kinase activity in the gel, probably autophosphorylation of protein kinases or active subunits thereof.

activity of the presumed protein kinase and perhaps even modifications of its regulatory state. A suitable tool is a specific assay for the lysophospholipid/Ca^{2+}-activated protein kinase and/or specific substrates for this enzyme and, later, specific antibodies. From this it is evident that studies of the isolated kinase are needed, as are the eventual purification and elucidation of its molecular properties. At present, we are studying the properties of the lysophospholipid/Ca^{2+}-activated protein kinase in isolated membranes (Martiny-Baron & Scherer, 1989), and initial attempts to solubilise the enzyme are being conducted.

One problem inherent in a lysophospholipid-activated protein kinase is that lysophospholipids are detergents. Hence, the distinction between detergent activation of a membrane-bound protein kinase and a more specific type of activation must be made. A detergent, by binding to the membrane, may simply expose more substrate hydroxyl groups for phosphorylation rather than directly activate the kinase. This may enhance phosphorylation but it would be mimicking genuine kinase activation by ligand binding.

We first investigated the lipid-specificity of activation by comparing different lysophospholipids and enantiomers of the chemically similar platelet-activating factor (Martiny-Baron & Scherer, 1989; Scherer, 1990*b*). It was found that the zwitterionic choline headgroup and a lysophospholipid-like lipid backbone are a requirement for activation of a membrane-associated protein kinase. More recently, it was found that the acidic lysophospholipids, lysophosphatidylglycerol, lysophosphatidylinositol and lysophosphatidic acid, but not lysophosphatidylserine, also weakly activated the membrane-associated protein kinase (A. Führ & G.F.E. Scherer, unpublished observation). Remarkably, none of the lipids known to activate protein kinase C had any activity in this assay, neither alone nor in combination (Martiny-Baron & Scherer, 1989; Scherer, 1990*b*).

When detergents as potential activators were tested in the membrane protein phosphorylation assay, they all activated to some extent the phosphorylation of the same proteins as did platelet-activating factor or lysophosphatidylcholine but at much higher concentrations than these lipids (Figs 3 and 4). In our assay, the phosphorylation of the most strongly phosphorylated proteins in zucchini microsomes (55–62 kDa) was quantified as the percentage of the total phosphorylation of all proteins. For a given concentration of platelet-activating factor this parameter was a constant value. Therefore, this value of relative phosphorylation can be used for comparison of different activators. In parallel, the solubilisation efficiency of the detergents was determined by centrifugation of the membranes for 45 min at 56000 *g* after detergent treatment. From these experiments it was evident that only platelet-activating factor and lysophosphatidylcholine activate phosphorylation better than they solubilise (Fig. 3). All other detergents tested solubilise membranes better at low concentrations than they activate protein phosphorylation of the 55–62 kDa proteins (Fig. 4). The best activating detergent was zwittergent 3–14, which, however, was roughly one order of magnitude weaker than platelet-activating factor (compare Fig. 3A with Fig. 4B). Another relatively potent detergent was Triton X-100, which contains ether bonds (Fig. 4A). Perhaps a zwitterionic head group and the lysolipid-like backbone of the zwittergents could be a structural requirement in detergents, leading them to be good activators of the membrane-associated protein kinase. The ether linkage in some detergents may also favour activation, as it is present also in the zwitterionic platelet-activating factor, which is more active than the plant lipid lysophosphatidylcholine (Fig. 3). Similarly, the alkane-ether detergents

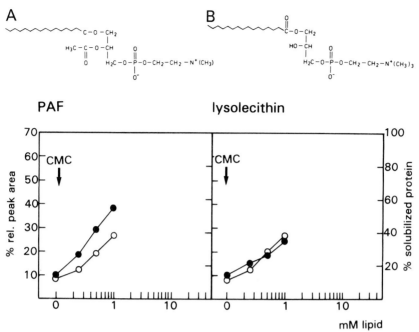

Fig. 3. Comparison of relative phosphorylation activation and of relative solubilisation by detergents. Shown are (A) platelet-activating factor (PAF) and (B) lysophosphatidylcholine. The structural formulae of the two lipids are given above the panels. For the determination of solubilisation (open circles) microsomes from zucchini hypocotyls (0.5 mg ml^{-1} protein) were treated with the lipids or detergents at the concentrations indicated. The arrows indicate the critical micelle concentrations (cmc). Membranes were centrifuged for 45 min at 36000 g, and protein was determined in the pellet and supernatant. Solubilisation is expressed as the percentage of protein in the supernatant. For the determination of relative phosphorylation (filled circles) microsomes (0.5 mg ml^{-1} protein) were phosphorylated and the relative phosphorylation was determined by laser-scan densitometry (Martiny-Baron & Scherer, 1989). In order to standardise the relative activities of different microsome preparations and of different exposures of autoradiograms, the phosphorylation of a group of three strongly phosphorylated proteins at 55–62 kDa was determined relative to the total phosphorylation of all phosphoproteins in the gel lane. These three phosphoproteins responded most strongly to lipids and detergents in phosphorylation assays.

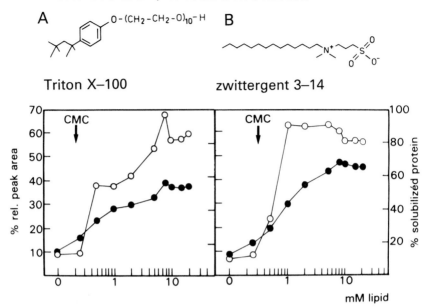

Fig. 4. Comparison of relative phosphorylation (filled circles) and relative solubilisation (open circles) of zucchini microsomal proteins by Triton X-100 (A) and by zwittergent 3–14 (B). The structural formulae of the detergents are given above the panels. Experimental procedures were as explained in the legend to Fig. 3. The arrows indicate the critical micelle concentrations (cmc).

thesit and isothesit were also good activators, as were the zwitterionic CHAPSO and CHAPS. However, the steran backbone was less effective than the lysolipid-like structure of the zwittergents. Rather weak activators were dodecylmaltoside and octylglucoside. Since these two detergents are mild and non-ionic, they seem to be ideal for the solubilisation of the lysophospholipid/Ca^{2+}-activated protein kinase from zucchini.

When the concentrations for half-maximal solubilisation and half-maximal activation of phosphorylation are compared, no obvious correlation exists for all detergents that were used. This may indicate that the activation was not due to increased substrate accessibility for the radioactive ATP by the permeabilisation of the membrane vesicles, but rather to activation by binding of detergents to the membranes. Again, platelet-activating factor and lysophosphatidylcholine were good activators at low concentrations, where hardly any solubilisation by them was observed (Fig. 3).

Solubilisation of the protein kinase was also determined by measuring phosphorylation activity in the supernatant and pellets after detergent treatment. Phosphorylation was determined by quantification of the 55–62 kDa phosphoprotein in autoradiograms after SDS–PAGE of supernatants and pellets (in both of which this group of phosphoproteins was found). With dodecylmaltoside as a detergent, 0.2 mM PAF activated protein phosphorylation in the presence of detergent. Since this detergent had the least activating properties it should be possible to use this experiment as a basis for the purification of the enzyme. It also demonstrated clearly the superior activating properties of PAF compared with detergent.

Another approach to studying the function of the lysophospholipid/Ca^{2+}-activated-protein kinase was to identify substrates of this kinase. We have identified two substrate proteins of the lysophospholipid/Ca^{2+}-activated protein kinase, the B subunit of tonoplast H^+-ATPase (Scherer, 1990b; Martiny-Baron et al., 1992) and the plasma membrane H^+-ATPase (Nickel et al., 1991). The latter substrate was first found as a 120 kDa phosphoprotein in plasma-membrane-enriched fractions (Martiny-Baron & Scherer, 1989) and later found to comigrate with a Western blot signal by an antibody against the plasma membrane H^+-ATPase. Usually, the molecular mass of this enzyme is found to be 100 kDa (Serrano, 1989) so that we were at first misled (Martiny-Baron & Scherer, 1989). It was known, however, that the plasma membrane H^+-ATPase was phosphorylated by protein kinase (Schaller & Sussman, 1988). Furthermore, PAF and lysophosphatidylcholine activate plasma membrane H^+-ATPase in isolated vesicles (Martiny-Baron & Scherer, 1989; Palmgren & Sommarin, 1989) so that regulatory phosphorylation may be involved in the activation mechanism. Several mechanisms for the activation of plasma membrane H^+-ATPase have been proposed, one of which is phosphorylation (Palmgren, 1991).

Since auxin is known to activate the plasma membrane H^+-ATPase and proton secretion by plant cells and tissues, it is an attractive hypothesis that auxin-stimulated phospholipase A_2 generates potential second messengers, lysophospholipids (André & Scherer, 1991; Scherer & André, 1989), which activate lysophospholipid/Ca^{2+}-activated protein kinase, which in turn may activate plasma membrane H^+-ATPase by phosphorylation (Scherer et al., 1988; Martiny-Baron & Scherer, 1989) so that proton secretion is stimulated (Scherer & Nickel, 1988; Nickel et al., 1991). Since both major H^+-ATPases of plant cells are phosphorylated in a lysophospholipid-dependent manner, and since they are known to be enzymes of central importance to all solute transport in plant cells (Sze, 1985), we have good reasons to believe that the lysophospholipid/Ca^{2+}-

activated protein kinase has important roles in regulation of cellular functions.

One question remains: why should a lipid-activated protein kinase in plants be different from the animal lipid-activated protein kinase C? The animal protein kinase C is activated by diacylglycerol, phosphatidylserine and Ca^{2+} ions (Nishizuka, 1986). Even though in some reports activation of plant protein kinase by diacylglycerol or by phosphatidylserine was described, this was not verified. In our test system these lipids had no effect (Martiny-Baron & Scherer, 1989; Scherer, 1990b). One answer could be that we are investigating a homologous enzyme with a different activator spectrum. Protein kinase C is also activated by short-chain phosphatidylcholines (Walker & Sando, 1988) or by lysophosphatidyl-choline (Oishi et al., 1988) or by cis-fatty acids (Murakami, Chan & Routtenberg, 1986) even though this seems to have no physiological function in animals. Hence, it is conceivable that a homologous plant enzyme exists but has evolved a different activator spectrum.

If one assumes that the lipid activator spectrum of a plant protein kinase could be different from that of animal protein kinase C, this has the obvious consequence that phospholipase C does not generate the activator for it but could instead be a phospholipase A_2, as we assume in our model. Another, less obvious consequence of our model would be that this may have as yet unknown, unprecedented roles in membrane structure. Diacylglycerol is very effectively down-regulated in the plasma membrane of animal cells (Berridge, 1987). It is a regulatory lipid in the plasma membrane but, at the same time, a precursor in lipid biosynthesis in the endoplasmic reticulum. This compartmentation may allow this dual function of the same lipid. We know very little about metabolic links and compartmentation of lysophospholipids in plants. Certainly, they are minor constituents of plant membranes. Plants have requirements quite different from those of animals for the biophysical functioning of their membranes as lipid bilayers. They are capable of tolerating rapid changes in the ambient temperature without losing the bilayer properties, which are recognised as essential for membrane function in all organisms. It will be interesting to see whether lysophospholipids as potential second messengers reflect specific biophysical properties of these lipids necessary for plant membrane structure.

Acknowledgements

This work was supported by a grant from the Bundesministerium für Forschung und Technologie to G.F.E.S.

References

André, B. & Scherer, G.F.E. (1991). Stimulation by auxin of phospholipase A in membrane vesicles from an auxin-sensitive tissue is mediated by an auxin receptor. *Planta* **185**, 209–14.

Berridge, M.J. (1987). Inositol trisphosphate and diacylglycerol: two interacting second messengers. *Annual Review of Biochemistry* **56**, 159–63.

Blowers, D.P. & Trewavas, A.J. (1987). Autophosphorylation of plasma membrane-bound calcium calmodulin-dependent protein kinase from pea seedlings and modification of catalytic activity by autophosphorylation. *Biochemical and Biophysical Research Communications* **143**, 691–6.

Bruce, W.B., Christensen, A.H., Klein, T., Fromm, M. & Quail, P.M. (1989). Photoregulation of a phytochrome gene promoter from oat transferred into rice by particle bombardment. *Proceedings of the National Academy of Sciences USA* **86**, 9692–6.

Cosio, E.G., Frey, T., Verduyn, R., van Boom, J. & Ebel, J. (1990). High-affinity binding of a synthetic heptaglucoside and fungal glucan phytoalexin elicitors to soybean membranes. *FEBS Letters* **271**, 223–6.

Dietrich, A., Mayer, J.E. & Hahlbrock, K. (1990). Fungal elicitor triggers rapid, transient, and specific phosphorylation in parsley cell suspension cultures. *Journal of Biological Chemistry* **265**, 6030–8.

Farmer, E.E., Moloshok, T.D. & Ryan, C.A. (1991). Oligosaccharide signaling in plants. Specificty of oligosaccharide-enhanced plasma membrane protein phosphorylation. *Journal of Biological Chemistry* **266**, 3140–5.

Felix, G., Grosskopf, D.G., Regenass, M. & Boller, T. (1991). Rapid changes of protein phosphorylation are involved in transduction of the elicitor signal in plant cells. *Proceedings of the National Academy of Sciences USA* **88**, 8831–4.

Gilman, A.G. (1984). G proteins and dual control of guanylate cyclase. *Cell* **36**, 577–9.

Hanahan, D. (1986). Platelet-activating factor: a biologically active phosphoglyceride. *Annual Review of Biochemistry* **55**, 483–509.

Harmon, A.C., Putnam-Evans, C. & Cormier, M.J. (1987). A calcium-dependent but calmodulin-independent protein kinase from soybean. *Plant Physiology* **83**, 830–7.

Harper, J.F., Sussman, M.R., Schaller, G.E., Putnam-Evans, C., Charbonneau, H. & Harmon, A.C. (1991). A calcium-dependent protein kinase with a regulatory domain similar to calmodulin. *Science* **252**, 951–4.

Hepler, P.K. & Wayne, R.O. (1985). Calcium and plant development. *Annual Review of Plant Physiology* **36**, 397–439.

Hooley, R., Beale, M.H. & Smith, S.J. (1991). Gibberellin perception at the plasma membrane of *Avena fatua* aleurone protoplasts. *Planta* **183**, 274–80.

Hornberg, C. & Weiler, E.W. (1984). High-affinity binding sites for abscisic acid on the plasmalemma of *Vicia faba* guard cells. *Nature (London)* **310**, 312–24.

Klämbt, D. (1990). A view about the function of auxin-binding proteins at the plasma membranes. *Plant Molecular Biology* **14**, 1045–50.

Lawton, M.A., Yamamoto, R.T., Hanks, S.K. & Lamb, C.J. (1989). Molecular cloning of plant transcripts encoding protein kinase homologs. *Proceedings of the National Academy of Sciences USA* **86**, 3140–4.

Ma, H., Yanovsky, M.F., & Meyerowitz, E.M. (1990). Molecular cloning and characterization of *GAP1*, a protein similar to a G protein α subunit gene from *Arabidopsis thaliana*. *Proceedings of the National Academy of Sciences USA* **87**, 3821–5.

Martiny-Baron, G., Manolson, M.F., Poole, R.J., Hecker, D. & Scherer, G.F.E. (1992). Proton transport and phosphorylation of tonoplast polypeptides from zucchini are stimulated by the phospholipid platelet-activating factor. *Plant Physiology*, in press.

Martiny-Baron, G. & Scherer, G.F.E. (1989). Phospholipid-stimulated protein kinase in plants. *Journal of Biological Chemistry* **264**, 18052–9.

Murakami, K., Chan, S.J. & Routtenberg, A. (1986). Protein kinase C activation by *cis*-fatty acids in the absence of Ca^{2+} and phospholipids. *Journal of Biological Chemistry* **261**, 15424–9.

Napier, R.M. & Venis, M.A. (1991). From auxin-binding protein to plant hormone receptor? *Trends in Biochemical Sciences* **16**, 72–5.

Nickel, R., Schütte, M., Hecker, D. & Scherer, G.F.E. (1991). The phospholipid platelet-activating factor stimulates proton extrusion in cultured soybean cells and protein phosphorylation and ATPase activity in plasma membranes. *Journal of Plant Physiology* **139**, 205–11.

Nishizuka, Y. (1986). Studies and perspectives of protein kinase C. *Science* **223**, 305–12.

Oishi, K., Raynor, R.L. Charp, P.A. & Ko, J.F. (1988). Regulation of protein kinase C by lysophospholipids. Potential role in signal transduction. *Journal of Biological Chemistry* **263**, 6865–71.

Palmgren, M.G. (1991). Regulation of plant plasma membrane H^+-ATPase activity. *Physiologia Plantarum* **83**, 314–23.

Palmgren, M.G. & Sommarin, M. (1989). Lysophosphatidylcholine stimulates ATP-dependent proton accumulation in isolated oat root plasma membrane vesicles. *Plant Physiology* **90**, 1009–14.

Ranjeva, R. & Boudet, A.M. (1987). Phosphorylation of proteins in plants: regulatory effects and potential involvements in stimulus/response coupling. *Annual Review of Plant Physiology* **38**, 73–93.

Schaller, G.E. & Sussman, M.G. (1988). Phosphorylation of the plasma membrane H$^+$-ATPase of oat roots by calcium-stimulated protein kinase. *Planta* **173**, 509–18.

Scherer, G.F.E. (1989). Ether phospholipid platelet-activating factor (PAF) and proton-transport-activating phospholipid (PAP): potential new signal transduction constituents for plants. In *Second Messengers in Plant Growth and Development* (ed. W.F. Boss & D.J. Morré), pp. 167–79. New York: Alan R. Liss.

Scherer, G.F.E. (1990*a*). Phospholipase A$_2$ and phospholipid-activated protein kinase in plant signal transduction. In *Hormone Perception and Signal Transduction in Animals and Plants* (ed. J. Roberts, C. Kirk & M. Venis), pp. 257–70. Cambridge, UK: The Company of Biologists.

Scherer, G.F.E. (1990*b*). Phospholipid-activated protein kinase in plants: coupled to phospholipase A$_2$? In *Signal Perception and Transduction in Higher Plants* (NATO-ASI Ser. H, vol. 47) (ed. R. Ranjeva & A.M. Boudet), pp. 69–82. Berlin: Springer.

Scherer, G.F.E. & André, B. (1989). A rapid response to a plant hormone: auxin stimulates phospholipase A$_2$ *in vivo* and *in vitro*. *Biochemical and Biophysical Research Communications* **163**, 111–17.

Scherer, G.F.E., André, B. & Martiny-Baron, G. (1990). Hormone-activated phospholipase A$_2$ and lysophospholipid-activated protein kinase: a new signal transduction chain and a new second messenger system in plants? *Current Topics in Plant Biochemistry and Physiology* **9**, 190–218.

Scherer, G.F.E., Martiny-Baron, G. & Stoffel, B. (1988). A new set of regulatory molecules in plants: a plant phospholipid similar to platelet-activating factor stimulates protein kinase and proton-translocating ATPase in membrane vesicles. *Planta* **175**, 241–53.

Scherer, G.F.E. & Nickel, R. (1988). The animal ether phospholipid platelet-activating factor stimulates acidification of the incubation medium by cultured soybean cells. *Plant Cell Reports* **7**, 575–8.

Serrano, R. (1989). Structure and function of plasma membrane ATPase. *Annual Review of Plant Physiology and Plant Molecular Biology* **40**, 61–94.

Soderling, T. (1990). Protein kinases. Regulation by autinhibitory domains. *Journal of Biological Chemistry* **265**, 1823–6.

Sze, H. (1985). H$^+$-translocating ATPases: advances using membrane vesicles. *Annual Review of Plant Physiology* **36**, 175–208.

Walker, J.M. & Sando, J.J. (1988). Activation of protein kinase C by short chain phosphatidylcholines. *Journal of Biological Chemistry* **263**, 4537–40.

N.H. BATTEY, S.M. RITCHIE
and H.D. BLACKBOURN

Distribution and function of Ca²⁺-dependent, calmodulin-independent protein kinases

I: Review of previous work

Background to plant Ca²⁺-dependent, calmodulin-independent protein kinases (Cpks)

A brief review of Ca^{2+}-regulated protein kinases in plants is useful because it places in context the smaller group of CPKs we are concerned with here. It also demonstrates that there is now considerable descriptive information available for CPKs, but that this precedes a clear understanding of their function(s) in plant cells.

By the beginning of the 1980s, researchers working on signal transduction in animals had described two main types of protein kinases, whose activity was controlled by cAMP and Ca^{2+}/calmodulin (Cohen, 1982). Therefore the first papers describing Ca^{2+}-regulated protein kinases from plants looked with particular interest for a role for calmodulin (Hetherington & Trewavas, 1982; Polya & Davies, 1982). During this time the evidence was accumulating for protein kinase C in animal cells (Takai et al., 1977, 1979; Nishizuka, 1984). There then followed several papers in which the focus of interest was the role of phospholipid and diacylglycerol in controlling Ca^{2+}-dependent protein kinase activity in plant cells (Schäfer et al., 1985; Muto & Shimogawara, 1985; Elliott & Skinner, 1986). However, there were anomalies in the work so far described; for example, the relatively high concentrations of calmodulin needed for activation effects, and interaction of calmodulin with the histone substrate (see Polya et al., 1990, for discussion); and the general absence of phorbol ester effects on putative protein kinase C activities in plant cells (see Hetherington, Battey & Millner, 1990).

Descriptions of Ca^{2+}-dependent, calmodulin (and phospholipid)-independent protein kinases (CPKs) then followed (Harmon, Putnam-

Society for Experimental Biology Seminar Series 53: *Post-translational modifications in plants*, ed. N.H. Battey, H.G. Dickinson & A.M. Hetherington. © Cambridge University Press 1993, pp. 123–147.

Table 1. *Characteristic properties of Ca²⁺-dependent, calmodulin-independent protein kinase*

Plant	Ca²⁺-dependent interaction on phenyl Sepharose	Inhibited by calmodulin antagonists: IC_{50} (µM)		Ca²⁺-dependent mobility shift: M_r of autophosphorylated kinase ($\times 10^{-3}$)		Native M_r (gel filtration) ($\times 10^{-3}$)	References
		TFP	W7	+ Ca²⁺	+ EGTA		
Soybean	✓	n.d.	110	43	46[a]	52	Harmon et al., 1987; Putnam-Evans et al., 1990
Apple	✓	45	15	45	56	n.d.	Battey & Venis, 1988b
Alfalfa	n.d.	n.d.	n.d.	50	56[b]	55	Bogre et al., 1988

[a] A band not showing a mobility shift was also detected at M_r = 50 000.
[b] Main autophosphorylated band.
n.d., Not determined.

Table 2. Species and tissue types containing Ca^{2+}-dependent, calmodulin-independent protein kinases (to 1992)

Plant	Tissue	Fraction			Reference
		Membrane	Soluble	Nuclear	
Soybean	Cell suspension		✓		Harmon et al., 1987
					Putnam-Evans et al., 1990
					Harper et al., 1991
Apple	Fruit	✓			Battey & Venis, 1988a, b
Alfalfa	Embryo		✓		Bogre et al., 1988
					Olah et al., 1989
Mougeotia			✓		Roberts, 1989
Wheat	Leaf		✓		Polya et al., 1990
Maize	Coleoptiles	✓	✓		Battey, 1990a, b
Dunaliella			✓		Guo & Roux, 1990[a]
Pea	Plumules			✓	Li et al., 1991
Soybean	Root nodules	✓	✓		Weaver et al., 1991
Oat	Roots	✓			Schaller et al., 1992
Very similar enzymes in which calmodulin-dependence was not specifically investigated					
Silver beet	Leaf	✓	✓		Klucis & Polya, 1987, 1988
					Polya et al., 1987
Wheat	Embryo		✓		Lucantoni & Polya, 1987
					Polya et al., 1989
Oat	Leaf	✓	✓		Minichiello et al., 1989
Barley	Leaf		✓		Klimczak & Hind, 1990

[a]Some significant differences from other CPKs; no Ca^{2+}-dependent mobility shift on SDS–PAGE.

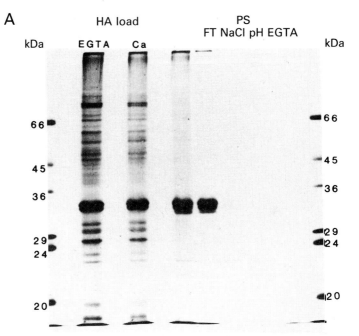

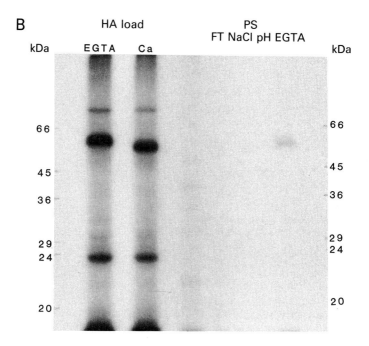

Evans & Cormier, 1987; Battey & Venis, 1988b; Bogre, Olah & Dudits, 1988). As the name suggests, these enzymes all appeared to be directly activated by Ca^{2+}, but also had a number of other common properties (Table 1). Subsequent papers have described similar (although not in all respects identical) CPKs from various plant tissues (Table 2). Of particular interest from the point of view of regulation were the effects of fatty acids on the Ca^{2+}-dependent protein kinases described by Polya and co-workers (see Polya & Chandra, 1990); and the characteristic Ca^{2+}-dependent mobility shift on SDS–PAGE described by others (see Fig. 1, for example), a property typical of calmodulin and other 'E-F hand' Ca^{2+}-binding proteins (Klee & Vanaman, 1982). Recently, the sequenced gene for CPK from soybean was shown to contain a region encoding a calmodulin-like (E-F hand) domain directly linked to the catalytic region (Harper *et al.*, 1991). This similarity in structure between CPK and the Ca^{2+}-dependent protease calpain (Ohno *et al.*, 1984) confirms the prediction of Battey & Venis (1988b).

This brief review suggests that CPKs, or related enzymes with the key property of direct Ca^{2+}-activation, are very widely distributed across a range of tissue types and plant species (Table 2). However, the most important question about these enzymes concerns their functions *in vivo*. Two types of evidence bear on this: the identity of endogenous substrates for CPKs, and the intracellular distribution of CPKs. These two topics are considered in more detail in the next section.

Proposed functions of CPKs

Evidence from protein substrates tested *in vitro*
Assays usually use histone H1, but although this is a good CPK substrate *in vitro* there is little positive evidence that it is a physiological substrate *in vivo*. Of the other proteins studied, casein, phosvitin and BSA can all be used to varying degrees by CPKs from different sources (see Polya & Chandra, 1990). Chicken gizzard myosin light chains approach histone as a substrate for soybean CPK (Putnam-Evans, Harmon & Cormier, 1990); a peptide with the sequence of the myosin light chain kinase site on the myosin light chains was used to identify and purify a CPK from *Mou-*

Fig. 1. Maize CPK (56–59 kDa), co-purifying with annexins (33 and 35 kDa), is separated by phenyl Sepharose chromatography. (A) SDS–PAGE of ^{32}P-labelled protein pre-hydroxyapatite (HA load), in the presence of EGTA or Ca^{2+}; and from the flowthrough (FT), 0.5 M NaCl (NaCl), pH 6.8 – 8.5 (pH) and EGTA at pH 8.5 (EGTA) fractions of a phenyl Sepharose (PS) column. (B) Autoradiograph of the same gel.

geotia and other algae (Roberts, 1989). A peptide based on the COOH-terminal sequence of nodulin-26 was phosphorylated *in vitro* by a soybean CPK, as was a symbiosome membrane protein of 28 kDa (Weaver *et al.*, 1991). The CPKs from wheat, silver beet and oats prefer synthetic substrate peptides with a Basic–X–X–Ser(Thr) motif; however, Basic–Basic–X–Ser(Thr), found for example in Kemptide (Kemp & Pearson, 1990), is not significantly phosphorylated (Polya, Morrice & Wettenhall, 1989; Polya & Chandra, 1990). Alfalfa CPK phosphorylated a peptide with the Ser–Phe–Lys motif thought to be specific for mammalian protein kinase C (Olah *et al.*, 1989). The phosphorylation of low molecular mass (10–12 kDa) basic proteins from wheat, barley and radish seeds has also been reported (Polya & Chandra, 1990). Some of these latter proteins are calmodulin antagonists.

Evidence from substrates phosphorylated *in vivo*
The use of [^{32}P]phosphate to label soybean nitrogen-fixing nodules led to the isolation of a phosphoprotein of 27 kDa, which cross-reacted with antibodies raised to the nodulin-26 peptide previously identified as an *in vitro* CPK substrate (Weaver *et al.*, 1991). This provides very persuasive evidence that the nodule protein is a substrate for CPK, and is the only evidence of this type available for plant CPK substrates.

Evidence from intracellular localisation
CPK activities have been extracted from both soluble and membrane (or particulate) fractions (Table 2). CPKs of very similar properties have been obtained from the soluble and particulate fractions of maize (Battey, 1990*b*) and soybean (Weaver *et al.*, 1991), while a Ca^{2+}-dependent protein kinase of M_r 37000 is also present in both fractions in barley (Klimczak & Hind, 1990). The Ca^{2+}-dependent protein kinases from silver beet and oat are localised to the plasma membrane (Klucis & Polya, 1988; Schaller, Harmon & Sussman, 1992), but membrane location has not been established in other work. The evidence from sub-cellular fractionation studies thus suggests that CPK may function in both soluble and membrane fractions, but there have been surprisingly few studies of the distribution of CPK *substrates* in the different cell fractions. Additionally, in pea nuclei a higher-M_r CPK (90000) has been reported to phosphorylate a polypeptide of M_r 43000, which may be identical with a red-light-stimulated phosphoprotein reported prevously (Li, Dauwalder & Roux, 1991).

Monoclonal antibodies to soybean CPK have been used to immunolocalise CPKs in onion root cells, *Tradescantia* pollen tubes (Putnam-Evans *et al.*, 1989) and in *Chara* internodal cells (Harmon & McCurdy, 1990). In onion and *Tradescantia*, CPK co-localises with actin, yet CPK does not bind (rabbit) F-actin in co-sedimentation assays (Putnam-Evans *et al.*, 1989). In *Chara*, CPK immunolocalises with actin cables *and* with small organelles and endoplasm, a staining pattern similar to that of myosin in *Chara* (Grolig *et al.*, 1988). This evidence, along with that for phosphorylation of myosin light chains by CPKs (see above), has led to the suggestion that CPK activation by increased cytosolic free Ca^{2+} leads to myosin phosphorylation and the subsequent inhibition of cytoplasmic streaming (Harmon & McCurdy, 1990). Paradoxically, in animals Ca^{2+}-dependent phosphorylation of myosin light chains *activates* the actin-dependent myosin ATPase (Adelstein & Eisenberg, 1980); however, *Physarum* actin-dependent myosin ATPase is inhibited by Ca^{2+}, and phosphorylation appears to be required for myosin sensitivity to Ca^{2+} (Kohama & Kendrick-Jones, 1986). Further work is needed to clarify the role of CPK in the regulation of actin–myosin interactions. The evidence reviewed in this section would suggest, however, that this is unlikely to be the only function of CPKs in plant cells.

II: Evidence from maize CPK

The preceding section has highlighted two important areas for our understanding of CPK function: first, the intracellular location of these enzymes, and the factors controlling this; and secondly, the physiological substrates of CPKs. In this section we discuss evidence for maize CPK on these two areas.

Intracellular location of CPKs

Ca^{2+}-dependent binding to lipids

Phenyl Sepharose chromatography can be used to purify Ca^{2+}-dependent protein kinase activity from the soluble fraction of dark-grown maize coleoptiles (Battey & Venis, 1992). The advantage of this method is that it separates calmodulin (which elutes from the phenyl Sepharose as Ca^{2+} is chelated with EGTA) from CPK (most of which elutes subsequently as the pH is raised from 6.8 to 8.5). A very similar CPK can be eluted from a maize microsomal membrane fraction with EGTA; this can then be chromatographed on phenyl Sepharose, eluting both as the pH is raised and as Ca^{2+} is chelated by EGTA at higher pH (Battey, 1990*b*). Neither of these CPK activities is activated by calmodulin, but they typically show 2–4-fold Ca^{2+} activation after phenyl Sepharose chromatography.

Confirmation of a maize CPK whose membrane association appears to be regulated by Ca^{2+} comes from the co-purification of CPK activity during the preparation of Ca^{2+}-dependent phospholipid-binding proteins (annexins) from maize. The method used involves tissue extraction in the presence of EDTA, the addition of excess Ca^{2+} and bovine brain lipid, centrifugation, and finally, elution of membrane/lipid-associated proteins with EDTA (see Blackbourn, Walker & Battey, 1991, for details of method). After overnight dialysis, the annexin fraction was incubated with 10 mM $MgCl_2$, 0.5 mM $CaCl_2$, 0.2 mM EGTA, 50 mM Hepes (pH 7), and carrier-free [γ-^{32}P]ATP (s.a. 3000 Ci mmol^{-1}; 5 μCi per assay; [ATP] = 2 nM), for 5 min at room temperature. TCA precipitation and SDS–PAGE in the presence of Ca^{2+} or EGTA were then carried out as described by Battey & Venis (1992). The distinctive Ca^{2+}-dependent mobility shift clearly demonstrated the presence of CPK (Fig. 1A,B).

These data imply that this CPK shows a Ca^{2+}-dependent association with membranes. However, it was necessary to test this critically using a well-characterised liposome-binding assay (Boustead, Walker & Geisow, 1988). After ^{32}P-labelling of the CPK as described above, the annexin/CPK fraction was incubated with liposomes, at a range of pCa at pH 6 or 7.4 (as described in Blackbourn et al., 1991). This was followed by centrifugation and separation of equal proportions of pellet and supernatant fractions by SDS–PAGE (Fig. 2A). Autoradiography revealed the distribution of CPK under these conditions (Fig. 2B). Clearly, the CPK associates only with liposomes containing acidic phospholipid, at the lower pH of 6, and there is a Ca^{2+}-dependent element to this association.

The striking similarity of this result to the binding shown by the annexin proteins p35 and p33 (Fig. 2A; see also Blackbourn et al., 1991) suggested that the lipid association of this form of CPK might be mediated by the annexins; some animal annexins, such as annexins I and II, are known to associate with cytoskeletal proteins as well as with membranes, in a Ca^{2+}-dependent manner (Glenney, 1986; Geisow et al., 1987). To test this, the post-hydroxyapatite annexin/CPK fraction was purified further by Ca^{2+}-dependent hydrophobic interaction chromatography on phenyl Sepharose as described by Battey & Venis (1988a, 1992). The annexins do not bind to the phenyl Sepharose column

---

Fig. 2. CPK shows a Ca^{2+}- and pH-dependent association with acidic phospholipids in the presence of annexins. (A) SDS–PAGE of ^{32}P-labelled proteins post-hydroxyapatite. After incubation with liposomes of equimolar phosphatidylserine : phosphatidylcholine (PS) or pure phosphatidylcholine (PC) in the presence of EGTA; or 5, 40 and 100 μM free Ca^{2+} (pH 6); or 5, 100 and 1000 μM free Ca^{2+} (pH 7.4), samples were centrifuged and equal proportions of pellet (left of pair) and supernatant (right of pair) loaded on the gel. (B) Autoradiograph of the same gel; CPK is the heavily labelled band of molecular mass 57 kDa.

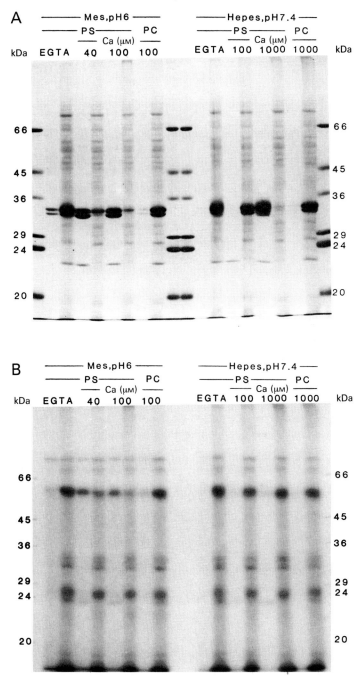

A

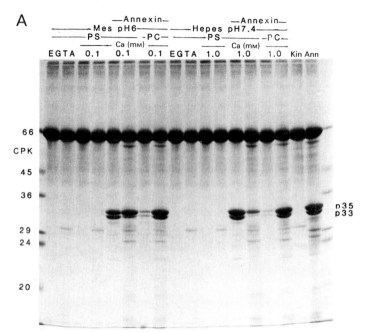

B

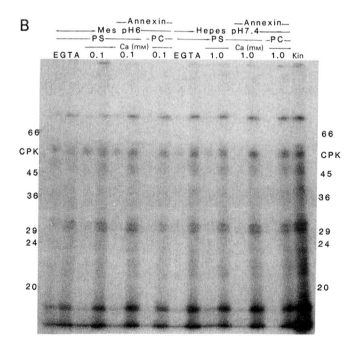

under these conditions, whereas CPK binds and is then eluted at high pH in the absence of Ca^{2+} (Fig. 1A,B). After concentration of the flow-through (annexin-rich) and the pooled high pH elution fractions from the phenyl Sepharose (CPK-rich), CPK was ^{32}P-labelled as described above. Both fractions were concentrated further and used in liposome binding assays (Fig. 3A,B). Again the result is clear: CPK associates with liposomes only at low pH, and the association is independent of annexins. The Ca^{2+} dependence of the low pH association of CPK is less pronounced in this experiment than in Fig. 2, but is nevertheless apparent.

This demonstration of a CPK with a Ca^{2+}- and pH-dependent capacity to associate with lipids raised the question of whether soluble CPK from maize has similar properties. Soluble CPK (purified on phenyl Sepharose as described by Battey & Venis, 1992) was concentrated, ^{32}P-labelled and tested for liposome association as described above (Fig. 4A,B). No association with liposomes could be detected under these conditions.

These results indicate that two distinct types of CPK exist in maize; the membrane-derived CPK shows a Ca^{2+} and pH-dependent interaction with liposomes, while the soluble type does not. It is not clear how these CPKs are related to each other; the membrane-derived form could be a precursor of the soluble form, as suggested for the oat CPK described by Schaller *et al.* (1992). However, since *ca.* 90% CPK activity is soluble in maize (N.H. Battey *et al.*, unpublished observation) it seems unlikely that it is all derived by proteolytic cleavage of a membrane-associated form, and there may be two distinct forms of CPK in this species.

Clearly, the association with *artificial* liposomes described here does not mimic *in vivo* membrane association, and there may be other factors, such as membrane lipids, fatty acids and proteins, which alter the pH and pCa dependence of CPK binding. Nevertheless, the presence of a pool of CPKs capable of reversible association with membranes is of considerable

---

Fig. 3. CPK association with liposomes at pH 6 is not dependent on the presence of annexins. (A) SDS–PAGE of ^{32}P-labelled post-phenyl Sepharose proteins. Liposome binding assays were carried out under the conditions indicated in the presence or absence of annexins (p35/p33) at 60 µg ml^{-1}. Equal proportions of pellet (left of pair) and supernatant (right of pair) were loaded on the gel. BSA at 50 µg ml^{-1} was added to supernatant fractions immediately before addition of TCA, as co-pre-cipitant. The same amount of BSA (25 µg) was added to the pellet fractions after resuspending in sample buffer, to prevent lane distortions on SDS–PAGE. 'Kin', aliquot of labelled post-phenyl Sepharose proteins not subjected to liposome binding assay. 'Ann', 30 µg of maize annexins from flowthrough of phenyl Sepharose column. (Both 'Kin' and 'Ann' have added BSA.) (B) Autoradiograph of the same gel.

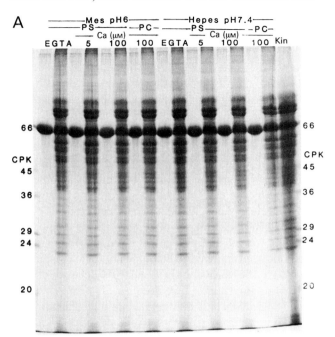

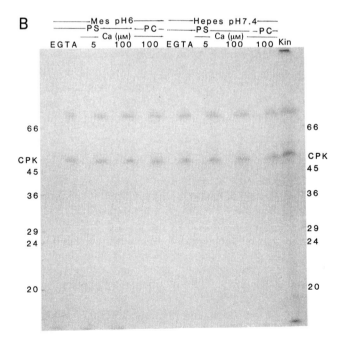

interest from the point of view of signal transduction. It is also noteworthy that such a pool has not been reported for either silver beet (Klucis & Polya, 1988) or oat root (Schaller *et al.*, 1992) where neither EGTA nor EDTA solubilised membrane-associated CPK.

Is CPK phosphorylated in plant cells?

Reasons to study phosphorylation *in vivo*

It is important to determine which proteins become phosphorylated within the cell and how this influences plant growth and development. The widely reported autophosphorylation of CPK is of potential significance. In other Ca^{2+}-regulated protein kinases, autophosphorylation can result in Ca^{2+}-independence (Saitoh & Schwartz, 1985; Lai, Nairn & Greengard, 1986); this has been proposed as a mechanism to allow molecular switching associated with learning in animals (Miller & Kennedy, 1986). Although there is little evidence that autophosphorylation influences Ca^{2+}-dependence of plant CPKs, this may be because the enzyme is already partially phosphorylated when extracted (see later). Even if this is not the case, it is possible that autophosphorylation acts as part of a signal transduction chain following elevation of intracellular Ca^{2+} levels. It would therefore be extremely useful to be able to detect changes in the phosphorylation state of CPK and other proteins during plant development, to discover their roles in this process.

Ways to study phosphorylation *in vivo*

These are summarised in Table 3. Labelling with $^{32}PO_4$ has been used very effectively to study phosphorylation of PEP carboxylase (Nimmo, this volume) and sucrose phosphate synthase (Huber & Huber, 1990), and also a 26 kDa protein from soybean root nodules as described in part I of this chapter. However, because this technique usually involves disturbing the plant in order to supply the $^{32}PO_4$, it may be more problematical

---

Fig. 4. CPK from the soluble fraction does not associate with liposomes. (A) SDS–PAGE of ^{32}P-labelled post-phenyl Sepharose proteins from the soluble fraction of maize. Liposome binding assays were carried out under the conditions indicated. Equal proportions of pellet (left of pair) and supernatant (right of pair) were loaded onto the gel. BSA at 50 µg ml^{-1} was added to supernatant fractions immediately before addition of TCA, as co-precipitant. The same amount of BSA (25 µg) was added to the pellet fractions after resuspending in sample buffer, to prevent lane distortions on SDS–PAGE. 'Kin', aliquot of labelled post-phenyl Sepharose proteins not subjected to liposome binding assay (with added BSA).

Table 3. *Ways to study phosphorylation in vivo*

Label with $^{32}PO_4$ and extract ^{32}P-phosphoproteins
Back-phosphorylation *in vitro*
Phosphoamino acid antibodies
Isoelectric focusing followed by immunoblotting
Antibodies that distinguish phosphorylated and non-phosphorylated forms of the same protein
Fe^{3+}–IDA–Sepharose chromatography?

for the study of phosphorylation events associated with developmental changes.

Back-phosphorylation is a method devised by Forn & Greengard (1978), and has been exploited to great effect to demonstrate the *in vivo* phosphorylation of two 54 kDa proteins from the insect central nervous system (Morton & Truman, 1986). The principle is that proteins that have been substrates during *in vitro* assays become unavailable for labelling following a developmental event, as a result of *in vivo* phosphorylation. This method has great potential for studying phosphorylation-dependent processes during plant development.

Phosphoamino acid antibodies (for example to detect changes in tyrosine phosphorylation of cdc2 during the cell cycle) (Morla *et al.*, 1989), and isoelectric focusing followed by immunoblotting (Maurides, Akkaraju & Jagus, 1989), both provide semi-quantitative methods of determining the phosphorylation state of unlabelled proteins. An even more powerful method is that provided by antibodies that recognise specifically either the phosphorylated or non-phosphorylated forms of a protein. Antibodies of this type have been used to demonstrate the preferential distribution of the phosphorylated form of microtubule-associated protein 1B in developing nerve axons, suggesting a role in axonal elongation (Sato-Yoshitake *et al.*, 1989).

Fe^{3+}–IDA–Sepharose affinity chromatography

Immobilised metal-ion affinity chromatography was introduced by Porath *et al.* (1975), and utilises the affinity between proteins and metal ions to allow separation of proteins. It has many applications, and both the metal ion (typically Zn^{2+}, Cu^{2+}, Hg^{2+}, Ni^{2+}, Fe^{3+}) and the chelating ligand (iminodiacetic acid (IDA), or N,N,N'-tris(carboxymethyl)ethylenediamine (TED)) influence the chromatographic behaviour of proteins (Porath & Olin, 1983; Lönnerdal & Keen, 1983). Recently, Ni^{2+} immobilised on a nitrilotriacetic acid (NTA) adsorbent has been shown to have great specificity for proteins with neighbouring histidine residues (Hochuli, Döbeli & Schacher, 1987), a finding that has found application in the purification of histidine-enriched fusion proteins from an expression vector (Abate *et al.*, 1990).

Fe^{3+}–IDA–Sepharose has been shown to bind proteins with some specificity (Porath & Olin, 1983). In keeping with the very high affinity of phosphoserine for Fe^{3+}, phosphoamino acids were found to be retarded on an Fe^{3+}–IDA chelate column when applied at acid pH, and could be eluted with phosphate (Andersson & Porath, 1986). Furthermore, adsorption of ovalbumin by the column was correlated with its degree of phosphorylation; elution of phospho-ovalbumin could be effected either by raising the pH or by 20 mM phosphate. The method has been used to purify phosphopeptides from photosystem II of spinach, where it proved its great selectivity for phosphorylated peptides (Michel & Bennett, 1987; Michel *et al.*, 1988).

In this context, it was of interest to develop a method for purifying CPK using Fe^{3+}–IDA–Sepharose affinity chromatography. IDA Sepharose (I4510) was purchased from Sigma (Dorset, UK) and a 3 × 1.5 cm column poured. After washing with nanopure water (Barnstead Nanopure II system, supplied by Fisons, UK), at least 10 column volumes of 30 mM $FeCl_3$ were applied to saturate the column with Fe^{3+}. Unbound Fe^{3+} was removed by washing with nanopure water, and the column was equilibrated with 20 mM Mes, pH 6.5.

Soluble maize protein was purified by hydrophobic interaction chromatography on phenyl Sepharose, as described previously (Battey & Venis, 1992). The CPK fraction was brought to pH 6.5 with Mes and applied to the Fe^{3+}–IDA–Sepharose column. The column was washed with 20 mM Mes (pH 6.5) until the A_{280} returned to baseline, and then developed with a gradient from pH 6.5 to 8.5 (limit buffer 20 mM Bicine, pH 8.5), or 0–500 mM potassium phosphate (in 20 mM Mes, pH 6.5). Fractions of volume 1 ml were collected and assayed in duplicate for CPK activity as follows. Aliquots of test samples were diluted to a final volume of 100 μl in a buffer of 10 mM $MgCl_2$, 0.2 mM EGTA, 0.5 mM $CaCl_2$, 50 mM Hepes (pH 7), 0.1 mg ml^{-1} histone (Sigma type IIIS), and [γ-^{32}P]ATP at a specific activity of *ca.* 1 nCi pmol^{-1} and at a concentration of 20 μM. After 10 min at room temperature, 40 μl aliquots were pipetted on to 2 cm filter squares, each of which had been impregnated with 100 μl of 25% (w/v) trichloroacetic acid (TCA), 20 mM sodium pyrophosphate, 10 mM EDTA. Filters were processed for Cerenkov counting as described for liquid scintillation counting by Battey & Venis (1992). For SDS–PAGE and autoradiography, 0.5 ml aliquots of each fraction were labelled and TCA precipitated as described in the previous section.

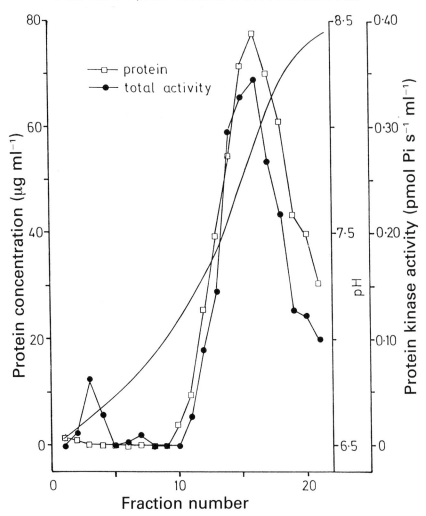

Fig. 5. Fe^{3+}–IDA–Sepharose chromatography of maize CPK: elution with a pH gradient. Post-phenyl Sepharose protein from 50 g (fresh mass) of maize coleoptiles was loaded onto the Fe^{3+}–IDA-Sepharose column as described in the text, and eluted with a gradient of pH. Protein was assayed by the method of Sedmak & Grossberg (1977).

Elution with a gradient of pH yields a single major peak of protein and CPK activity (Fig. 5). When pH is used to elute CPK, recovery is usually about 40%. These results were interesting, particularly in view of the relatively high loading pH compared with that used previously (typically around pH 5) (see Andersson & Porath, 1986; Michel & Bennett, 1987).

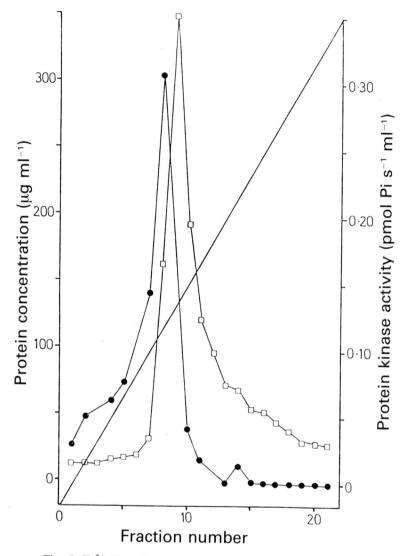

Fig. 6. Fe^{3+}–IDA–Sepharose chromatography of maize CPK: elution with a KPO_4 gradient. Method as for Fig. 5, but elution was with a gradient of 0–500 mM phosphate. Symbols as Fig. 5.

This was necessary because loading at pH less than 6 resulted in reduction of CPK activity.

Elution with a gradient of phosphate also recovers most of the activity as a single peak (Fig. 6); autoradiography of the labelled fractions clearly shows the distribution of CPK activity (Fig. 7A,B). After purification the

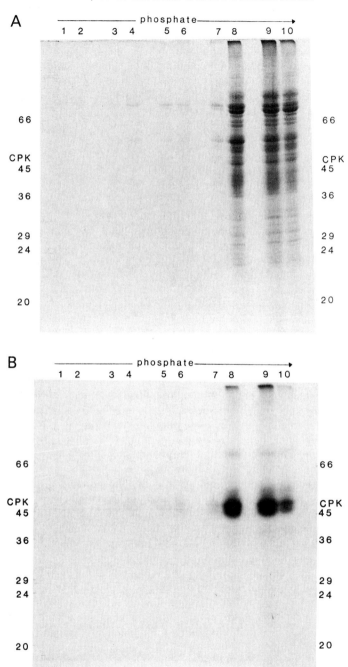

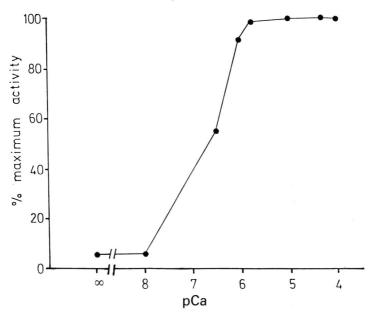

Fig. 8. The pCa optimum of maize CPK after Fe^{3+}–IDA–Sepharose chromatography. Peak fractions from a KPO_4 elution were pooled and assayed for CPK activity as described in the text; pCa levels were set by using the method of Robertson & Potter (1984).

enzyme is almost completely Ca^{2+}-dependent, and shows a pCa optimum of about 2 μM (Fig. 8).

The question that remains is whether CPK binds to Fe^{3+}–IDA–Sepharose because it is phosphorylated. Previous work would certainly suggest that this is so, as would the almost complete binding of the enzyme; in eight experiments, on average only 10% of CPK was recovered in the flowthrough. However, the purification factor (typically only 4-fold) and the number of other polypeptides present after chromatography (Fig. 7A) suggest a lack of specificity for phosphoproteins; either that, or a surprising proportion of proteins co-purifying with CPK on phenyl Sepharose are phosphorylated. We are currently testing ways to dephosphorylate CPK to establish unequivocally whether

---

Fig. 7. Protein and CPK activity from the chromatographic separation illustrated in Fig. 6. (A) 0.5 ml of each fraction was labelled with carrier free [^{32}P]ATP (10 μCi per fraction), TCA precipitated and subjected to SDS–PAGE. (B) Autoradiograph of the same gel.

Fe^{3+}-affinity chromatography can be used to determine the phosphorylation status of CPK during plant development.

III: Summary and general conclusions

1. CPKs are apparently widely distributed in plants. However, there is clearly heterogeneity within this overall grouping, and much work needs to be done to discover the relationship between the CPKs described here.

2. Studies of substrate proteins *in vitro* suggest that myosin may be an important CPK substrate. Immunolocalisation data from several species are consistent with this, and suggest a role for CPK in regulating cytoplasmic streaming. However, CPK is present in both soluble and membrane fractions, and clearly phosphorylates other proteins as well. There is particularly good evidence to suggest that a 26 kDa symbiosome protein from soybean is a substrate.

3. CPK from maize exists in at least two forms. One, membrane-derived, form associates with liposomes in a Ca^{2+}- and pH-dependent manner; the other, derived from the soluble fraction, does not. Chromatography of the soluble CPK on Fe^{3+}–IDA–Sepharose has potential because of good recovery and reasonable purification, and because binding may be a result of phosphorylation. We intend to develop this method to allow the *in vivo* phosphorylation state of CPK and other proteins to be reported.

Acknowledgements

We are grateful to the AFRC and the Royal Society for financial support. After this manuscript had been prepared, Alice Harmon was kind enough to let us see a preprint of the following highly relevant article: Roberts, D.M. & Harmon, A.C. (1992). Calcium-modulated proteins: targets of intracellular calcium signals in higher plants. *Annual Review of Plant Physiology and Plant Molecular Biology* **43**, 375–414.

References

Abate, C., Luk, D., Gentz, R., Ranscher, F.J. III & Curran, T. (1990). Expression and purification of the leucine zipper and DNA-binding domains of Fos and Jun: both Fos and Jun contact DNA directly. *Proceedings of the National Academy of Sciences USA* **87**, 1032–6.

Adelstein, R.S. & Eisenberg, E. (1980). Regulation and kinetics of the actin-myosin-ATP interaction. *Annual Review of Biochemistry* **49**, 921–56.

Andersson, L. & Porath, J. (1986). Isolation of phosphoproteins by immobilized metal (Fe^{3+}) affinity chromatography. *Analytical Biochemistry* **154**, 250–4.

Battey, N.H. (1990*a*). Dephosphorylation of calcium-activated protein kinase from maize. *Current Topics in Plant Biochemistry and Physiology* **9**, 413.

Battey, N.H. (1990*b*). Calcium-activated protein kinase from soluble and membrane fractions of maize coleoptiles. *Biochemical and Biophysical Research Communications* **170**, 17–22.

Battey, N.H. & Venis, M.A. (1988*a*). Separation of calmodulin from calcium-activated protein kinase using calcium-dependent hydrophobic interaction chromatography. *Analytical Biochemistry* **170**, 116–22.

Battey, N.H. & Venis, M.A. (1988*b*). Calcium-dependent protein kinase from apple fruit membranes is calmodulin-independent but has calmodulin-like properties. *Planta* **176**, 91–7.

Battey, N.H. & Venis, M.A. (1992). Calcium-dependent hydrophobic interaction chromatography. In *Methods in Molecular Biology* vol. 11 (*Practical Protein Chromatography*) (ed. A. Kenney & S. Fowell), pp. 73–80. New Jersey: Humana.

Blackbourn, H.D., Walker, J.H. & Battey, N.H. (1991). Calcium-dependent phospholipid-binding proteins in plants: their characterisation and potential for regulating cell growth. *Planta* **184**, 67–73.

Bogre, L., Olah, Z. & Dudits, D. (1988). Ca^{2+}-dependent protein kinase from alfalfa (*Medicago varia*): partial purification and autophosphorylation. *Plant Science* **58**, 135–44.

Boustead, C.M., Walker, J.H. & Geisow, M.J. (1988). Isolation and characterization of two novel calcium-dependent phospholipid-binding proteins from bovine lung. *FEBS Letters* **233**, 233–8.

Cohen, P. (1982). The role of protein phosphorylation in neural and hormonal control of cellular activity. *Nature (London)* **296**, 613–20.

Elliott, D.C. & Skinner, J.D. (1986). Calcium-dependent, phospholipid-activated protein kinase in plants. *Phytochemistry* **25**, 39–44.

Forn, J. & Greengard, P. (1978). Depolarizing agents and cyclic nucleotides regulate the phosphorylation of specific neuronal proteins in rat cerebral cortex slices. *Proceedings of the National Academy of Sciences USA* **75**, 5195–9.

Geisow, M.J., Walker, J.H., Boustead, C. & Taylor, W. (1987). Annexins – new family of Ca^{2+}-regulated phospholipid-binding proteins. *Bioscience Reports* **7**, 289–98.

Glenney, J.R. (1986). Two related but distinct forms of the M_r 36,000 tyrosine kinase substrate (calpactin) that interact with phospholipid and actin in a Ca^{2+}-dependent manner. *Proceedings of the National Academy of Sciences USA* **83**, 4258–62.

Grolig, F., Williamson, R.E., Parker, J., Miller, C. & Anderton, B.H. (1988). Myosin and Ca^{2+}-sensitive streaming in the alga *Chara*: detection of two polypeptides reacting with a monoclonal anti-myosin and their localization in the streaming endoplasm. *European Journal of Cell Biology* **47**, 22–31.

Guo, Y.-L. & Roux, S.J. (1990). Partial purification and characterization of a Ca^{2+}-dependent protein kinase from the green alga, *Dunaliella salina. Plant Physiology* **94**, 143–50.

Harmon, A.C. & McCurdy, D.W. (1990). Calcium-dependent protein kinase and its possible role in the regulation of the cytoskeleton. *Current Topics in Plant Biochemistry and Physiology* **9**, 119–28.

Harmon, A.C., Putnam-Evans, C. & Cormier, M.J. (1987). A calcium-dependent but calmodulin-independent protein kinase from soybean. *Plant Physiology* **83**, 830–7.

Harper, J.F., Sussman, M.R., Schaller, G.E., Putnam-Evans, C., Charbonneau, H. & Harmon, A.C. (1991). A calcium-dependent protein kinase with a regulatory domain similar to calmodulin. *Science* **252**, 951–4.

Hetherington, A.M., Battey, N.H. & Millner, P.A. (1990). Protein kinase. In *Methods in Plant Biochemistry*, vol. 3 (ed. P.J. Lea), pp. 371–83. London: Academic Press.

Hetherington, A.M. & Trewavas, A.J. (1982). Calcium-dependent protein kinase in pea shoot membranes. *FEBS Letters* **145**, 67–71.

Hochuli, E., Döbeli, H. & Schacher, A. (1987). New metal chelate adsorbent selective for proteins and peptides containing neighbouring histidine residues. *Journal of Chromatography* **411**, 177–84.

Huber, S.C. & Huber, J.L.A. (1990). Regulation of spinach leaf sucrose-phosphate synthase by multisite phosphorylation. *Current Topics in Plant Biochemistry and Physiology* **9**, 329–43.

Kemp, B.E. & Pearson, R.B. (1990). Protein kinase recognition sequence motifs. *Trends in Biochemical Sciences* **15**, 342–6.

Klee, C.B. & Vanaman, T.C. (1982). Calmodulin. *Advances in Protein Chemistry* **35**, 213–321.

Klimczak, L.J. & Hind, G. (1990). Biochemical similarities between soluble and membrane-bound calcium-dependent protein kinases of barley. *Plant Physiology* **92**, 919–23.

Klucis, E. & Polya, G.M. (1987). Calcium-independent activation of two plant leaf calcium-regulated protein kinases by unsaturated fatty acids. *Biochemical and Biophysical Research Communications* **147**, 1041–7.

Klucis, E. & Polya, G.M. (1988). Localization, solubilization and characterization of plant membrane-associated calcium-dependent protein kinases. *Plant Physiology* **88**, 164–71.

Kohama, K. & Kendrick-Jones, J. (1986). The inhibitory Ca^{2+}-regulation of the actin-activated Mg-ATPase activity of myosin from *Physarum polycephalum* plasmodia. *Journal of Biochemistry* **99**, 1433–46.

Lai, Y., Nairn, A.C. & Greengard, P. (1986). Autophosphorylation reversibly regulates the Ca^{2+}/calmodulin-dependence of Ca^{2+}/calmodulin-dependent protein kinase II. *Proceedings of the National Academy of Sciences USA* **83**, 4253–7.

Li, H., Dauwalder, M. & Roux, S.J. (1991). Partial purification and characterization of a Ca^{2+}-dependent protein kinase from pea nuclei. *Plant Physiology* **96**, 720–7.

Lönnerdal, B. & Keen, C.L. (1983). Metal chelate affinity chromatography of proteins. *Journal of Applied Biochemistry* **4**, 203–8.

Lucantoni, A. & Polya, G.M. (1987). Activation of wheat embryo calcium-regulated protein kinase by unsaturated fatty acids in the presence and absence of calcium. *FEBS Letters* **221**, 33–6.

Maurides, P.A., Akkaraju, G.R. & Jagus, R. (1989). Evaluation of protein phosphorylation state by a combination of vertical slab gel isoelectric focusing and immunoblotting. *Analytical Biochemistry* **183**, 144–51.

Michel, H.P. & Bennett, J. (1987). Identification of the phosphorylation site of an 8.3kDa protein from photosystem II of spinach. *FEBS Letters* **212**, 103–8.

Michel, H.P., Hunt, D.F., Shabanowitz, J. & Bennett, J. (1988). Tandem mass spectrometry reveals that three photosystem II proteins of spinach chloroplasts contain *N*-acetyl-*O*-phosphothreonine at their NH_2 termini. *Journal of Biological Chemistry* **263**, 1123–30.

Miller, S.G. & Kennedy, M.B. (1986). Regulation of brain type II $Ca^{2+}/$ calmodulin-dependent protein kinase by autophosphorylation: a Ca^{2+}-triggered molecular switch. *Cell* **44**, 861–70.

Minichiello, J., Polya, G.M. & Keane, P.J. (1989). Inhibition and activation of oat leaf calcium-dependent protein kinase by fatty acids. *Plant Science* **65**, 143–52.

Morla, A.O., Draetta, G., Beach, D. & Wang, J.Y.J. (1989). Reversible tyrosine phosphorylation of cdc2: dephosphorylation accompanies activation during entry into mitosis. *Cell* **58**, 193–203.

Morton, D.B. & Truman, J.W. (1986). Substrate phosphoprotein availability regulates eclosion hormone sensitivity in an insect CNS. *Nature* **323**, 264–7.

Muto, S. & Shimogawara, K. (1985). Calcium- and phospholipid-dependent phosphorylation of ribulose-1,5-bisphosphate carboxylase/ oxygenase small subunit by a chloroplast envelope-bound protein kinase *in situ*. *FEBS Letters* **193**, 88–92.

Nishizuka, Y. (1984). The role of protein kinase C in cell surface signal transduction and tumour promotion. *Nature (London)* **308**, 693–8.

Ohno, S., Encori, Y., Imajoh, S., Kawasaki, H., Kisaragi, M. & Suzuki, K. (1984). Evolutionary origin of a calcium-dependent protease by fusion of genes for a thiol protease and a calcium-binding protein. *Nature (London)* **312**, 566–70.

Olah, Z., Bogre, L., Lehel, C., Farago, A., Seprodi, J. & Dudits, D. (1989). The phosphorylation site of Ca^{2+}-dependent protein kinase from alfalfa. *Plant Molecular Biology* **12**, 453–61.

Polya, G.M. & Chandra, S. (1990). Ca^{2+}-dependent protein phosphorylation in plants: regulation, protein substrate specificity and product dephosphorylation. *Current Topics in Plant Biochemistry and Physiology* **9**, 164–80.

Polya, G.M. & Davies, J.R. (1982). Resolution of Ca^{2+}-calmodulin-activated protein kinase from wheatgerm. *FEBS Letters* **150**, 167–71.

Polya, G.M., Morrice, N. & Wettenhall, R.E.H. (1989). Substrate specificity of wheat embryo calcium-dependent protein kinase. *FEBS Letters* **253**, 137–40.

Polya, G.M., Nott, R., Klucis, E., Minichiello, J. & Chandra, S. (1990). Inhibition of plant calcium-dependent protein kinases by basic polypeptides. *Biochimica et Biophysica Acta* **1037**, 259–62.

Porath, J., Carlsson, J., Olsson, I. & Belfrage, G. (1975). Metal chelate affinity chromatography, a new approach to protein fractionation. *Nature (London)* **258**, 598–9.

Porath, J. & Olin, B. (1983). Immobilized metal ion affinity adsorption and immobilized metal ion affinity chromatography of biomaterials. Serum protein affinities for gel-immobilized iron and nickel ions. *Biochemistry* **22**, 1621–30.

Putnam-Evans, C., Harmon, A.C. & Cormier, M.J. (1990). Purification and characterization of a novel calcium-dependent protein kinase from soybean. *Biochemistry* **29**, 2488–95.

Putnam-Evans, C., Harmon, A.C., Palevitz, B.A., Fechheimer, M. & Cormier, M.J. (1989). Calcium-dependent protein kinase is localized with F-actin in plant cells. *Cell Motility and the Cytoskeleton* **12**, 12–22.

Roberts, D.M. (1989). Detection of a calcium-activated protein kinase in *Mougeotia* by using synthetic peptide substrates. *Plant Physiology* **91**, 1613–19.

Robertson, S. & Potter, J.D. (1984). The regulation of free Ca^{2+} ion concentration by metal chelators. *Methods in Pharmacology* **5**, 63–75.

Saitoh, T. & Schwartz, J.H. (1985). Phosphorylation-dependent subcellular translocation of a Ca^{2+}/calmodulin-dependent protein kinase produces an autonomous enzyme in *Aplysia* neurons. *Journal of Cell Biology* **100**, 835–42.

Sato-Yoshitake, R., Shiomura, Y., Miyasaka, H. & Hirokawa, N. (1989). Microtubule-associated protein 1B: molecular structure, localization, and phosphorylation-dependent expression in developing neurons. *Neuron* **3**, 229–38.

Schäfer, A., Bygrave, F., Matzenauer, S. & Marmé, D. (1985). Identification of a calcium- and phospholipid-dependent protein kinase in plant tissue. *FEBS Letters* **187**, 25–8.

Schaller, G.E., Harmon, A.C. & Sussman, M.R. (1992). Characterization of a calcium- and lipid-dependent protein kinase associated with the plasma membrane of oat. *Biochemistry* **31**, 1721–7.

Sedmak, J.J. & Grossberg, S.E. (1977). A rapid, sensitive and versatile assay for protein using Coomassie Brilliant Blue G250. *Analytical Biochemistry* **79**, 544–52.

Takai, Y., Kishimoto, A., Iwasa, Y., Kawahara, Y., Mori, T. & Nishizuka, Y. (1979). Calcium-dependent activation of a multifunctional protein kinase by membrane phospholipids. *Journal of Biological Chemistry* **254**, 3692–5.

Takai, Y., Yamamoto, M., Inoue, M., Kishimoto, A. & Nishizuka, Y. (1977). A proenzyme of cyclic nucleotide-independent protein kinase and its activation by calcium-dependent neutral protease from rat liver. *Biochemical and Biophysical Research Communications* **77**, 542–50.

Weaver, C.D., Crombie, B., Stacey, G. & Roberts, D.M. (1991). Calcium-dependent phosphorylation of symbiosome membrane proteins from nitrogen-fixing soybean nodules. *Plant Physiology* **95**, 222–7.

M.R. SUSSMAN

Phosphorylation of the plasma membrane proton pump

Introduction

The plasma membrane proton pump (H^+-ATPase) is the primary means by which metabolic energy is coupled to solute transport in plants and fungi. In animal cells, the plasma membrane sodium pump (Na^+,K^+-ATPase) plays a similar role. Thus, animal cells utilise sodium-coupled carriers for solute movement, whereas in plants and fungi these carriers are proton-coupled. Electrophysiological measurements indicate that in actively transporting cells, such as those of plant root hairs or rapidly dividing fungi, the proton pump may be the single greatest consumer of cellular ATP (Felle, 1982). The reliance of plants and fungi on the ever-present proton, and their rather large resting electric potential (-160 to -240 mV, interior negative), may account for their ability to survive in media with little nutritional value. These organisms are capable of scavenging low concentrations of nutrients; the proton pump is essential for this process. Genetic disruption experiments with a gene encoding the plasma membrane proton pump of yeast has provided conclusive evidence that this enzyme is essential for normal growth. The enzyme contains a single polypeptide of *ca.* $M_r = 100\,000$ with conserved sequences and topology that place it within the P-type family of ion pumps (Sussman & Harper, 1989). Other members of this family include the animal plasma membrane Na^+,K^+-ATPase and Ca^{2+}-ATPase.

Most higher plant cells grow at a much slower rate than laboratory-grown yeast. It seems likely that, in place of a rapid doubling time, plants emphasise a precise regulation of cellular growth rates, to ensure that the tissues and organs of this complex multicellular eukaryote form and develop in an orderly fashion. Several decades of physiological research in green plants have indicated that the secretion of protons is clearly tied to growth rates of individual cells. Although earlier studies focused on

Society for Experimental Biology Seminar Series 53: *Post-translational modifications in plants*, ed. N.H. Battey, H.G. Dickinson & A.M. Hetherington. © Cambridge University Press 1993, pp. 149–159.

changes occurring within hours of treatment with plant growth regulators, modern electrophysiological and biochemical experiments focus on the very early, rapid events occurring within the first few seconds or minutes after chemical application. As will be described below, the complex multicellular structure of most plant tissues creates problems for performing critical biochemical experiments with isolated plant plasma membrane vesicles. Although recombinant DNA techniques offer the potential to perform genetic studies, gene disruption experiments are still difficult or impossible to perform, even in *Arabidopsis thaliana*. Fortunately, yeast and plants rely on the same enzyme, a proton pump, for similar transport functions at the plasma membrane. Although yeast is unresponsive to plant growth regulators, a rapid effect of glucose on activation of the yeast plasma membrane proton pump (Serrano, 1983) could represent a useful paradigm to help future efforts to understand how the plant enzyme's activity is likewise affected by growth regulators and other efficacious agents.

Activation of the yeast plasma membrane proton pump by glucose

When yeast cells are starved of glucose by centrifugation and resuspension in media lacking sugar, the plasma membrane proton pump becomes catalytically less active. This phenomenon is easily recorded by homogenising yeast cells after glucose depletion or addition, and then assaying for vanadate-sensitive H^+-ATPase activity in plasma membrane vesicles *in vitro*. The K_m for ATP is 2–5 times lower, and the V_{max} 2–5 times higher when cells are metabolising glucose than when they are starved for glucose. Glucose is without effect on isolated vesicles, indicating that glucose metabolism is required, or that the enzymes mediating the glucose effect are unstable or removed from the membrane during homogenisation.

The effect of glucose on extractable ATPase activity is very rapid, requiring 10 s or less for maximal activation. The rapidity of this response indicates that changes in covalent modification (e.g. kinase-mediated phosphorylation), associated lipids or a 'regulatory' polypeptide could be involved. Recent biochemical and genetic studies indicate that ATPase phosphorylation, possibly at the carboxy-terminus, could account for ATPase catalytic activity regulation. The evidence for this is summarised below.

First, when the amino-terminal 11 residues of the yeast ATPase are genetically altered, the catalytic activity is stimulated (Portillo, de Larrinoa & Serrano, 1989). Proteolytic degradation of the yeast and plant enzyme likewise indicates that the carboxy terminus is an 'inhibitory' domain whose normal function is to restrain catalytic activity (Palmgren, Larsson & Sommarin, 1990). When synthetic peptides with sequences corresponding to this domain are added back to proteolytically activated enzyme, the enzyme is inhibited, indicating that the carboxy-terminus contains an autoinhibitory domain (Palmgren *et al.*, 1991). Since the activation of ATPase by lysolecithin (Pedchenko, Nasirova & Palladina, 1990) is not additive to that observed by proteolytic activation, it is likely that the two activate by similar mechanisms.

According to this model, phosphorylation of an amino acid residue in the carboxy terminus would stimulate activity by interfering with the autoinhibitory domain. A consensus sequence for calmodulin-dependent protein kinases (Arg/Lys–X–X–Ser/Thr) exists at residues Arg^{909} to Thr^{912}, and site-directed mutagenesis at this site reduces the ability of glucose to elicit ATPase activation. It is interesting to note that when either of these residues is mutated, the mutant enzyme no longer displays glucose-induced changes in K_m and V_{max}, but a glucose-induced change in vanadate sensitivity is retained. This result is important since it indicates that protein phosphorylation, at least at Thr^{912}, is not the whole story. These results indicate that although kinase-mediated phosphorylation could explain some of the rapid effects of glucose, there may be additional mechanisms operating as well. Glucose could be altering the lipid composition of plasma membranes, and changes in vanadate sensitivity may thus reflect lipid-induced changes in ATPase conformation.

Direct evidence that glucose alters the phosphorylation status was provided recently by Chang & Slayman (1991). Earlier studies by Serrano (1985) with extracted ATPase indicated that glucose did not produce a measurable change in ^{32}P incorporated from phosphate supplied *in vivo*. Chang & Slayman (1991) followed up this question by proteolysing the extracted ATPase and measuring ^{32}P incorporation at discrete locations in the polypeptide. Two-dimensional peptide maps demonstrated that at least seven different proteolytic fragments were labelled when ^{32}P was provided *in vivo*. Of these multiple fragments, only one showed reproducible changes in ^{32}P incorporation after glucose treatment of whole cells. Another important conclusion of the studies by Chang & Slayman (1991) is that as the ATPase moves from its site of synthesis, the endoplasmic reticulum, to the plasma membrane, it becomes multiply phosphorylated. Of course, these experiments are correlative in nature, and indicate that kinase-mediated phosphorylation of the ATPase could play a role in ATPase biogenesis or in the regulation of catalytic activity at the cell surface. Amino acid sequence of the phosphorylated peptides would provide the location of important serine and threonine residues and genetic alterations of such residues should determine whether they are required for biogenesis and/or glucose activation.

The plant plasma membrane H$^+$-ATPase gene family in *Arabidopsis thaliana*

Using oat roots, Schaller & Sussman (1988) observed that the 100 kDa H$^+$-ATPase polypeptide is readily phosphorylated *in vitro* by a protein kinase that remains associated with plasma membrane vesicles after homogenisation. This kinase required low, micromolar calcium concentrations for maximal activity. In order to characterise the interaction of the plant ATPase with this calcium-dependent protein kinase, we embarked on studies to clone and sequence cDNA clones encoding each of these two enzymes, and these results are summarised below.

Molecular biology studies were performed with *Arabidopsis thaliana*, a model higher plant with a small genome. Initially, three genes encoding different isoforms of the *Arabidopsis* H$^+$-ATPase (AHA) were cloned and sequenced (Sussman & Harper, 1989; Harper, Surowy & Sussman, 1989; Pardo & Serrano, 1989; Harper *et al.*, 1990). The three proteins predicted from the DNA sequences showed greater than 85% identity in amino acid sequence. Using low-stringency hybridisation and the polymerase chain reaction, seven other AHA gene clones have now been isolated. Although all show a high degree of amino acid sequence identity, one (AHA 10) shows a great deal of divergence in the carboxy-terminus. This indicates that the isoforms may not share all of their regulatory properties.

Whole-cell studies with green plants indicate that there are two clear instances where the plasma membrane proton pump is regulated in as rapid a fashion as glucose acts in yeast. Thus, electrophysiological studies with root hairs of mustard plants indicate that, in these cells, fusicoccin activates the plasma membrane proton pump within 10–20 s after application (Felle, 1982). A second example is in guard cells, where Assmann, Simoncini & Schroeder (1985) showed by means of patch-clamp studies that blue light induces a dramatic increase in ATPase activity. Blue-light activation of the proton pump could lead to hyperpolarisation and concomitant opening of a voltage-gated potassium channel that allows a massive potassium influx and increase in turgor.

Although we have demonstrated that the oat root plasma membrane H$^+$-ATPase incorporates ^{32}P from [γ-^{32}P]ATP at serine and theonine residues via a calcium-dependent protein kinase, there is no direct evidence that phosphorylation regulates the ATPase. In comparison with yeast, these studies with the plant ATPase are complicated by the lower amounts of enzyme present in extracts, a barely detectable *in vitro* fusiococcin response, the difficulty of performing biochemical experiments with blue light responsive guard cells, and confusion as to which of the ten ATPase isoforms is present in purified vesicles isolated from multicellular tissue.

Plasma membrane proton pump kinases

In order to study the interaction of the proton pump with endogenous protein kinases, experiments were initiated to purify the kinases and characterise their molecular structure and regulatory properties. Initial experiments with plasma membrane vesicles isolated from oat roots demonstrated that a calcium-dependent protein kinase, which phosphorylates the pump, is tightly associated with the membrane. This enzyme was not removed by osmotic shock or by washes with 0.5 M KCl or 4 mM EDTA. However, after the addition of 0.9% (v/v) Triton X-100, 90% of the total kinase activity was solubilised. Utilising ion exchange and size-exclusion chromatography, this solubilised kinase was partially purified *ca.* 80-fold from the starting membranes. Western blots were performed with a set of four monoclonal antibodies isolated by Professor Alice Harmon (University of Florida), directed against a calcium-dependent protein kinase purified from the soluble fraction of soybean extracts (Putnam-Evans, Harmon & Cormier, 1990). All four antibodies gave a dark immunoreactive band of $M_r=61000$ (61 kDa) and a weaker one of $M_r=79000$ (79 kDa) in the partially purified oat plasma membrane kinase preparation. Upon storage, or after treatment with low concentrations of trypsin, the 61 kDa polypeptide is transformed into a 58 kDa species that is no longer capable of associating with Triton X-114 micelles. This observation suggests that the 55 kDa kinase purified from the soluble fraction of soybean cells may be a proteolytic artefact and that *in situ* the enzyme could be membrane-associated. When precautions against proteolysis are taken, there is no detectable immunoreactive kinase in the soluble fraction of oat root extracts (Schaller, Harmon & Sussman, 1992).

Immunocytochemical experiments with the same four monoclonal antibodies identify microfilaments in the cytoskeleton as a site of location for the calcium-dependent protein kinase (Putnam-Evans *et al.*, 1989). There is thus a discrepancy between localisation results determined microscopically and those determined biochemically. There are two ways to resolve this conflict. First, the membrane-bound kinase may in fact be bound to membranes via microfilaments attached to the plasma membrane (Traas,

1990). A second possibility is that there are different isoforms encoding the kinase and these may be located at different cellular locations. This latter explanation accounts for the observation of protein kinase C at several intracellular sites in animal cells (Masmoudi *et al.*, 1989; Mochly-Rosen *et al.*, 1990, 1991*a,b*).

In order to clarify whether the 'soluble' soybean and 'membrane-bound' oat kinases represent different isoforms of the same kinase, cDNA clones encoding the plant calcium-dependent protein kinase were isolated and sequenced. These studies are focused on characterising the entire gene family encoding this enzyme in one plant species, *Arabidopsis thaliana*. At least two different *A. thaliana* gene isoforms have been found that contain the hallmark structure of plant calcium dependent protein kinases, a 35 kDa amino-terminal kinase catalytic domain (Hanks, Quinn & Hunter, 1988) fused to a carboxy-terminal 'regulatory' domain resembling calmodulin and containing four E-F hands (Harper *et al.*, 1991). These clones have now been expressed in *Escherichia coli*; as expected, the purified proteins catalyse calcium-dependent protein kinase activity (M.R. Sussman, unpublished results).

Most interesting is the observation that the protein expressed by the *A. thaliana* calcium-dependent protein kinase cDNA clone shows activation by lipids. Earlier studies had demonstrated that the partially purified protein kinase from oat root plasma membranes was activated by lipids. The activation showed specificity in that lysophosphatidylcholine and phosphatidylinositol were active, but phosphatidylcholine and phosphatidylserine were inactive. A similar specificity is observed with the recombinant *A. thaliana* enzyme. This result is provocative, since this kinase represents the predominant calcium-dependent protein kinase in plant extracts, and may be the focus for diverse signal transduction pathways. As yet, protein kinase C has not been conclusively identified in plant extracts, nor has a plant clone containing protein kinase C-like sequence been identified. Further speculation on the significance for lipid activation of the plant calcium-dependent kinase seems unwarranted until the effect of lipids with endogenous protein substrates is measured. At present, histone IIIS is the best substrate for the plant kinase and studies to identify endogenous substrates are in progress. Purified 100 kDa H^+-ATPase polypeptide from oat roots is an excellent substrate, either attached to the membrane or as a solubilised protein. However, there is still no data indicating which residue(s) is (are) labelled or whether the phosphorylation affects activity.

Concluding remarks

In contrast to the situation with plants, there is no biochemical evidence for a calcium-dependent protein kinase that phosphorylates, or is associated with, the yeast plasma membranes. However, Lew (1989) presented electrophysiological data which indicate that higher cytoplasmic calcium concentrations activate the *Neurospora* plasma membrane proton pump (H^+-ATPase).

Recent 'genomic' project results indicate that Hunter (1987) was probably correct in predicting that eukaryotic cells may contain many hundreds, or even thousands, of different protein kinases. For example, yeast chromosome III has now been completely sequenced; this 400 kilobasepair chromosome has 200 genes, of which two are protein kinases. Similarly, a protein kinase has been observed about every 100–200 kilobasepairs in plants. Although the sample size for these experiments still remains small, it appears possible that *Arabidopsis thaliana*, which contains *ca.* 15 000 genes, could contain 150–300 different protein kinases. The key question, of course, is what are their substrates and how are they regulated?

Since the plasma membrane proton pump of yeast is phosphorylated at multiple sites, the yeast and plant enzymes may be regulated by several protein kinases, with varying effects. For example, there are reports of several other plant plasma membrane kinase activities that are not calcium-dependent (Gallagher *et al.*, 1988; Farmer, Pearce & Ryan, 1989; Felix *et al.*, 1991). Changes in membrane lipid composition may also affect the pump. For example phosphatidylinositol 4,5-bisphosphate (PIP_2) and other phosphoinositides can activate the plant plasma membrane H^+-ATPase *in vitro*, and there are *in vivo* changes in phosphoinositide concentrations which correlate with changes in ATPase activity (Memon & Boss, 1990). Until we learn whether the *in vitro* lipid effect on the oat and *Arabidopsis* calcium-dependent protein kinase reflects a similar activation to the ATPase *in situ*, it will be unclear whether lipids are acting directly on the pump, or via a closely associated protein kinase.

Kolarov *et al.* (1988) reported on a protein kinase that co-purifies with the yeast proton pump even after detergent solubilisation from membranes. This kinase prefers casein over histone, and is unaffected by calcium or other known regulators. In light of recent reports that a plasma membrane casein kinase from animal cells is regulated by PIP_2 (Brockman & Anderson, 1990), serious consideration should be given to the suggestion of Memon & Boss (1990) that a phospholipase C-mediated signal transduction pathway is less important than other lipid effectors (see also Palmgren *et al.*, 1988).

Finally, we should not lose sight of other possible modes of regulation during biogenesis. For example, Hager *et al.* (1991) has reported immunochemical data indicating that the plasma membrane H^+-ATPase in coleoptiles has a remarkably short half-life of *ca.* 10 min. Since the cytoplasmic calcium concentration has been implicated in controlling the secretory pathway of yeast (Rudolph *et al.*, 1989), it may be necessary to consider rapid effector-induced changes in ATPase flow between membrane compartments as a means of regulating the plasma membrane protonmotive force.

In conclusion, the higher plant is a complex mixture of individual cells growing and differentiating into well-organised tissues and organs. The plasma membrane proton pump (H^+-ATPase) plays a central role in nutrition, and possibly in signal transduction as well. The recent observation that plants contain many multiple genes encoding slightly different enzymes, expressed in different tissues, makes our job of determining what is going on in a single cell very difficult. However, this reductionist approach is necessary to first identify the important players in the plant's signal transduction system. We should not assume that the predominant systems working in yeast or animals are operating to the same extent or in the same manner in green plants. The discovery of a new family of calcium-dependent protein kinases in soybean and *Arabidopsis thaliana* (see Harper *et al.*, 1991) supports this notion. As we gather a greater understanding of the molecular structure of important plant plasma membrane proteins and their regulators, we can turn our attention to asking how these proteins interact to form the signal transduction pathway.

References

Assmann, S.M., Simoncini, L. & Schroeder, J.I. (1985). Blue light activates electrogenic ion pumping in guard cell protoplasts of *Vicia faba*. *Nature (London)* **318**, 285–7.

Brockman, J.L. & Anderson, R.A. (1990). Casein kinase IIs regulated by phosphatidylinositol 4,5-bisphosphate in native membranes. *Journal of Biological Chemistry* **266**, 1–5.

Chang, A. & Slayman, C.W. (1991). Maturation of the yeast plasma membrane H^+-ATPase involves phosphorylation during intracellular transport. *Journal of Cell Biology* **115**, 289–96.

Farmer, E.E., Pearce, G. & Ryan, C.A. (1989). *In vitro* phosphorylation of plant plasma membrane proteins in response to the proteinase inhibitor inducing factor. *Proceedings of the National Academy of Sciences USA* **86**, 1539–42.

Felix, G., Grosskopf, D.G., Regenass, M. & Boller, T. (1991). Rapid changes of protein phosphorylation are involved in transduction of the elicitor signal in plant cells. *Proceedings of the National Academy of Sciences USA* **88**, 8831–4.

Felle, H. (1982). Effects of fusicoccin upon membrane potential, resistance and current-voltage characteristics in root hairs of *Sinapis alba*. *Plant Science Letters* **25**, 219–25.

Gallagher, S., Short, T.W., Ray, P.M., Pratt, L.H. & Briggs, W.R. (1988). Light-mediated changes in two proteins found associated with plasma membrane fractions from pea stem sections. *Proceedings of the National Academy of Sciences USA* **85**, 8003–7.

Hager, A., Debus, G., Edel, H.-G., Stransky, H. & Serrano, R. (1991). Auxin induces exocytosis and the rapid synthesis of a high-turnover pool of plasma membrane H^+-ATPase. *Planta* **185**, 527–37.

Hanks, S.K., Quinn, A.M. & Hunter, T. (1988). The protein kinase family: conserved features and deduced phylogeny of the catalytic domains. *Science* **241**, 42–52.

Harper, J.F., Manney, L., DeWitt, N., Yoo, M.H. & Sussman, M.R. (1990). The *Arabidopsis thaliana* plasma membrane H^+-ATPase multigene family. *Journal of Biological Chemistry* **265**, 13601–8.

Harper, J.F., Surowy, T.K. & Sussman, M.R. (1989). Molecular cloning and sequence of cDNA encoding the plasma membrane proton pump (H^+-ATPase) of *Arabidopsis thaliana*. *Proceedings of the National Academy of Sciences USA* **86**, 1234–8.

Harper, J.F., Sussman, M.R., Schaller, G.E., Putnam-Evans, C., Charbonneau, H. & Harmon, A.C. (1991). A calcium-dependent protein kinase with a regulatory domain similar to calmodulin. *Science* **252**, 951–4.

Hunter, T. (1987). A thousand and one protein kinases. *Cell* **50**, 823–30.

Kolarov, J., Kulpa, J., Baijot, M. & Goffeau, A. (1988). Characterization of a protein serine kinase from yeast plasma membrane. *Journal of Biological Chemistry* **263**, 10613–19.

Lew, R. (1989). Calcium activates an electrogenic proton pump in *Neurospora* plasma membrane. *Plant Physiology* **91**, 213–16.

Masmoudi, A., Labourdette, G., Mersel, M., Huang, F.L., Huang, K.-P., Vincendon, G. & Malviya, A.N. (1989). PKC located in rat liver nuclei. *Journal of Biological Chemistry* **264**, 1172–9.

Memon, A.R. & Boss, W.F. (1990). Rapid light-induced changes in phosphoinositide kinases and H^+-ATPase in plasma membrane of sunflower hypocotyls. *Journal of Biological Chemistry* **265**, 14817–21.

Mochly-Rosen, D., Henrich, C.J., Cheever, L., Khaner, H. & Simpson, P.C. (1990). A protein kinase C isozyme is translocated to cytoskeletal elements on activation. *Cell Regulation* **1**, 693–706.

Mochly-Rosen, D., Khaner, H. & Lopez, J. (1991*a*). Identification of intracellular receptor proteins for activated protein kinase C. *Proceedings of the National Academy of Sciences USA* **88**, 3997–4000.

Mochly-Rosen, D., Khaner, H., Lopez, J. & Smith, B.L. (1991*b*). Intracellular receptors for activated protein kinase C: identification of a binding site for the enzyme. *Journal of Biological Chemistry* **266**, 14866–8.

Palmgren, M.G., Larsson, C. & Sommarin, M. (1990). Proteolytic activation of the plant plasma membrane ATPase by removal of a carboxy terminal domain. *Journal of Biological Chemistry* **265**, 13423–6.

Palmgren, M.G., Sommarin, M., Serrano, R. & Larsson, C. (1991). Identification of an autoinhibitory domain in the C-terminal region of the plant plasma membrane H⁺-ATPase. *Journal of Biological Chemistry* **266**, 20470–5.

Palmgren, M.G., Sommarin, M., Ulvskov, P. & Jorgensen, P.L. (1988). Modulation of plasma membrane H⁺-ATPase from oat roots by lysophosphatidylcholine, free fatty acids and phospholipase A₂. *Physiologia Plantarum* **74**, 11–19.

Pardo, J.M. & Serrano, R. (1989). Structure of a plasma membrane H⁺-ATPase gene from the plant *Arabidopsis thaliana*. *Journal of Biological Chemistry* **264**, 8557–62.

Pedchenko, V.K., Nasirova, G.F. & Palladina, T.A. (1990). Lysophosphatidylcholine specifically stimulates plasma membrane H⁺-ATPase from corn roots. *FEBS Letters* **275**, 205–8.

Portillo, F., de Larrinoa, I.F. & Serrano, R. (1989). Deletion analysis of yeast plasma membrane H⁺-ATPase and identification of a regulatory domain at the carboxy terminus. *FEBS Letters* **247**, 381–5.

Portillo, F., Eraso, P. & Serrano, R. (1991). Analysis of the regulatory domain of yeast plasma membrane H⁺-ATPase by directed mutagenesis and intergenic suppression. *FEBS Letters* **287**, 71–4.

Putnam-Evans, C.L., Harmon, A.C. & Cormier, M.J. (1990). Purification and characterization of a novel calcium-dependent protein kinase from soybean. *Biochemistry* **29**, 2488–95.

Putnam-Evans, C., Harmon, A.C., Palevitz, B.H., Fechheimer, M. & Cormier, M. (1989). Calcium-dependent protein kinase is localized with F-actin in plant cells. *Cell Motility and the Cytoskeleton* **12**, 12–22.

Rudolph, H.K., Antebi, A., Fink, G.R., Buckley, C.M., Dorman, T.E., LeVitre, J., Davidow, L.S., Mao, J. & Moir, D.T. (1989). The yeast secretory pathway is perturbed by mutations in *PMR1*, a member of a Ca²⁺ ATPase family. *Cell* **58**, 133–45.

Schaller, G.E., Harmon, A.C. & Sussman, M.R. (1992). Characterization of a calcium- and lipid-dependent protein kinase associated with the plasma membrane of oat. *Biochemistry* **31**, 1721–7.

Schaller, G.E. & Sussman, M.R. (1988). Phosphorylation of the plasma membrane H⁺-ATPase of oat roots by a calcium-stimulated protein kinase. *Planta* **173**, 509–18.

Serrano, R. (1983). In vivo glucose activation of the yeast plasma membrane ATPase. *FEBS Letters* **156**, 11–14.

Serrano, R. (1985). *Plasma Membrane ATPase of Plants and Fungi.* Boca Raton, Florida: CRC Press.

Sussman, M.R. & Harper, J.F. (1989). Molecular biology of the plant plasma membrane. *Plant Cell* **1**, 953–60.

Traas, J.A. (1990). The plasma membrane associated cytoskeleton. In *The Plant Plasma Membrane* (ed. C. Larsson & I.M. Moller), pp. 269–92. Berlin: Springer-Verlag.

H.G. NIMMO

The regulation of phosphoenolpyruvate carboxylase by reversible phosphorylation

Introduction

There is increasing interest in the possible roles of reversible protein phosphorylation in intracellular signalling mechanisms in higher plants. Phosphorylation is an extremely effective and adaptable regulatory mechanism because the conformation and biological function of the target protein can be altered in a number of different ways (e.g. Barford, 1991). It is involved in the regulation of a plethora of processes in animal cells. Typically the activity of a preexisting protein kinase or phosphatase is altered either by non-covalent binding of a second messenger, such as cyclic nucleotides, Ca^{2+} ions or diacylglycerol, or by covalent modification, for example via a cascade of phosphorylation reactions. There have been a number of speculations that this animal paradigm may also be applicable to plant cell signalling. In this chapter I review our understanding of the regulation of phosphoenolpyruvate carboxylase (PEPc) in higher plants and suggest that in this case the activity of a protein kinase is controlled by a different mechanism, one that involves protein synthesis.

PEPc catalyses the fixation of CO_2 (as HCO_3^-) into oxaloacetate. It plays an anaplerotic role in non-photosynthetic tissues and in the leaves of C_3 plants (e.g. Andreo, Gonzalez & Iglesias, 1987). However, the leaves of Crassulacean acid metabolism (CAM) and C_4 plants contain distinct isoenzymes of PEPc with specialised roles in CO_2 fixation. In CAM plants external CO_2 is fixed into malate at night via PEPc and malate dehydrogenase, and the malate is stored in the vacuole. During the day the malate is released from the vacuole and decarboxylated, and the resulting CO_2 is re-fixed photosynthetically. In C_4 plants PEPc is located in leaf mesophyll cells and catalyses the first committed step in the C_4 pathway of CO_2 fixation. Consideration of the pathways of carbohydrate metabolism suggests that mechanisms must exist to reduce or

Society for Experimental Biology Seminar Series 53: *Post-translational modifications in plants*, ed. N.H. Battey, H.G. Dickinson & A.M. Hetherington. © Cambridge University Press 1993, pp. 161–170.

eliminate the flux through PEPc at night in C_4 plants and during the day in CAM plants.

Higher plant PEPc is an allosteric enzyme, subject to activation by glucose 6-phosphate and to feedback inhibition by malate (e.g. Andreo *et al.*, 1987). It is widely thought that fluctuations in the cytosolic concentration of malate may play a role in regulating flux through PEPc. Several years ago, however, it became clear from the work of several groups that CAM leaf PEPc must be controlled at an additional level, in that the enzyme is more sensitive to inhibition by malate (i.e. less active under physiological conditions) during the day than at night (Kluge, Brulfert & Queiroz, 1981; Winter, 1982; Buchanan-Bollig & Smith 1984; Nimmo *et al.*, 1984). Further investigations led to the discovery that CAM PEPc is regulated by reversible phosphorylation, and this conclusion was subsequently extended to C_4 plants.

Phosphorylation of PEPc in intact tissue

Diurnal changes in the activity of PEPc were studied systematically by examining the properties of the enzyme in freshly prepared and desalted extracts of the CAM plant *Bryophyllum fedtschenkoi*. This work (Nimmo *et al.*, 1984) revealed four important points. First, PEPc was significantly more sensitive to inhibition by malate during the light period (apparent K_i for malate 0.3 mM) than in the middle of the dark period (apparent K_i 3.0 mM). Secondly, the specific activity of PEPc measured under V_{max} conditions did not change significantly throughout the diurnal cycle. Immunochemical studies subsequently confirmed the suggestion that synthesis–degradation of PEPc was not involved. Thirdly, the conversions of the 'night' to the 'day' form of PEPc and vice versa coincided with the cessation and onset of malate accumulation, respectively, suggesting that these conversions are very important in regulating the flux through PEPc. Finally, both conversions occurred during the dark period, about 1–2 h before the lights came on in the morning and about 4–6 h after the lights went off at night. This information suggested that the conversions between the two kinetically distinct forms of PEPc are controlled by a circadian rhythm rather than by light–dark transitions.

The molecular mechanism responsible for these conversions was then investigated, first by immunoprecipitation of PEPc from detached leaves that had been prelabelled with $^{32}P_i$. This showed that the 'night' form of the enzyme contained ^{32}P whereas the 'day' form did not (Nimmo *et al.*, 1984). On purification of the two forms of the enzyme it was found that the 'night' form was phosphorylated on serine residues (Nimmo *et al.*, 1986). Moreover, removal of the phosphate groups by treatment with alkaline phosphatase increased the malate sensitivity of the purified 'night' form of PEPc to that of the 'day' form. These data indicated that the conversion of the 'day' to the 'night' form of *Bryophyllum* PEPc correlated with phosphorylation of the enzyme. A similar conclusion was reported for the PEPc of *Kalanchoë* species (Brulfert *et al.*, 1986).

Detached *Bryophyllum* leaves can show persistent circadian rhythms of CO_2 metabolism in continuous darkness or continuous light (e.g. Wilkins, 1959, 1984). In continuous darkness and CO_2-free air at 15 °C, leaves exhibit a rhythm of CO_2 output that is directly attributable to changes in flux through PEPc (Warren & Wilkins, 1961); CO_2 that would otherwise be released is periodically re-fixed into malate. Hence the question arose as to whether phosphorylation of PEPc could be responsible for this circadian rhythm in flux through the enzyme. Kinetic studies and immunoprecipitation from $^{32}P_i$-labelled leaves showed that, under these conditions (continuous darkness, CO_2-free air, 15 °C), PEPc exhibited a persistent circadian rhythm of interconversions between a malate-sensitive, dephosphorylated form and a less sensitive, phosphorylated form. Importantly, the changes in the K_i of the PEPc for malate were exactly in phase with the rhythm of CO_2 output (Nimmo *et al.*, 1987*b*). These results significantly strengthened the conclusions that phosphorylation of *Bryophyllum* PEPc (a) regulates flux through the enzyme and (b) is controlled by an endogenous circadian rhythm rather than by changes in illumination.

In parallel with these studies of CAM leaf PEPc, several groups reported that the PEPc of C_4 plant leaf tissue could be interconverted between two kinetically distinct forms (e.g. Karabourniotis, Manetas & Gavalas, 1983; Huber & Sugiyama, 1986; Doncaster & Leegood, 1987; Nimmo *et al.*, 1987*a*; Jiao & Chollet, 1988). In our hands (Nimmo *et al.*, 1987*a*) the K_i of maize PEPc for malate changes some 2- to 3-fold, from about 0.4 mM to 1.0 mM, upon illumination. The mechanism responsible for this reduction in malate sensitivity of maize leaf PEPc upon illumination was investigated both by immunoprecipitation of PEPc from $^{32}P_i$-labelled tissue (Nimmo *et al.*, 1987*a*) and by purification of both forms of the enzyme (Jiao & Chollet, 1988). Both studies showed that illumination led to a significant increase in the phosphorylation state of PEPc, though it is not yet clear to what extent this affects metabolic flux through the enzyme.

In summary, this work showed that the phosphorylation state of PEPc exhibited a circadian rhythm in *Bryophyllum* and was controlled by light–darkness in maize. In order to understand the underlying mechanisms, the properties of the enzymes responsible for phosphorylation and dephosphorylation of PEPc were investigated.

Properties and regulation of PEPc kinase and phosphatase

Recently, a significant insight into plant protein phosphatases has been obtained by comparing them with their mammalian counterparts. Mammalian tissues contain four main types of protein phosphatase catalytic subunit (termed 1, 2A, 2B and 2C) that dephosphorylate protein phosphoserine or phosphothreonine residues. These enzymes can be distinguished by their substrate specificities, sensitivities to certain inhibitors and requirements for activity (e.g. Cohen, 1989). It is now clear that higher plants contain activities that are very similar to the mammalian phosphatases 1, 2A and 2C (MacKintosh & Cohen, 1989; Carter et al., 1990; MacKintosh, Coggins & Cohen, 1991; MacKintosh & MacKintosh, this volume); they must also contain additional phosphatases, for example in chloroplasts (Mackintosh et al., 1991).

Following up this work, we showed that the phosphorylated forms of *Bryophyllum* and maize PEPc could be dephosphorylated by the purified catalytic subunit of protein phosphatase 2A from rabbit skeletal muscle, but not by the type 1 enzyme. Dephosphorylation of PEPc completely reversed the effects of phosphorylation on its sensitivity to malate (Carter et al., 1990; McNaughton et al., 1991). We partially purified the type 2A catalytic subunit from *Bryophyllum* and showed that it, too, could dephosphorylate PEPc (Carter et al., 1990). The activity of protein phosphatase 2A does not change significantly during the normal diurnal cycle in *Bryophyllum*, nor is it affected by illumination in maize (Carter et al., 1991; McNaughton et al., 1991), which suggests that the phosphorylation state of PEPc might be regulated largely by changes in the activity of PEPc kinase.

PEPc kinase has been partially purified from *Bryophyllum* (Carter et al., 1991), maize (Jiao & Chollet, 1989; McNaughton et al., 1991) and sorghum (Bakrim et al., 1992). This enzyme can phosphorylate the dephosphorylated form of PEPc to a stoichiometry approaching 1 per subunit and causes essentially the same changes in kinetic properties of PEPc as are observed with intact tissue (Carter et al., 1991; McNaughton et al., 1991). In addition, sorghum contains a Ca^{2+}-calmodulin dependent protein kinase that can phosphorylate PEPc (Echevarria et al., 1988). However, this kinase does not appear to affect the kinetic properties of PEPc, presumably because it phosphorylates a distinct site (Bakrim et al., 1992). The Ca^{2+}/calmodulin-independent enzyme which does affect the kinetic properties of PEPc is termed PEPc kinase; it phosphorylates *in vitro* the same serine residue as is labelled in intact tissue, which has now been identified as Ser15 in the maize enzyme (Jiao & Chollet, 1990; Carter et al., 1991; Jiao et al., 1991b). The activity of PEPc kinase *in vitro* was not greatly affected by a number of potential effectors (Jiao & Chollet, 1989; Carter *et al.*, 1991; McNaughton *et al.*, 1991) and this work did not yield much information about how the kinase activity might be regulated. However, studies of changes in PEPc kinase activity during the diurnal cycle in *Bryophyllum* and in response to illumination in maize and sorghum proved much more informative (Echevarria *et al.*, 1990; Carter *et al.*, 1991; McNaughton *et al.*, 1991; Jiao *et al.*, 1991a; Bakrim *et al.*, 1992).

To measure PEPc kinase activity, freshly prepared and desalted leaf extracts are incubated with [γ-^{32}P]ATP either with or without the purified dephosphorylated form of PEPc (Carter *et al.*, 1991). Samples are denatured and resolved by SDS–polyacrylamide gel electrophoresis and the incorporation of ^{32}P into the exogenous PEPc is monitored by autoradiography. Application of this technique to *Bryophyllum* leaf extracts showed that PEPc kinase activity appeared some 4–6 h after the lights went off at night and disappeared some 2 h before they went on in the morning. Leaf extracts that contained PEPc kinase activity also contained PEPc in the phosphorylated form with a high K_i for malate, whereas in those extracts lacking PEPc kinase activity, PEPc was in the dephosphorylated form (Carter *et al.*, 1991). These results, allied to the fact that the activity of protein phosphatase 2A did not change significantly during the diurnal cycle, indicated that the major factor that determines the phosphorylation state of PEPc is the presence or absence of PEPc kinase activity.

The mechanism underlying the nocturnal appearance of PEPc kinase was investigated by allowing leaves to take up inhibitors of protein or RNA synthesis. Treatment of leaves with puromycin, cycloheximide, rifampicin or actinomycin D all prevented the nocturnal appearance of PEPc kinase (Carter *et al.*, 1991, and unpublished data). This indicates that *de novo* synthesis of a protein is required for the appearance of PEPc kinase. This protein could be the kinase itself, or another component that is required to activate the kinase. *Bryophyllum* leaves kept in constant environmental conditions show a persistent rhythm of appearance and disappearance of PEPc kinase activity (Carter *et al.*, 1991), indicating that the protein synthesis step exhibits a circadian rhythm.

Darkened maize leaves contained little detectable PEPc kinase activity, but illumination caused a marked increase in activity after 20–60 min (Echevarria *et al.*, 1990; McNaughton *et al.*, 1991). This increase in activity was prevented by prior treatment of the leaves with cycloheximide, whereas the light activations of NADP-dependent malate dehydrogenase and pyruvate, phosphate dikinase were not affected (Jiao *et al.*, 1991a). Similar data have recently been reported for sorghum

(Bakrim *et al.*, 1992). Hence it can be concluded that, in C_4 plants as in the CAM plant *Bryophyllum*, protein synthesis is required for the activation of PEPc kinase. In C_4 plants, however, this activation is triggered by light. Inhibitors of non-cyclic electron flow (e.g. methyl viologen, DCMU) and of the Calvin cycle (DL-glyceraldehyde) can block the light activation of PEPc in C_4 plants (Karabourniotis *et al.*, 1983; Samaras, Manetas & Gavalas, 1988; McNaughton *et al.*, 1991). It has recently been shown that these inhibitors can all reduce the light-triggered activation of PEPc kinase (Jiao & Chollet, 1992). Non-cyclic electron flow from H_2O to NADP is largely, if not entirely, restricted to the granal mesophyll chloroplasts in maize (e.g. Hatch, 1987), so the electron transport inhibitors presumably exert their effects in the mesophyll tissue. However, the effects of DL-glyceraldehyde and the failure to detect light activation of PEPc in maize mesophyll protoplasts (Jiao & Chollet, 1992) imply that a signal derived from Calvin cycle metabolism in the bundle sheath tissue interacts with the protein synthesis event in the mesophyll tissue in such a way that it is essential for the activation of PEPc kinase.

Fig. 1 shows a simplified diagram of the regulation of PEPc in CAM and C_4 leaf tissue that illustrates the similarity between the two systems. Comparison of the phosphorylation state of *Bryophyllum* PEPc with

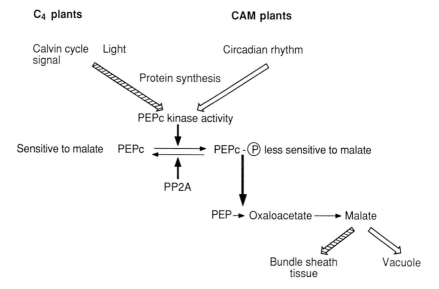

Fig. 1. Phosphorylation of PEPc in CAM and C_4 plants. Hatched arrows, C_4 plants; open arrows, CAM plants; solid arrows, common to both; PP2A, protein phosphatase 2A.

malate accumulation or the rate of CO_2 output by detached leaves suggests that the phosphorylated form of the enzyme is largely or solely responsible for CO_2 fixation into malate. Phosphorylation of PEPc may be essential to allow flux through the enzyme in the face of the cytoplasmic malate concentration. In contrast, in C_4 plants the increase in the phosphorylation state of PEPc is much slower than the onset of the C_4 pathway. Hence, at least in the first 20–30 min of illumination, the dephosphorylated form of PEPc must be capable of CO_2 fixation *in vivo*. (The diagram in Fig. 1 is therefore an over-simplification.) However, it has been argued that the phosphorylation of C_4 leaf PEPc is needed to allow the enzyme to function in the presence of high malate concentrations after longer periods of illumination (Jiao & Chollet, 1992). The functions of the phosphorylation of PEPc in the two systems are therefore similar, and the mechanisms involved (protein synthesis leading to an increase in PEPc kinase activity) also seem similar, at least at this stage in our understanding. The signalling mechanisms through which *Bryophyllum* responds to an endogenous oscillator whereas C_4 plants respond to light may, however, be quite different.

Conclusions

Perhaps the most interesting point to emerge from recent work in the phosphorylation of PEPc in CAM and C_4 plants is that the system does not follow the typical animal paradigm. Changes in the activity of PEPc kinase reflect protein synthesis–degradation (possibly of the protein kinase itself) rather than the effects of a second messenger or a protein phosphorylation cascade. Clearly, purification of PEPc kinase to homogeneity is an important objective that, once achieved, will allow a more detailed analysis of the mechanism involved to be made.

Recent work suggests that the phosphorylation of PEPc is not restricted to CAM and C_4 plants. The enzyme becomes phosphorylated in C_3 plant leaf tissue in response to light and/or nitrogen supply (Van Quy, Foyer & Champigny, 1991) and in stomatal guard cells in response to fusicoccin, which promotes stomatal opening (J.P.S. Nelson, C.A. Fewson & H.G. Nimmo, unpublished). The fact that PEPc becomes phosphorylated in response to different signals in different types of higher plant cell makes the PEPc system an extremely attractive one for studies of the involvement of protein phosphorylation in intracellular signalling in higher plants.

Acknowledgements

Work from this laboratory has been supported by the Agricultural and Food Research Council and carried out in collaboration with Profs. M.B. Wilkins and C.A. Fewson and Drs. G.A. Nimmo, G.A.L. McNaughton and P.J. Carter, to whom I express my thanks.

References

Andreo, C.S., Gonzalez, D.H. & Iglesias, A.A. (1987). Higher plant phosphoenolpyruvate carboxylase. *FEBS Letters* **213**, 1–8.

Bakrim, N., Echevarria, C., Cretin, C., Arrio-Dupont, M., Pierre, J.N., Vidal, J., Chollet, R. & Gadal, P. (1992). Regulatory phosphorylation of *Sorghum* leaf phosphoenolpyruvate carboxylase. *European Journal of Biochemistry* **204**, 821–30.

Barford, D. (1991). Molecular mechanisms for the control of enzymatic activity by phosphorylation. *Biochimica et Biophysica Acta* **1133**, 55–62.

Brulfert, J., Vidal, J., LeMarechal, P., Gadal, P. & Queiroz, O. (1986). Phosphorylation–dephosphorylation process as a probable mechanism for the diurnal regulatory changes of phosphoenolpyruvate carboxylase in CAM plants. *Biochemical and Biophysical Research Communications* **136**, 151–9.

Buchanan-Bollig, I.C. & Smith, J.A.C. (1984). Circadian rhythms in Crassulacean acid metabolism: phase relationships between gas exchange, leaf water relations and malate metabolism in *Kalanchoë daigremontiana*. *Planta* **161**, 314–19.

Carter, P.J., Nimmo, H.G., Fewson, C.A. & Wilkins, M.B. (1990). *Bryophyllum fedtschenkoi* protein phosphatase 2A can dephosphorylate phosphoenolpyruvate carboxylase. *FEBS Letters* **263**, 233–6.

Carter, P.J., Nimmo, H.G., Fewson, C.A. & Wilkins, M.B. (1991). Circadian rhythms in the activity of a plant protein kinase. *EMBO Journal* **10**, 2063–8.

Cohen, P. (1989). The structure and regulation of protein phosphatases. *Annual Review of Biochemistry* **58**, 453–508.

Doncaster, H.D. & Leegood, R.C. (1987). Regulation of phosphoenolpyruvate carboxylase activity in maize leaves. *Plant Physiology* **84**, 82–7.

Echevarria, C., Vidal, J., Jiao, J.A. & Chollet, R. (1990). Reversible light activation of the phosphoenolpyruvate carboxylase protein-serine kinase in maize leaves. *FEBS Letters* **275**, 25–8.

Echevarria, C., Vidal, J., LeMarechal, P., Brulfert, J., Ranjeva, R. & Gadal, P. (1988). The phosphorylation of *Sorghum* leaf phosphoenolpyruvate carboxylase is a Ca^{2+}-calmodulin dependent process. *Biochemical and Biophysical Research Communications* **155**, 835–40.

Hatch, M.D. (1987). C_4 photosynthesis: a unique blend of modified biochemistry, anatomy and ultrastructure. *Biochimica et Biophysica Acta* **895**, 81–106.

Huber, S.C. & Sugiyama, T. (1986). Changes in sensitivity to effectors of maize leaf phosphoenolpyruvate carboxylase during light/dark transitions. *Plant Physiology* **81**, 674–7.

Jiao, J.A. & Chollet, R. (1988). Light/dark regulation of maize leaf phosphoenolpyruvate carboxylase by *in vivo* phosphorylation. *Archives of Biochemistry and Biophysics* **261**, 409–17.

Jiao, J.A. & Chollet, R. (1989). Regulatory seryl-phosphorylation of C_4 phosphoenolpyruvate carboxylase by a soluble protein kinase from maize leaves. *Archives of Biochemistry and Biophysics* **269**, 526–35.

Jiao, J.A. & Chollet, R. (1990). Regulatory phosphorylation of serine-15 in maize phosphoenolpyruvate carboxylase by a C_4-leaf protein-serine kinase. *Archives of Biochemistry and Biophysics* **283**, 300–5.

Jiao, J.A. & Chollet, R. (1992). Light activation of maize phosphoenolpyruvate carboxylase protein-serine kinase is inhibited by mesophyll and bundle sheath-directed photosynthesis inhibitors. *Plant Physiology* **98**, 152–6.

Jiao, J.A., Echevarria, C., Vidal, J. & Chollet, R. (1991*a*). Protein turnover as a component in the light/dark regulation of phosphoenolpyruvate carboxylase protein-serine kinase activity in C_4 plants. *Proceedings of the National Academy of Sciences USA* **88**, 2712–15.

Jiao, J.A., Vidal, J., Echevarria, C. & Chollet, R. (1991*b*). *In vivo* regulatory phosphorylation site in C_4-leaf phosphoenolpyruvate carboxylase from maize and sorghum. *Plant Physiology* **96**, 297–301.

Karabourniotis, G., Manetas, Y. & Gavalas, N.A. (1983). Photoregulation of phosphoenolpyruvate carboxylase in *Salsola soda* L. and other C_4 plants. *Plant Physiology* **73**, 735–9.

Kluge, M., Brulfert, J. & Queiroz, O. (1981). Diurnal changes in the regulatory properties of PEP-carboxylase in Crassulacean Acid Metabolism (CAM). *Plant Cell and Environment* **4**, 251–6.

MacKintosh, C., Coggins, J. & Cohen, P. (1991). Plant protein phosphatases. *Biochemical Journal* **273**, 733–8.

MacKintosh, C. & Cohen, P. (1989). Identification of high levels of type 1 and type 2A protein phosphatases in higher plants. *Biochemical Journal* **262**, 335–9.

McNaughton, G.A.L., MacKintosh, C., Fewson, C.A., Wilkins, M.B. & Nimmo, H.G. (1991). Illumination increases the phosphorylation state of maize leaf phosphoenolpyruvate carboxylase by causing an increase in the activity of a protein kinase. *Biochimica et Biophysica Acta* **1093**, 189–95.

Nimmo, G.A., McNaughton, G.A.L., Fewson, C.A., Wilkins, M.B. & Nimmo, H.G. (1987*a*). Changes in the kinetic properties and phosphorylation state of phosphoenolpyruvate carboxlyase in *Zea mays* leaves in response to light and dark. *FEBS Letters* **213**, 18–22.

Nimmo, G.A., Nimmo, H.G., Fewson, C.A. & Wilkins, M.B. (1984). Diurnal changes in the properties of phosphoenolpyruvate carboxylase in *Bryophyllum* leaves: a possible covalent modification. *FEBS Letters* **178**, 199–203.

Nimmo, G.A., Nimmo, H.G., Hamilton, I.D., Fewson, C.A. & Wilkins, M.B. (1986). Purification of the phosphorylated night form and dephosphorylated day form of phosphoenolpyruvate carboxylase from *Bryophyllum fedtschenkoi*. *Biochemical Journal* **239**, 213–20.

Nimmo, G.A., Wilkins, M.B., Fewson, C.A. & Nimmo, H.G. (1987*b*). Persistent circadian rhythms in the phosphorylation state of phosphoenolpyruvate carboxylase from *Bryophyllum fedtschenkoi* leaves and in its sensitivity to inhibition by malate. *Planta* **170**, 408–15.

Samaras, Y., Manetas, Y. & Gavalas, N.A. (1988). Effects of temperature and photosynthetic inhibitors on light activation of C_4-phosphoenolpyruvate carboxylase. *Photosynthesis Research* **16**, 233–42.

Van Quy, L., Foyer, C. & Champigny, M.L. (1991). Effect of light and NO_3^- on wheat leaf phosphoenolpyruvate carboxylase activity. *Plant Physiology* **97**, 1476–82.

Warren, D.M. & Wilkins, M.B. (1961). An endogenous rhythm in the rate of dark-fixation of carbon dioxide in leaves of *Bryophyllum fedtschenkoi*. *Nature (London)* **191**, 686–8.

Wilkins, M.B. (1959). An endogenous rhythm in the rate of carbon dioxide output of Bryophyllum. I. Some preliminary experiments. *Journal of Experimental Botany* **10**, 377–90.

Wilkins, M.B. (1984). A rapid circadian rhythm of carbon dioxide in *Bryophyllum fedtschenkoi*. *Planta* **161**, 381–4.

Winter, K. (1982). Properties of phosphoenolpyruvate carboxylase in rapidly prepared, desalted leaf extracts of the Crassulacean acid metabolism plant *Mesembryanthemum crystallinum* L. *Planta* **154**, 298–309.

L. RENSING, W. KOHLER, G. GEBAUER
and A. KALLIES

Protein phosphorylation and circadian rhythms

Introduction

The basic cellular oscillator which controls the various circadian rhythms of cells and organisms is often called the 'circadian clock'. Knowledge about the mechanism of this 'clock' is emerging only slowly, progress being made particularly in the analysis of the period (per) gene and its products in *Drosophila* and other animal organisms (reviews: Rosbash & Hall, 1989; Young *et al.*, 1989; Hall, 1990). Earlier approaches to the analysis of the clock mechanism were based mainly on phase shifting experiments with perturbing pulses or on biochemical evidence of oscillatory changes. These data indicated that protein synthesis, membrane potential, Ca^{2+}-calmodulin and possibly other second messenger systems had roles in the clock mechanism or in the input pathways (reviews: Edmunds, 1988; Rensing and Hardeland, 1990). Evidence for the involvement of protein phosphorylation in this mechanism has also been indirectly derived from treatments with agents affecting kinase or phosphatase activities and from measurements of circadian rhythmic protein phosphorylation (Schroeder-Lorenz & Rensing, 1987; Techel *et al.*, 1990). In addition, phase shifting pulses of light and elevated temperatures cause – among other effects – changes of protein phosphorylation within the cell (Lauter & Russo, 1990; Nover, 1990).

There are numerous 'downstream' processes, so called 'hands' of the clock, which also oscillate and cannot easily be distinguished from oscillating parts of the mechanism. Circadian rhythms in the phosphorylation state of phosphoenolpyruvate carboxylase (Carter *et al.*, 1991; Nimmo, this volume), for example, may represent such a hand. Similarly, the circadian rhythm in the translational control of the luciferin binding protein in *Gonyaulax* (Morse, Fritz & Hastings, 1990) can be regarded as a hand even though translational control may also play a role in the clock mechanism. In *Neurospora*, upon which organism this article will

Society for Experimental Biology Seminar Series 53: *Post-translational modifications in plants*, ed. N.H. Battey, H.G. Dickinson & A.M. Hetherington. © Cambridge University Press 1993, pp. 171–185.

concentrate, the circadian clock controls the differentiation of aerial hyphae and conidia and is involved in the morphogenesis of concentric bands of these differentiated forms (Rensing, 1992). Protein phosphorylation changes are likely to be involved in this differentiation process.

Is there a role for protein phosphorylation in the clock mechanism?

There is considerable evidence that pulsatile or constant changes of second messenger levels, such as cyclic nucleotides or Ca^{2+}, affect the function of the circadian clock. A role of protein phosphorylation can be deduced from these results, since these second messengers act mainly on protein kinases. However, there are considerable differences among different species as to the extent of the response.

Role of cyclic nucleotides

In several animal cells, phase shifting of the circadian clock has been achieved by manipulating the cAMP level (Eskin & Takahashi, 1983; Gillette & Prosser, 1988). This has not been possible in the avian pineal (Zatz & Mullen, 1988; Nikaido & Takahashi, 1989). The role of cAMP in higher plants is also still ambiguous in this respect even though methylxanthines were reported to have an influence on phase and period length (review: Engelmann & Schrempf, 1980). In *Neurospora*, quinidine, caffeine, aminophylline, theophylline, cyclic nucleotide derivatives and antagonists were moderately effective in phase shifting the clock (Feldman & Dunlap, 1983; Techel *et al.*, 1990); however, these compounds are not as effective as light and temperature pulses. cAMP may thus play a role in the input pathway of the clock rather than being a part of the mechanism (Edmunds, 1988). In the *Aplysia* eye the effect of light on the rhythm is mimicked by the addition of 8 bromo-cGMP (Eskin *et al.*, 1984), which may influence the clock through changes in membrane permeability.

Role of calcium – calmodulin

When the Ca^{2+}-content of cells was changed by applying the ionophore A23187, by different external Ca^{2+}-concentrations or by incubation with Ca^{2+}-channel blockers, phase shifts of the circadian rhythm were observed in a considerable number of organisms (Khalsa & Block, 1990; Edmunds, 1992) including *Neurospora* (Nakashima, 1984; Techel *et al.*, 1990). Calmodulin inhibitors caused even larger phase shifts in *Neurospora* (Nakashima, 1986; Techel *et al.*, 1990), thus indicating a particular role of calmodulin-dependent enzymes, especially protein kinases, in the input pathway or in the mechanism of the clock.

Role of inositol phosphates

In animal cells the second messenger pathway of Ca^{2+} release involves phospholipase C and inositol phosphates (reviews: Berridge & Irvine, 1989; Rasmussen, 1990). Similar pathways may also exist in plants (review: Lehle 1990). In *Neurospora* a considerable Ca^{2+} release was induced in isolated vacuoles by IP_3, a release which could be inhibited by adding dantrolene (Cornelius, Gebauer & Techel, 1989). Dantrolene, on the other hand, was shown to act as a phase shifting agent on the circadian rhythm of conidiation (Techel *et al.*, 1990). At present, there is too little evidence available to ascertain what roles inositol phosphates might play as inputs or parts of the clock mechanism.

Are phase shifts of the clock by light and temperature pulses mediated by protein phosphorylation?

Light and temperature pulses are the most powerful phase shifting signals ('Zeitgeber') of the circadian clock. It is not clear yet how these signals are transmitted from the receptor to the clock within a cell. There is some indirect evidence that cyclic nucleotides, calcium and inositol phosphates as well as protein synthesis and phosphorylation may be involved (Edmunds, 1988). Because of the enormous amount of data on the general effects of light and temperature on cellular processes, only a few recent results in *Neurospora* with regard to the possible effect on the clock will be discussed here.

Light

Light pulses induce transient changes in the IP_3 level in cells of *Neurospora* (Fig. 1). A pulse of 90 s (2.6 W/m^2), for example, induced an IP_3 increase of about 50%, whereas longer pulses (10–60 min) lead to considerable decreases. The kinetics of a short light pulse of 2 s also showed an IP_3 increase after 90 s and after 5 min. Whether or not IP_3 elicits calcium oscillations with periods in the range of seconds (see Berridge & Irvine, 1989) is not clear. A role for IP_3 in light perception of *Neurospora* is also suggested by experiments with inositol-requiring mutants, which showed a higher light sensitivity at low external inositol levels (P.L. Lakin-Thomas, personal communication). Light perception and transmission of the signal to the clock mechanism seem to require the activity of H^+-ATPases, possibly coupled to Ca^{2+} transport (Nakashima

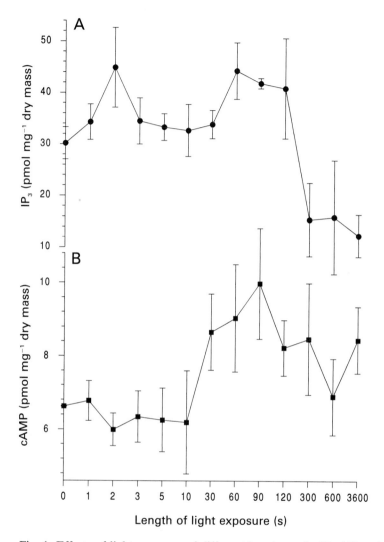

Fig. 1. Effects of light exposure of different length on the IP_3 (A) and cAMP (B) content. The light intensity was 2.6 W m^{-2}. The treated mycelia were immediately cooled to 0 °C after light exposure. The IP_3 and cAMP content were determined by binding proteins with labelled IP_3 and cAMP, respectively. Each point represents the mean of 4–9 independent determinations (bars: standard deviation). Note increasing time intervals on abscissa.

1982; Nakashima & Fujimura, 1982). Light perceived by the blue light receptor also seems to influence the intracellular cAMP concentration (Kritzky *et al.*, 1984; Hasunuma *et al.*, 1987). Recent experiments have corroborated these results in principle, and showed an increase of up to 50% after 30–90 s of light (Fig. 1). After exposure to 2–10 min of light the cAMP level dropped. The kinetics of a short (2 s) pulse again showed an increase in cAMP after 5 min in the dark.

Lauter & Russo (1990) reported that protein phosphorylation in *Neurospora* was altered after a light pulse. They mainly observed a dephosphorylation, particularly in a protein of 33 kDa. This result was corroborated by us when light treated mycelia were homogenised and tested as to phosphorylation changes. In addition, a two dimensional electrophoretic separation of a 45–52 kDa region of proteins revealed two spots of approximately similar M_r but different isoelectric points (Fig. 2). Both spots were dephosphorylated after a 2 min *in vivo* illumination. The response of these phosphorylated bands of M_r 45–52 to *in vivo* illumination was also demonstrated on 1-D gels (Fig. 2). This is of special interest, because the 52 kDa band is also sensitive to heat shock and

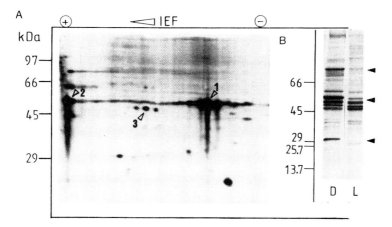

Fig. 2. Light-sensitive protein phosphorylation. (A) Two-dimensional gel electrophoresis of *in vitro* [γ-^{35}S]ATP-labelled proteins (11% acrylamide). Arrows indicate spots which were determined for labelling after light exposure (2 min, 2.6 W m^{-2}) *in vivo* or during incubation with labelled ATP. The phosphorylation of spot 1 was decreased after illumination *in vivo*, whereas spot 2 was only slightly dephosphorylated after illumination *in vivo*. Spot 3 was unchanged. (B) One-dimensional PAGE of *in vitro* [γ-^{32}P]ATP-labelled proteins after 5 min (2.6 W m^{-2}) light exposure *in vivo*. Arrows indicate proteins dephosphorylated by light (L). D, dark control.

shows circadian rhythmicity (see below). Kinases have been shown to exist in this M_r range. Trevillyan & Pall (1982) described a 46 kDa, cAMP-binding protein in *Neurospora*, which resembles the regulatory subunit of mammalian cells and which is autophosphorylated. The isoelectric point (IP) of the regulatory subunit was determined to be pH 5.5 whereas the IP of the catalytic subunit is pH 6.7.

The phosphorylation of proteins within this area (47, 52 kDa) is, however, strongly inhibited by trifluoperazine, an inhibitor of calmodulin and protein kinase C (Techel *et al.*, 1990). The 52 kDa protein may be identical with a protein known to be phosphorylated in a Ca^{2+}-calmodulin-dependent way (Van Tuinen *et al.*, 1984). The proteins of this M_r may in fact be Ca^{2+}-dependent kinases; there is a report concerning two autophosphorylating protein kinases of similar molecular mass in *Alfalfa* (Bögre, Olah & Dudits, 1988) and in other plants (see Battey, Ritchie & Blackbourn, this volume).

Light, furthermore, induces the synthesis of different translatable mRNA species in *Neurospora* after time lags of 2–45 min (Nawrath & Russo, 1990). The mechanism of gene control is not yet clear. There is evidence that light-induced proteins may play a role in the phase-shifting of the clock (Johnson pers. comm.; Raju, Yeung & Eskin, 1990).

Elevated temperature

There is an enormous literature on the effects of elevated temperature (heat shock) on cells, including phosphorylation changes of proteins (reviews: Morimoto, Tissieres & Georgopoulos, 1990; Nover, 1991). In *Neurospora*, a heat shock of 44 °C changed the IP_3 as well as the cAMP level (Fig. 3). With increasing duration of the elevated temperature treatment, the IP_3 level increased to about 50% above that of the controls (after 5 min) and then dropped and increased again after longer exposures. A similar response can be observed in the cAMP level: an initial maximum is reached after 2 min at 44 °C and a second maximum after 45 min.

The Ca^{2+} leakage through the membranes of isolated vacuoles was also enhanced after raising the temperature (Cornelius *et al.*, 1989). A dephosphorylation was observed including the 52 kDa and the 33 kDa proteins after heat shock in most of the cases. The degree of dephosphorylation seemed to vary with the phase of the circadian period (Techel, Kohler & Rensing, unpubl.). Thus, the 52 kDa proteins seem to represent a target for both light and temperature signals.

It is interesting to look for other common targets since light and temperature signals most often show similar phase shifting effects on the

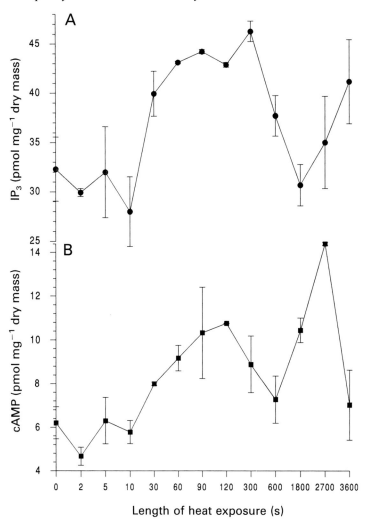

Fig. 3. Effects of heat (44 °C) exposure for different lengths of time on the IP$_3$ (A) and cAMP (B) contents. Details as for Fig. 1.

circadian rhythm. Light- and heat shock-inducible genes in *Neurospora* are apparently different, even though similarities in M_r and IP exist in the 38 kDa region (Kohler & Rensing, 1992). The role of heat shock proteins in the resetting or functioning of the circadian clock is not clear (Rensing *et al.*, 1987).

Circadian changes of second messengers and protein phosphorylation

Up to this point the possible input pathways to the clock have been discussed (Fig. 6). These processes should be sensitive to light and temperature changes but need not oscillate. Parts of the clock, or its 'hands', should, on the other hand, show a circadian oscillation.

The IP_3 level in *Neurospora* seems to fluctuate with a short period length of about 3–4 h (Fig. 4). No obvious circadian rhythmicity was detected, except for some cases in which a semicircadian oscillation

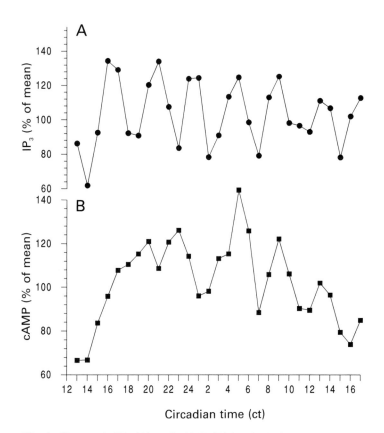

Fig. 4. Changes in IP_3 (A) and cAMP (B) levels during the first circadian period in constant darkness (mean of two experiments). Data were processed by using a moving average ($n=2$). Ordinate, percentage deviation from the mean; abscissa, circadian time (ct); one ct is the period length of one circadian period (21.5 h) divided by 24.

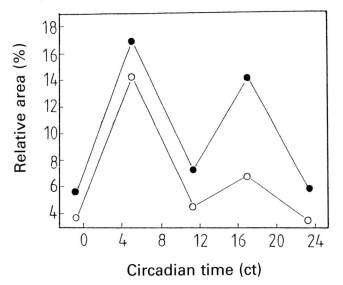

Fig. 5. Changes of *in vivo* phosphorylation of proteins of about 50 kDa during a circadian period (see Fig. 2). Densitometric measurements of one-dimensional gels revealing the area of the band as a percentage of the total gel area. (From Techel *et al.*, 1990.)

appeared. To our knowledge, there are no reports in the literature on rhythmic IP_3 changes.

In contrast, circadian rhythms in cAMP levels are well known (Nikaido & Takahashi, 1989; Prosser & Gillette, 1991; Edmunds, 1992). In *Neurospora*, more or less well defined circadian variations have been observed (Hasunuma *et al.*, 1987; X. Perlman & J. Feldman, personal communication). Most of the *Neurospora* data show two peaks per period. In recent experiments (Fig. 4), a circadian rhythm was measured with a maximum cAMP level at ct 5 and a minimum at about ct 13–16. This rhythm also showed ultradian components of about $\frac{1}{6}$ circadian period length. Previous studies suggested a circadian rhythmicity in the cAMP-diesterase activity as the basis for the cAMP rhythmicity (Techel *et al.*, 1990; Prosser & Gillette, 1991).

The phosphorylation of the proteins of 47–52 kDa described above also showed a bimodal rhythmicity with peaks at around 5 and 17 ct (Fig. 5). Similar bimodal changes were observed in the activity of membrane ATPases in *Neurospora* (G. Lysek, personal communication).

It is difficult to decide whether or not this phosphorylation/dephosphorylation process is part of the clock mechanism or part of the 'hands'.

The same uncertainty applies to phosphorylation changes observed in non-circadian sporulation rhythms of other fungi (review: Jerebzoff & Jerebzoff-Quintin, 1992).

At present, there are no genetic data available connecting kinases or phosphatases to the function of the circadian clock. This point contrasts with the cell cycle oscillator, whose mechanism is much better understood and which definitely consists of protein phosphorylation changes, possibly coupled to Ca^{2+} and H^+-oscillations (Grandin & Charbonneau, 1991).

Circadian clock-controlled differentiation

In *Neurospora*, differentiation of vegetative hyphae into aerial hyphae and conidia is induced by external factors, such as starvation, oxygen and light (Nelson, Selitrennikoff & Siegel, 1975). Heat shock treatment is also able to induce differentiation (Rensing, 1992). Independent of these external inducing factors, differentiation is controlled by the endogenous circadian clock; this is especially clear in the 'band' (bd) mutant (review: Feldman & Dunlap, 1983). The external factors as well as the circadian clock may trigger differentiation via second messenger systems, protein phosphorylation and other signals (Fig. 6).

There is considerable evidence that changes in mammalian cells, including cAMP, diacylglycerol and IP_3 and corresponding synthesis or phosphorylation of transcription factors, play important roles in triggering differentiation. Mammalian *c*AMP *r*esponsive *e*lements (CRE) have been found to activate transcription in yeast and to bind a yeast factor similar to mammalian ATF (Jones & Jones, 1989). Down regulation of protein kinase C activity and c-myc expression are early events in the induction of differentiation, for example, in *neuroblastoma* cells (Wada *et al.*, 1989; Thiele, Reynolds & Israel, 1985).

In *Neurospora*, differentiation-inducing signals, such as light and temperature, affect the levels of cAMP and IP_3 as discussed above. The evidence for circadian rhythmic changes of cAMP concentrations has also been reviewed. Both cAMP and calcium affect growth patterns, branching and morphology (Pall & Robertson 1986; Reissig & Kinney, 1983; Scott & Solomon, 1975). There is also evidence for a role of cAMP in the formation of aerial hyphae and conidia (Hasunuma, 1985). Whether or not the circadian triggering of the differentiation process is elicited by second messenger molecules or by signals which exert a circadian control of gene activity in plants (Otto *et al.*, 1988; Nagy *et al.*, 1988) is not known. In *Neurospora* there is evidence for a clock-control of two genes (Loros, Denome & Dunlap, 1989).

Conidia are formed about 10–12 h after the differentiation-inducing

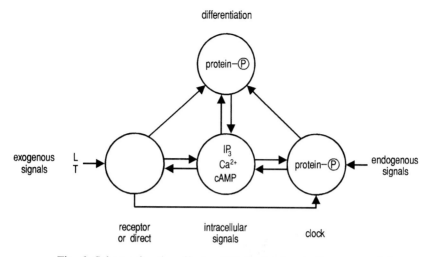

Fig. 6. Scheme for the effects of 'Zeitgeber' and other external factors on the circadian clock and aerial hyphae and conidia formation (differentiation). 'Zeitgeber', such as light and temperature signals, function as control inputs and can reset the circadian clock. The clock, in turn, controls a number of processes, some of which lead to differentiation (endogenous control). Light and temperature – as well as other external factors – are, furthermore, able to induce differentiation directly (exogenous control).

external signal (or the putative signal from the clock). A differential expression of genes is observed along with differentiation (Berlin & Yanofsky, 1985; Roberts *et al.*, 1988). Conidia seem to express stress genes at a higher constitutive level since conidia are exposed to light, higher temperature and oxygen and exist under starving conditions (Scholle & Rensing, unpubl.).

Acknowledgement

We thank Dr M. Vicker for critically reading the manuscript.

References

Berlin, V. & Yanofsky, C. (1985). Isolation and characterization of genes differentially expressed during conidiation of *Neurospora crassa*. *Molecular and Cellular Biology* **5**, 849–55.

Berridge, M.J. & Irvine, R.F. (1989). Inositol phosphates and cell signalling. *Nature (London)* **341**, 197–205.

Bögre, L., Olah, Z. & Dudits, D. (1988). Ca²⁺-dependent protein kinase from *Alfalfa* (*Medicago varia*): partial purification and auto-phosphorylation. *Plant Science* **58**, 135–44.

Carter, P.J., Nimmo, H.G., Fewson, C.A. & Wilkins, M.B. (1991). Circadian rhythms in the activity of a plant protein kinase. *EMBO Journal* **10**, 2063–8.

Cornelius, G., Gebauer, G. & Techel, D. (1989). Inositol trisphosphate induces calcium release from *Neurospora crassa* vacuoles. *Biochemical and Biophysical Research Communications* **162**, 852–6.

Edmunds, L.N. (1988). *Cellular and Molecular Bases of Biological Clocks*. Heidelberg: Springer-Verlag.

Edmunds, L.N. (1992). Regulation of cell division cycles by circadian oscillators: Role of the cAMP pathway. In *Oscillations and Morphogenesis* (ed. L. Rensing), pp. 227–48. New York: Marcel Dekker.

Engelmann, W. & Schrempf, M. (1980). Membrane models for circadian rhythms. *Photochemical and Photobiological Reviews* **5**, 49–86.

Eskin, A. & Takahashi, J.S. (1983). Adenylate cyclase activation shifts the phase of a circadian pacemaker. *Science* **220**, 82–4.

Eskin, A., Takahashi, J.S., Zatz, M. & Block, G.D. (1984). Cyclic guanosine 3′,5′-monophosphate mimics the effects of light on a circadian pacemaker in the eye of *Aplysia*. *Journal of Neuroscience* **4**, 2466–71.

Feldman, J.F. & Dunlap, J.C. (1983). *Neurospora crassa*: A unique system for studying circadian rhythms. *Photochemical and Photobiological Reviews* **7**, 319–68.

Gillette, M.U. & Prosser, R.A. (1988). Circadian rhythm of rat suprachiasmatic brain slice is rapidly reset by daytime application of cAMP analogs. *Brain Research* **474**, 348–53.

Grandin, N. & Charbonneau, M. (1991). Cycling of intracellular free calcium and intracellular pH in *Xenopus* embryos: a possible role in the control of the cell cycle. *Journal of Cell Science* **99**, 5–11.

Hall, J. (1990). Genetics of circadian rhythms. *Annual Review of Genetics* **24**, 659–97.

Hasunuma, K. (1985). Formation of aerial hyphae and conidia under orthophosphate limited conditions in *Neurospora crassa*: Role of repressible cyclic phosphodiesterase. *Botanical Magazine Tokyo* **98**, 203–17.

Hasunuma, K., Funadera, K., Shinohara, Y., Furikawa, K. & Watanabe, M. (1987). Circadian oscillations and light-induced changes in the concentration of cyclic nucleotides in *Neurospora*. *Current Genetics* **12**, 127–33.

Jerebzoff, S. & Jerebzoff-Quintin, S. (1992). Oscillators and morphogenesis in fungi. In *Oscillations and Morphogenesis* (ed. L. Rensing), pp. 311–26. New York: Marcel Dekker.

Jones, R. & Jones, N. (1989). Mammalian cAMP responsive element can activate transcription in yeast and binds a yeast factor(s) that resembles the mammalian transcription factor ATF. *Proceedings of the National Academy of Sciences USA* **86**, 2176–80.

Khalsa, S.B.S. & Block, G.D. (1990). Calcium in phase control of the *Bulla* circadian pacemaker. *Brain Research* **506**, 40–5.

Kohler, W. & Rensing, L. (1992). Light and the recovery from heat shock induce the synthesis of 38 kDa mitochondrial proteins in *Neurospora crassa*. *Archives of Microbiology*, **158**, 5–8.

Kritzky, M.S., Afanasieva, T.P., Belozerskaya, T.A., Chailakhian, L.M., Chernysheva, E.K., Filippovich, S.Y., Levina, N.N., Potapova, T.V. & Sokolovsky, V.Y. (1984). Photoreceptor mechanism of *Neurospora crassa*: Control over the electrophysiological properties of cell membrane and over the level of nucleotide regulators. In *Blue Light Effects in Biological Systems* (ed. H. Senger), pp. 207–21. Heidelberg: Springer-Verlag.

Lauter, F. & Russo, V.E.A. (1990). Light-induced dephosphorylation of a 33 kDa protein in the wild-type strain of *Neurospora crassa*: the regulatory mutants wc-1 and wc-2 are abnormal. *Journal of Photochemistry and Photobiology B* **5**, 95–103.

Lehle, L. (1990). Phosphatidylinositol metabolism and its role in signal transduction in growing plants. *Plant Molecular Biology* **15**, 647–58.

Loros, J.J., Denome, S.A. & Dunlap, J.C. (1989). Molecular cloning of genes under control of the circadian clock in *Neurospora*. *Science* **243**, 385–8.

Morimoto, R.I., Tissieres, A. & Georgopolous, C. (eds) (1990). *Stress Proteins in Biology and Medicine*. Cold Spring Harbor Laboratory Press.

Morse, D.S., Fritz, L. & Hastings, J.W. (1990). What is the clock? Translational regulation of circadian bioluminescence. *Trends in Biochemical Sciences* **15**, 262–5.

Nagy, F., Kay, S.A. & Chua, N.H. (1988). A circadian clock regulates transcription of the wheat Cab-1 gene. *Genes and Development* **2**, 376–82.

Nakashima, H. (1982). Effects of membrane ATPase inhibitors on light-induced phase shifting of the circadian clock in *Neurospora crassa*. *Plant Physiology* **69**, 619–23.

Nakashima, H. (1984). Calcium inhibits phase shifting of the circadian conidiation rhythm of *Neurospora crassa* by the calcium ionophore A23187. *Plant Physiology* **76**, 612–14.

Nakashima, H. (1986). Phase shifting the circadian conidiation rhythm in *Neurospora crassa* by calmodulin antagonists. *Journal of Biological Rhythms* **1**, 163–9.

Nakashima, H. & Fujimura, Y. (1982). Light-induced phase shifting of the circadian clock in *Neurospora crassa* requires ammonium salt at high pH. *Planta* **155**, 431–6.

Nawrath, C. & Russo, V.E.A. (1990). Fast induction of translatable mRNA by blue light in *Neurospora crassa*: the wc-1 and wc-2 mutants are blind. *Journal of Photochemistry and Photobiology* B **4**, 261–71.

Nelson, R.E., Selitrennikoff, C.P. & Siegel, R.W. (1975). Cell changes in *Neurospora*. In *Cell Cycle and Cell Differentiation* (ed. J. Reinert & H. Holtzer), pp. 291–310. Heidelberg: Springer-Verlag.

Nikaido, S.S. & Takahashi, J.S. (1989). Twenty-four hour oscillation of cAMP in chick pineal cells: role of cAMP in acute and circadian regulation of melatonin production. *Neuron* **3**, 609–19.

Nover, L. (ed.) (1991). *Heat Shock Response*. Boca Raton, Florida: CRC Press.

Otto, B., Grimm, G., Ottersbach, P. & Kloppstech, K. (1988). Circadian control of accumulation of mRNAs for light- and heat-inducible chloroplast proteins in pea (*Pisum sativum*). *Plant Physiology* **88**, 21–5.

Pall, M.L. & Robertson, C.K. (1986). Cyclic AMP control of hierarchical growth pattern of hyphae in *Neurospora crassa*. *Experimental Mycology* **10**, 161–5.

Prosser, R.A. & Gillette, M.U. (1991). Cyclic changes in cAMP concentration and phosphodiesterase activity in mammalian circadian clock studied in vitro. *Brain Research* **568**, 185–92.

Raju, U., Yeung, S.J. & Eskin, A. (1990). Involvement of proteins in light resetting ocular circadian oscillators of *Aplysia*. *American Journal of Physiology* **258**, R256–R262.

Rasmussen, H. (1990). The complexities of intracellular Ca^{2+} signalling. *Biological Chemistry Hoppe Seyler* **371**, 191–206.

Reissig, J.L. & Kinney, S.G. (1983). Calcium as a branching signal in *Neurospora crassa*. *Journal of Bacteriology* **154**, 1397–402.

Rensing, L. (1992). Morphogenesis of periodic conidiation pattern in *Neurospora crassa*, its controls by circadian rhythm and daily light and temperature signals. In '*Oscillations and Morphogenesis*' (ed. L. Rensing), pp. 327–41. New York: Marcel Dekker.

Rensing, L., Bos, A., Kroeger, J. & Cornelius, G. (1987). Possible link between circadian rhythm and heat shock response in *Neurospora crassa*. *Chronobiology International* **4**, 543–9.

Rensing, L. & Hardeland, R. (1990). The cellular mechanism of circadian rhythms – a view on evidence, hypotheses and problems. *Chronobiology International* **7**, 353–70.

Roberts, A.N., Berlin, V., Hager, K.M. & Yanofsky, X. (1988). Molecular analysis of a *Neurospora crassa* gene expressed during conidiation. *Molecular and Cellular Biology* **8**, 2411–18.

Rosbash, M. & Hall, J.C. (1989). The molecular biology of circadian rhythms. *Neuron* **3**, 387–98.

Schroeder-Lorenz, A. & Rensing, L. (1987). Circadian change in protein synthesis rate and protein phosphorylation in cell free extracts of *Gonyaulax polyedra*. *Planta* **170**, 7–13.

Scott, W.A. & Solomon, B. (1975). Adenosine 3′,5′-cyclic monophosphate and morphology in *Neurospora crassa*: drug induced alterations. *Journal of Bacteriology* **113**, 1015–25.

Techel, D., Gebauer, G., Kohler, W., Brauman, T., Jastorff, B. & Rensing, L. (1990). On the role of Ca^{2+}-calmodulin-dependent and cAMP-dependent protein phosphorylation in the circadian rhythm of *Neurospora crassa*. *Journal of Comparative Physiology* **159**B, 695–706.

Thiele, C.J., Reynolds, C.P. & Israel, M.A. (1985). Decreased expression of N-myc precedes retinoic acid-induced morphological differentiation of human *neuroblastoma*. *Nature* **313**, 404–6.

Trevillyan, J.M. & Pall, M.L. (1982). Isolation and properties of a cyclic AMP-binding protein from *Neurospora*. *Journal of Biological Chemistry* **257**, 3978–86.

Van Tuinen, D., Ortega-Perez, R., Marme, D. & Turian, G. (1984). Calcium calmodulin-dependent protein phosphorylation in *Neurospora crassa*. *FEBS Letters* **176**, 317–20.

Wada, H., Ohno, S., Kubo, K., Taya, C., Tsuji, S., Yonehara, S. & Suzuki, K. (1989). Down regulation of mRNAs for PKCα and nPKC upon *in vitro* differentiation of a mouse *neuroblastoma* cell line *Neuro 2a*. *Biochemical and Biophysical Research Communications* **165**, 523–8.

Young, M.W., Bargiello, T.A., Baylies, M.K., Saez, L. & Spray, D.C. (1989). Molecular Biology of the *Drosophila* clock. In *Neuronal and Cellular Oscillators* (ed. J.W. Jacklet), pp. 529–42. New York: Marcel Dekker.

Zatz, M. & Mullen, D.A. (1988). Norepinephrine, acting via adenylate cyclase, inhibits melatonin output but does not phase shift the pacemaker in cultured chick pineal cells. *Brain Research* **450**, 137–43.

A.D. SHIRRAS, J. SOMMERVILLE,
A.K. MASTERS and A.M. HETHERINGTON

Control of translation by phosphorylation of mRNP proteins in *Fucus* and *Xenopus*

Introduction

The study of plant gametogenesis, fertilisation and embryonic development has been hindered by the inaccessibility of the gametes and embryo in higher plants. It has therefore been necessary to turn to lower plants as model systems for studying these early phases of development. Large brown marine algae of the genus *Fucus* have been used extensively for this purpose (Quatrano, 1990). This and another recent study (Masters, Shirras & Hetherington, 1992) provides evidence that the eggs of *Fucus serratus* contain maternal mRNA and that this mRNA may be sequestered in mRNP particles in a manner similar to that found in several animal systems. Further, as previous work has demonstrated that phosphorylation–dephosphorylation of messenger ribonucleoproteins plays a role in the control of the translational availability of *Xenopus* stored mRNA, we have sought evidence that a similar strategy is employed in *Fucus* eggs.

Maternal mRNA

The unfertilised *Fucus* egg shows no apparent asymmetry, an axis of symmetry forming after fertilisation in response to external stimuli such as light. When unidirectional light is shone on *Fucus* zygotes the shaded side forms a protuberance. The first cell division is asymmetric, dividing the protuberance from the larger cell mass of the original zygote. The polarisation of the embryo can be divided into two phases: a reversible axis formation and an irreversible axis fixation. Axis fixation depends on a reorganisation of actin filaments and an interaction between cell surface molecules and the forming cell wall (Kropf, Kloareg & Quatrano, 1988; Kropf, Berge & Quatrano, 1989*a*).

Results of experiments using inhibitors of RNA and protein synthesis

Society for Experimental Biology Seminar Series 53: *Post-translational modifications in plants*, ed. N.H. Battey, H.G. Dickinson & A.M. Hetherington. © Cambridge University Press 1993, pp. 187–196.

have shown that early development depends on RNA synthesised during the first 6 h following fertilisation (Kropf, Hopkins & Quatrano, 1989*b*). This suggests that the role played by any maternal mRNA in *Fucus* is minimal. However, when the synthesis of particular proteins during the first 24 h of development was examined, a number of proteins were shown to be synthesised in the presence of actinomycin D, applied immediately after fertilisation, indicating that they are translated from pre-formed mRNA (Kropf *et al.*, 1989*b*). Two of these proteins were identified as actin and β-tubulin.

The existence of maternal mRNA has been well documented in animal species (Davidson, 1986) where it is a common strategy to accumulate mRNAs during oogenesis which are maintained in an untranslated state until fertilisation, when they are used to programme early development. The mechanism of translational control has been studied extensively in the African clawed toad, *Xenopus laevis* (Sommerville, 1990). In this species mRNA accumulates during the early pre-vitellogenic stage of oogenesis and becomes complexed with specific proteins to form messenger ribonucleoprotein (mRNP) particles (Fig. 1). There are four major

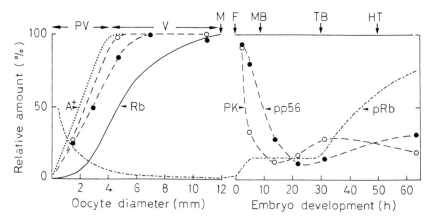

Fig. 1. Changes in pool size of RNP particles during oogenesis and early embryogenesis of *Xenopus laevis*. Relative amounts plotted are: mRNA molecules as estimated from poly (A) titration (A⁺); mRNP-bound casein kinase II activity (PK); non-translating mRNP particles as assayed by immunoblotting with anti-pp56 (pp56); ribosomes as measured by absorbance in sucrose gradients (Rb); ribosomes in polysomes as measured by absorbance in gradients containing high salt and sodium deoxycholate (pRb). (PV) (V) indicate the extents of pre-vitellogenesis and vitellogenesis, respectively. The timing of fertilisation (F), maturation (M), mid-blastula (MB), tail-bud stage (TB) and hatching tadpole (HT) are indicated by arrows. (From Sommerville (1990); with permission.)

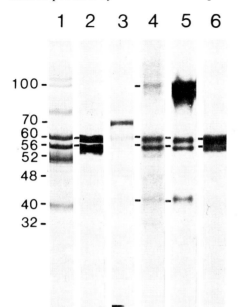

Fig. 2. Proteins of *Xenopus* mRNP particles. Polypeptides are separated by SDS–PAGE and molecular masses (kDa) shown. Track 1, total protein; track 2, proteins selectively bound to oligo (dT)-cellulose; track 3, protein selectively bound to poly(A)-Sepharose; track 4, autoradiograph after phospholabelling *in vivo*; track 5, autoradiograph after phospholabelling *in vitro*; track 6, as track 5 but after phospholabelling the particles were UV cross-linked, denatured in 0.5% SDS, selectively bound to oligo (dT)-cellulose and eluted by treatment with ribonuclease. (From Sommerville (1990); with permission.)

polypeptide components of the mRNP particles with apparent molecular masses of 50, 54, 56 and 60 kDa (Fig. 2). The 56 and 60 kDa proteins have been shown to be heavily phosphorylated *in vivo* (Dearsly *et al.*, 1985) (Fig. 2) and have thus been termed pp56 and pp60. Binding of these proteins to mRNA inhibits translation and this binding depends on the proteins being in the phosphorylated state (Kick *et al.*, 1987). The 56 and 60 kDa proteins become phosphorylated when purified mRNPs are incubated *in vitro* with [^{32}P]ATP, indicating that mRNPs possess endogenous kinase activity (Cummings & Sommerville, 1988). This kinase is of the casein kinase II type (J. Sommerville, unpublished observations). The suggested model for translational control of maternal mRNA in *Xenopus* is therefore that, during early oogenesis, the mRNP-associated kinase is active and phosphorylates the pp56 and pp60 mRNP proteins, allowing them to bind to mRNA and inhibit translation. Kinase activity is seen to rise along with numbers of mRNP particles during the pre-vitello-genic stage of oogenesis (Fig. 1). Following fertilisation there is a sharp decrease in kinase activity, presumably due to the release of kinase inhibitors. At the same time the number of intact mRNP particles falls (Fig. 1). This is thought to be due to an increase in phosphoprotein phosphatase activity, which dephosphorylates the pp56 and pp60 proteins, releasing mRNA and making it available for translation.

Fucus maternal mRNA

To determine whether unfertilised eggs of *F. serratus* contain functional mRNA, RNA was extracted from unfertilised eggs and translated in the wheatgerm or reticulocyte lysate cell-free systems.

Attempts to isolate translatable RNA from *Fucus* eggs by standard methods were unsuccessful owing to the large amounts of sulphated polysaccharides and polyphenols which formed an insoluble aggregate with RNA following ethanol precipitation. Precipitation with 2 M LiCl immediately following phenol–chloroform extraction of the egg homo-

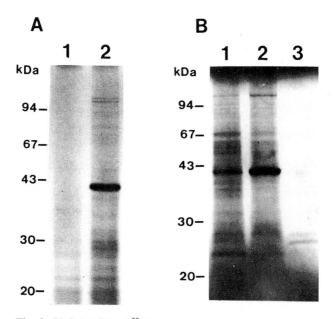

Fig. 3. SDS–PAGE of ^{35}S-labelled *in vitro* translation products of RNA from unfertilised *Fucus* eggs. (A) Wheatgerm lysate: lane 1, no added RNA; lane 2, 100 μg ml^{-1} total RNA. (B) Reticulocyte lysate: lane 1, 40 μg ml^{-1} poly(A)+RNA; lane 2, 100 μg ml^{-1} total RNA; lane 3, no added RNA. (From Masters *et al.* (1992); with permission.)

genate removed most of the contaminating polysaccharide and yielded RNA that could be translated *in vitro*.

Fig. 3 shows the results of translating total RNA in the wheat germ system and total or poly (A)+ RNA in the reticulocyte system. In both cases a range of polypeptides are produced, showing that the unfertilised eggs of *Fucus* contain functional mRNA. The major *in vitro* translation product has a molecular mass of approximately 42 kDa. This is close to the molecular mass of actin monomers; since actin has an important role in axis formation, the identity of this protein was further investigated. An anti-actin antibody was used to immunoprecipitate *in vitro* translation products from the reticulocyte lysate system programmed with RNA from unfertilised *Fucus* eggs. Fig. 4 shows that the immunoprecipitated

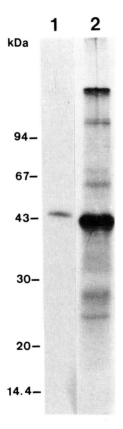

Fig. 4. SDS–PAGE of *in vitro* translation products immunoprecipitated using anti-actin antibody. Lane 1: immunoprecipitated translation products. Lane 2: total translation products programmed by 100 µg ml^{-1} total RNA from *Fucus* eggs. (From Masters *et al.* (1992); with permission.)

polypeptide co-migrates with the prominent 42 kDa *in vitro* translation product, suggesting that it is indeed actin.

In vivo labelling experiments using $Na_2[^{14}C]O_3$ have shown that there is negligible synthesis of actin in unfertilised *Fucus* eggs, its synthesis increasing sharply after fertilisation and then continuing at a constant rate during the first day of development (Kropf *et al.*, 1989*b*). This early synthesis of actin would therefore appear to be programmed by a large store of maternal actin mRNA, the translation of which is blocked until after fertilisation.

Do *Fucus* eggs contain mRNP particles?

The existence of maternal mRNA in *Fucus* eggs, which is apparently translationally blocked, prompted a study to determine whether similarities exist between *Fucus* and the well-characterised *Xenopus* system.

The large amounts of polysaccharide present in *Fucus* eggs did not allow the purification of putative mRNP particles by centrifugation through glycerol or sucrose gradients. Accordingly an alternative strategy was adopted. After homogenisation of *Fucus* eggs the homogenate was centrifuged at 10 000 *g* for 20 min to remove cell debris. The resulting supernatant was then centrifuged at 90 000 *g* for 6 h. The pellet was dissolved and applied to an oligo (dT) cellulose column. After extensive washing with loading buffer, bound material was eluted with 20% or 60% formamide. Formamide was removed by dialysis and the putative mRNP particles concentrated by ethanol precipitation (for full details of the method, see Hetherington *et al.*, 1990).

Previous results have shown there to be around four major polypeptides in material eluted from the oligo (dT) column, with molecular masses in the region 55–75 kDa (Hetherington *et al.*, 1990). To determine whether there is similarity between any of these proteins and the pp56 mRNP protein of *Xenopus*, a Western blot of material eluted from the oligo (dT) column was probed with an antiserum raised against the pp56 protein. The results of this experiment are shown in Fig. 5A. The antibody recognises a protein of approximately 68 kDa in the track containing material from the oligo (dT) column, whereas there is no band detectable in the track containing unbound material even though large amounts of protein are present as indicated by Coomassie Blue staining of a replica sample (data not shown). The 68 kDa protein is therefore clearly enriched in the putative mRNP preparation. To determine whether the cross-reaction observed between the anti-pp56 antiserum and the 68 kDa proteins is genuinely due to epitopes shared between the two proteins, monospecific antibodies were prepared by elution from

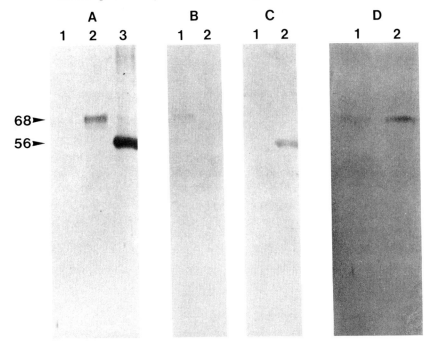

Fig. 5. Western blot analysis of mRNP proteins. To purify *Fucus* mRNP proteins, a 90 000 *g* supernatant from a *Fucus* egg extract was applied to an oligo (dT) cellulose column. After extensive washing with loading buffer, bound proteins were eluted with 60% (v/v) formamide, dialysed and ethanol precipitated. The arrows indicate the positions of the 56 and 68 kDa bands. (A) anti-pp56 antiserum. Track 1: *Fucus* unbound proteins (110 μg). Track 2: *Fucus* bound proteins (5 μg). Track 3: *Xenopus* mRNP proteins (10 μg). (B) Anti-*Fucus* 68 kDa-specific antibodies. Track 1: *Fucus* bound proteins (5 μg). Track 2: *Xenopus* mRNP proteins (10 μg). (C) Anti-*Xenopus* 56 kDa-specific antibodies. Track 1: *Fucus* bound proteins (5 μg). Track 2: *Xenopus* mRNP proteins (10 μg). (D) Anti-*Fucus* 68 kDa-specific antibodies. Track 1: *Fucus* bound proteins (5 μg). Track 2: *Xenopus* mRNP proteins (100 μg).

either the *Fucus* 68 kDa band or the *Xenopus* 56 kDa band and used to re-probe Western blots of *Fucus* and *Xenopus* mRNP proteins. Fig. 5B shows that the *Fucus* 68 kDa-specific antibodies recognise the *Fucus* 68 kDa protein but not the *Xenopus* 56 kDa protein. Conversely, the *Xenopus* 56 kDa-specific antibodies recognise the *Xenopus* 56 kDa protein but not the *Fucus* 68 kDa protein (Fig. 5C). The *Fucus* and *Xenopus* proteins therefore appear to be immunologically unrelated. The observed cross-reaction between the anti-pp56 antiserum and the 68 kDa *Fucus* protein is presumably due to a second population of antibodies within the antiserum. A Western blot of a 10-fold greater amount of *Xenopus* mRNP protein relative to Fig. 5A and probed with *Fucus* 68 kDa-specific antibodies reveals a band of 68 kDa (Fig. 5D). The pp56 used to prepare the antiserum may therefore have been contaminated with this protein. The identity of this protein is at present unknown.

Endogenous protein kinase activity of the putative mRNP fraction was investigated by using *Fucus* mRNP proteins as substrate. Fig. 6 shows that when mRNP proteins are incubated with [^{32}P]ATP the major phosphorylated protein has a molecular mass of approximately 17 kDa. Other labelled polypeptides of molecular mass 42 kDa and 79 kDa are also observed. None of these proteins corresponds to polypeptides observed

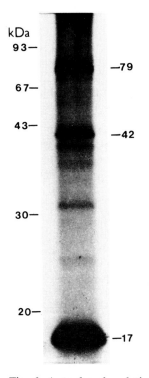

Fig. 6. Autophosphorylation of *Fucus* proteins eluted from oligo (dT) cellulose. Eluted proteins (5 µg) were incubated with 1.85 MBq [γ-^{32}P]ATP (74 TBq mmol^{-1}) for 30 min at 18 °C. Labelled proteins were subjected to SDS–PAGE and visualised by autoradiography. The 17, 42 and 79 kDa phosphorylated products are indicated.

by silver staining in the 55–75 kDa range. Preliminary experiments to investigate the substrate-specificity of the putative mRNP kinase suggest that it belongs to the histone kinase class rather than to the casein kinase II class as described for the *Xenopus* mRNP kinase (J. Sommerville, unpublished observations).

Conclusions and perspectives

Unfertilised eggs of *F. serratus* have been shown to contain maternal mRNA, a large fraction of which apparently codes for actin. Since there is little or no synthesis of actin in the unfertilised egg, the translation of this maternal mRNA must be blocked. Preliminary experiments designed to investigate the nature of this translational control suggest that *Fucus* eggs may share some features in common with *Xenopus* eggs. *Fucus* eggs contain material which can be enriched by oligo (dT) cellulose chromatography and which contains protein and RNA. One of the proteins enriched in the material eluted from the oligo (dT) column cross-reacts with an antiserum raised against a major *Xenopus* mRNP protein. At this stage it is not clear whether the material purified by oligo (dT) cellulose chromatography represents *bona fide* mRNP particles, that is complexes of mRNA and protein which have a functional significance, or whether a subset of proteins and RNA have coincidentally co-purified. UV cross-linking studies will help to resolve this question.

The putative mRNP particles clearly have associated kinase activity but, unlike the *Xenopus* mRNP kinase, do not phosphorylate the major mRNP proteins *in vitro*. The role of protein phosphorylation in translational control must await the purification of larger amounts of mRNP particles so that their translational capacity in the phosphorylated and unphosphorylated states can be determined.

Acknowledgements

A.M.H. is grateful to the Marine Biological Association of the UK and to the Royal Society for providing support for the investigations described in this chapter. We are also grateful to Dr Clive Lloyd (John Innes Centre for Plant Science Research) for the gift of the anti-actin antibody.

References

Cummings, A. & Sommerville, J. (1988). Protein kinase activity associated with stored messenger ribonucleoprotein particles of *Xenopus* oocytes. *Journal of Cell Biology* **107**, 45–56.

Davidson, E.H. (1986). *Gene Activity in Early Development*, 3rd edn. New York: Academic Press.

Dearsly, A.L., Johnson, R.M., Barrett, P. & Sommerville, J. (1985). Identification of a 60 kDa phosphoprotein that binds stored messenger RNA of *Xenopus* oocytes. *European Journal of Biochemistry* **150**, 95–103.

Hetherington, A.M., Sommerville, J., Masters, A.K. & Mitchell, A.G. (1990). Evidence which supports the presence of stored messenger ribonucleoprotein (mRNP) in the unfertilized eggs of *Fucus serratus*. In *Mechanism of Fertilization: Plant to Human*. (NATO ASI, Vol. H45) (ed. B. Dale), pp. 654–61. Berlin: Springer-Verlag.

Kick, D., Barrett, P., Cummings, A. & Sommerville, J. (1987). Phosphorylation of a 60 kDa polypeptide from *Xenopus* oocytes blocks messenger RNA translation. *Nucleic Acids Research* **15**, 4099–109.

Kropf, D.L., Berge, S.K. & Quatrano, R.S. (1989*a*). Actin localization during *Fucus* embryogenesis. *Plant Cell* **1**, 191–200.

Kropf, D.L., Hopkins, R. & Quatrano, R.S. (1989*b*). Protein synthesis and morphogenesis are not tightly linked during embryogenesis in *Fucus*. *Developmental Biology* **134**, 452–61.

Kropf, D.L., Kloareg, B. & Quatrano, R.S. (1988). Cell wall is required for fixation of the embryonic axis in *Fucus* zygotes. *Science* **239**, 187–90.

Masters, A.K., Shirras, A.D. & Hetherington, A.M. (1992). Maternal mRNA and early development in *Fucus serratus*. *Plant Journal*, **2**, 619–22.

Quatrano, R.S. (1990). Model algal systems to study plant development. In *Experimental Embryology in Aquatic Plants and Animals* (ed. H.J. Marthy), pp. 41–55. New York: Plenum Press.

Sommerville, J. (1990). RNA-binding phosphoproteins and the regulation of maternal mRNA in *Xenopus*. *Journal of Reproduction and Fertilization* Supplement 42, pp. 225–33.

R.W. MacKINTOSH and C. MacKINTOSH

Regulation of plant metabolism by reversible protein (serine/threonine) phosphorylation

Introduction: intracellular signalling in plant cells

Plants are able to detect and respond quickly to environmental stimuli. For example, dramatic changes in intracellular metabolism and leaf orientation occur in plants, in response to changing light conditions, that allow them to optimise their ability to harvest energy from the sun at all times of the day. Many plants are disease-resistant because they are able to detect fungal attack and respond by mounting a battery of chemical defences.

Other potential routes for signalling pathways in plant cells are those emanating from specialised internal regulators of timed events such as cell cycle or circadian rhythms (often called 'internal clocks'). There may also be regulatory networks controlling exchange of information between different cellular compartments (e.g. between chloroplast and cytoplasm) which might be mediated by metabolites or by specialised signal molecules.

It is becoming increasingly clear that, as in animal cells (see Hardie, this volume), changes in intracellular protein (serine/threonine) phosphorylation in plants are central in transduction of extracellular signals, in control of the cell cycle and circadian rhythms and in internal metabolic control of cells (Fig. 1).

It is the critical property of the *reversibility* of intracellular protein phosphorylation which suits this mechanism for its role in 'switching' activities of target proteins from one steady state to another and allows for great sensitivity of control and integration of cellular regulation. Therefore, control of protein phosphatase activity is just as important as control of the protein kinases.

Society for Experimental Biology Seminar Series 53: *Post-translational modifications in plants*, ed. N.H. Battey, H.G. Dickinson & A.M. Hetherington. © Cambridge University Press 1993, pp. 197–212.

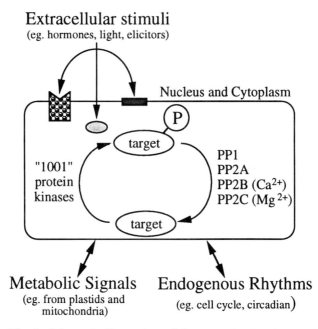

Fig. 1. Schematic illustration of the central role of reversible protein (serine/threonine) phosphorylation in regulation of cytoplasmic and nuclear processes in eukaryotic cells.

Protein phosphatases 1, 2A and 2C in plants

The phosphatases responsible for dephosphorylation of serine and threonine residues of proteins in the cytoplasmic and nuclear compartments of mammalian cells have been classified into two groups; namely, type 1 (protein phosphatase 1) and type 2 (protein phosphatases 2A, 2B and 2C) (for review see Cohen, 1989). Each of these enzymes can be measured individually in cell-free extracts by exploiting differences in their substrate specificities, divalent cation dependencies and response to specific inhibitors and activators (Cohen, Klumpp & Schelling, 1989). Upon applying these procedures to plant extracts, surprisingly high levels of protein phosphatase 1 (PP1), protein phosphatase 2A (PP2A) and protein phosphatase 2C (PP2C) were found in a variety of plant extracts from both mono- and dicotyledonous plants. Moreover, their properties (substrate specificity, sensitivity to inhibitors) were remarkably similar to those of the corresponding enzymes in mammalian cells (MacKintosh & Cohen, 1989; MacKintosh, Coggins & Cohen, 1991). The molecular explanation for these similarities became clear when plant-derived

```
Rabbit PP1α    M S D S E K L N L D S I I G R L L E V Q G S R P G K N V Q L T E N E I R G L C L K S R E I F L S   48
Rabbit PP2Aα                           M D E K V F T K E L D Q W I E Q L N E C K Q L S E S Q V K S L C E K A K E I L T K   41
Brassica PP2A                          S I P T D A T L D L D E Q I S Q L M Q C K P L S E Q Q V R A L C E K A K E I L M D   41

Rabbit PP1α    Q P I L L E L E A P L K I C G D I H G Q Y Y D L L R L F E Y G G F P P E S N Y L F L G D Y V D R   96
Brassica PP1                                   C G D I H G Q Y H D L L R L F E Y G G Y Y P E S N Y L F L G D Y V D R   34
Rabbit PP2Aα   E S N V Q E V R C P V T Y C G D Y H G Q F H D L M E L F R I G G K S P D T N Y L F M G D Y V D R   89
Brassica PP2A  E S N V Q P V K S P V T C G D Y H G Q F H D L M E L F R I G G M C P D T N Y L F M G D Y V D R   89

Rabbit PP1α    G K Q S L E T I C L L L A Y K I K Y P E N F F L L R G N H E C A S I N R I Y G F Y D E C K R R Y   144
Brassica PP1   G K Q S L E T I C L L L A Y K I R Y P S K I X L L R G N H E D A S I N R I Y G F Y D E C K R R E   82
Rabbit PP2Aα   G Y Y S V E T V T L L V A L K V R Y R E R I T I L R G N H E S R Q I T Q V V Y G F Y D E C L R K Y   137
Brassica PP2A  G Y Y S V E T V T L L V G L K V R Y P Q R I T I L R G N H E S R Q I T Q V V Y G F Y D E C L R K Y   137

Rabbit PP1α    - N I K L W K T F T D C F N C L P I A A A I V D E K I F C C H G G L S P D L Q S M E Q I R R I M R   191
Brassica PP1   - N Y R L W K I E T D C F N C L P V A A A I D D K L I C M H G G L S P E L D N L Q I R E I Q R   129
Rabbit PP2Aα   G N A N V W K Y F T D L F D Y L P L T A L V D G Q I F C L H G G L S P S I D T L D H I R A L D R   185
Brassica PP2A  G N A N V W K Y F T D L F D Y L P L T A L V D S E I F C L H G G L S P S I E T L D N I R N F D R   185

Rabbit PP1α    P T D V P D Q G L L C D L L W S D P D K D K G W G E N D R G V S F T F G A E V V A K F L H K H   239
Brassica PP1   P T E I P D S G L L C D L L W S D P D D Q K I E G W G E N D R G V S C T F G A D K V A E F L D K N   177
Rabbit PP2Aα   L Q E V P H E G P M C D L L W S D P D D R G W G I S P R G A G Y T F G Q D I S E T F N H A N   232
Brassica PP2A  V Q E V P H G G P M C D L L W S D P D - D R C G W G I S P R G A G Y T F G Q D I S N Q F N H S N   232

Rabbit PP1α    D L D L I C R A H Q V V E D G Y E F F A K R Q L V T L F S A P N Y C G E F D N A G A M M S V D E   287
Brassica PP1   D L D L I C R G H Q V V E D G Y E F F A K R Q L V T L F S A P N Y G G E F D N A G A L L S V D E   225
Rabbit PP2Aα   G L T L V S R A H Q L V M E G Y N W C H D R N V V T I F S A P N Y C Y R C G N Q A A I M E L D D   280
Brassica PP2A  S L K L I S R A H Q L V M E G Y N W A H E Q K G T I F S A P N Y C Y R C G N M A S I L E L D D   280

Rabbit PP1α    T L M C S F Q I L K P A D K N K G K Y G Q F S G L N P G G R P I T P P R N S A K A K K   330
Brassica PP1   S L V C S F E I M K P A L A S S - - - - H P L K K V P K M G K S   255
Rabbit PP2Aα   T L K Y S F L Q F D P A P R R G E P H V T R R T P D Y F L   309
Brassica PP2A  C R N H T E I Q F L Q F E P A P R R G E P D V T R R T P D Y F L   309
```

Fig. 2. Comparison of the deduced amino acid sequences of plant and mammalian protein phosphatases 1 and 2A. Boxed residues indicate identities and underlines the conservative replacements. These data were originally reported by MacKintosh et al. (1990b).

cDNA clones were isolated whose deduced amino acid sequences were more than 70% identical to the catalytic subunits of mammalian PP1 and PP2A (Fig. 2) (MacKintosh *et al.*, 1990*b*).

Like their mammalian counterparts, PP1 and PP2A in plants are potently inhibited by okadaic acid (the toxin from marine dinoflagellates which causes diarrhetic shellfish poisoning), microcystin (the liver toxin from freshwater blue-green algae) and tautomycin (a polyketide from species of the soil bacterium *Streptomyces*). Okadaic acid, microcystin and tautomycin do not seem to be taken up by intact seeds or plant roots (C. MacKintosh, unpublished). However, these toxins can enter plant cells in culture or through cut surfaces of seeds, roots, stems or leaves of plants (e.g. Siegl, MacKintosh & Stitt, 1990; and C. MacKintosh, unpublished). This means that these compounds are powerful new pharmacological agents for identifying processes in plants that are controlled by reversible phosphorylation. Testing whether application of the toxins to plants mimics, potentiates or prevents any of the controlled responses of plants to environmental stimuli gives clues about which enzymes and cellular processes are regulated by PP1 and/or PP2A. Experiments can then be performed in crude extracts and using purified proteins to confirm (or reject) the idea that target proteins implicated from *in vivo* studies with the toxins are indeed controlled by reversible phosphorylation.

Regulation of light-coupled cytoplasmic enzymes by reversible protein phosphorylation

Sucrose-phosphate synthase

Upon illumination of darkened spinach leaves sucrose-phosphate synthase (SPS) is converted from a low-activity form (sensitive to P_i inhibition, with a high K_m for Fru6P) to a high activity form (less sensitive to inhibition by P_i, with a lower K_m for Fru6P). The effect of light on SPS activity is a consequence of increased rates of photosynthesis rather than a direct effect. Other factors which change photosynthetic rates such as changing CO_2 levels or feeding mannose (a phosphate-sequestering agent) to leaves also affect the activity of SPS (Huber & Huber, 1992).

Experiments performed with purified enzymes and in spinach leaf extracts have demonstrated that the low-activity form of SPS is phosphorylated and can be activated (dephosphorylated) by PP2A but not by PP1 or PP2C (Siegl *et al.*, 1990). Furthermore, *in vivo* (in discs cut from leaves) okadaic acid and microcystin-LR both prevent the light-induced activation of SPS and decrease the rate of sucrose biosynthesis (Siegl *et al.*, 1990).

How the relevant SPS kinase(s) and/or PP2A are controlled in response to changing rates of photosynthesis is not yet understood.

SPS is a cytoplasmic enzyme whose activity is dependent upon substrates generated by light-dependent reactions in the chloroplast (see Fig. 3). Therefore, these studies on SPS highlight the interplay of regulation between the chloroplast and cytoplasm and raise the question of whether other light-coupled cytoplasmic processes (such as nitrate assimilation) might also be regulated by reversible protein phosphorylation.

Nitrate reductase

Nitrate reductase (NR) in leaves is a cytoplasmic NADH-dependent enzyme, which reduces nitrate to nitrite. Nitrite is then converted to ammonia by reduced-ferredoxin-dependent nitrite reductase (NiR) in the chloroplast. The expression of NR is highly regulated by both light and external nitrate at the level of gene transcription and possibly protein turnover (reviewed in Campbell, 1989). However, there are reasons for suspecting that this enzyme might also be a candidate for short-term post-translational regulation during the normal dark–light cycle. Firstly, some mechanism to ensure tight coordination of the activities of NR and NiR might be expected so that nitrite (which is very toxic) is never allowed to build up under conditions (e.g. low light levels) where reduced ferredoxin supply occurs and hence NiR activity is low. Secondly, there are several reports that the activity of NR in extracts does not always match the amount of NR protein measured by immunological methods (Lillo, 1991). In particular, short-term changes in NR activity (within minutes) are sometimes observed with changing light conditions (Remmler & Campbell, 1986) or CO_2 levels (Kaiser & Brendle-Behnisch, 1991), i.e. conditions which alter the rate of photosynthesis. Recently, Kaiser & Spill (1991) and Kaiser, Spill & Brendle-Behnisch (1992) have found that NR activity in extracts of spinach leaves harvested after one hour of illumination could be inactivated in a time-dependent manner in the presence of MgATP.

One of us has performed experiments, *in vivo*, in spinach leaf extracts and using purified enzymes and specific protein phosphatase inhibitors (okadaic acid, microcystin and inhibitor 2) which strongly suggest that nitrate reductase is activated upon illumination due to dephosphorylation by PP2A. *In vitro*, nitrate reductase from illuminated plants can be inactivated by a protein kinase which can be separated from NR by Blue-Sepharose chromatography (MacKintosh, 1992).

When spinach leaves incubated in the dark were illuminated for one hour, subsequent assay revealed that NR activity was increased 8-fold compared with leaves that remained in the dark. In leaves that had been incubated with okadaic acid or microcystin-LR before illumination, this activation of NR was prevented. Methylated okadaic acid and methylated microcystin, which do not inhibit PP1 or PP2A (Holmes *et al.*, 1990; C. MacKintosh, unpublished data), had no effect. None of the toxins had any effect on NR activity in leaves that had been kept in the dark throughout the experiment (MacKintosh, 1992). These experiments suggest (but do not prove) that NR is phosphorylated in the dark and is activated upon illumination due to dephosphorylation by either PP1 or PP2A. However, one must be cautious because the effects of the toxins could be the indirect consequence of other effects. For example, in these experiments it was noticed that leaves treated with toxin took up less water than the control leaves. However, in this case, control of NR activity by reversible phosphorylation was confirmed by experiments performed in cell extracts. Extracts of leaves harvested in the dark had low NR activity, which could be increased by incubation at 30 °C and this activation was prevented by okadaic acid or microcystin but not by inhibitor 2 (a specific inhibitor of PP1) (MacKintosh, 1992). These experiments suggested that PP2A is the major phosphatase in spinach leaf extracts which dephosphorylates and activates NR. NR activity in extracts of leaves that had been illuminated was high and was decreased in a time-dependent manner at 30 °C in the presence of 5 mM $MgCl_2$ plus 1 mM ATP (MacKintosh, 1992).

Taken together, these results demonstrate that NR is regulated by phosphorylation–dephosphorylation in a very similar way to the regulation of SPS. Both enzymes are dephosphorylated and activated by PP2A and it is tempting to speculate that inactivation of both NR and SPS may be catalysed by the same protein kinase (Figure 3). However, interrelationships between carbon and nitrogen assimilation in leaves are very complex and vary enormously under different physiological situations (for example, as discussed by Quy, Foyer & Champigny, 1991). Sometimes, there is *competition* between carbon and nitrogen assimilation for photolytically-derived reducing power; under other circumstances, nitrate assimilation actually *stimulates* photosynthetic carbon metabolism. In order to answer the question of how reversible phosphorylation of NR and SPS contributes to these different patterns of regulation, it will be crucial to determine how the activities of NR and SPS kinases (or kinase) and PP2A are controlled *in vivo*.

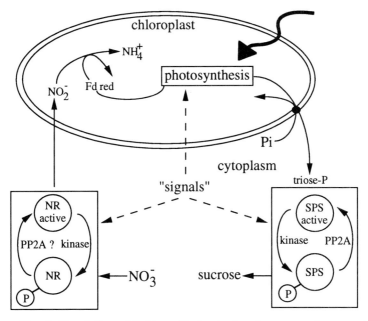

Fig. 3. Regulation of light-coupled cytoplasmic enzymes by reversible protein phosphorylation. Sucrose-phosphate synthase (SPS) and nitrate reductase (NR) are both cytoplasmic enzymes provided with substrates generated by light-dependent reactions in the chloroplast. (NADH is provided to NR via the malate/oxaloacetate shuttle.) The nature of the 'signals' that regulate phosphorylation/dephosphorylation of SPS and NR, to coordinate the activities of these enzymes with photosynthesis, is unknown.

Other functions of PP1 and PP2A in plants

Although our results to date suggest roles for PP2A in the regulation of cytosolic metabolism, it seems likely that as in animal cells PP1 and PP2A will turn out to be involved in regulating many other plant functions such as regulation of gene expression and the cell cycle. For example, PP2A is involved in preventing the activation of p34/cdc 2, the protein kinase which plays a central role in driving cell cycle events in other eukaryotic cells and PP1 activity is required for chromosome separation in insects, fungi and yeast (reviewed in Clarke & Karsenti, 1991).

A highly conserved protein kinase cascade in plants

Bearing in mind the striking similarities observed between the animal and plant protein phosphatases, we decided to investigate whether there were homologues in higher plants of any mammalian protein kinases. Our colleagues in Dundee have been studying a protein kinase cascade in animal cells, the central enzyme of which is the AMP-activated protein kinase (Hardie, Carling & Sim, 1989; Hardie, 1992) which inactivates three regulatory enzymes of mammalian lipid metabolism *in vivo*, i.e. HMG-CoA reductase (cholesterol/isoprenoid biosynthesis), acetyl-CoA carboxylase (fatty acid synthesis), and hormone-sensitive lipase (triglyceride and cholesterol ester breakdown) (Hardie, 1992). The protein kinase can be activated up to 5-fold by 5'-AMP (Carling *et al.*, 1989), but it is also activated a further 10-fold by phosphorylation, and this reaction is catalysed by a distinct kinase kinase (Moore, Weekes & Hardie 1991) (Fig. 4).

Several observations suggested to us that plants might possess a homologue of the mammalian AMP-activated protein kinase. The sequence around the regulatory site on hamster HMG-CoA reductase (serine-871),

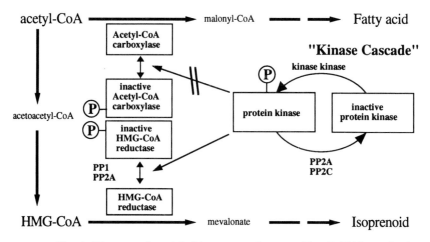

Fig. 4. Diagram of protein kinase cascade controlling lipid biosynthesis. In animal cells the kinase that phosphorylates and inactivates HMG-CoA reductase and acetyl-CoA carboxylase is the AMP-activated protein kinase. In plants this kinase is not activated by AMP and it appears only to inactivate HMG-CoA reductase but not acetyl-CoA carboxylase. In both plants and animals the kinase is itself regulated by phosphorylation, being activated by a kinase kinase and inactivated by PP2A or PP2C.

phosphorylated by the AMP-activated protein kinase (Clarke & Hardie, 1990), is conserved in the deduced amino acid sequences of HMG-CoA reductase from many species, including plants such as *Arabidopsis thaliana* (Learned & Fink, 1989), potato (Stermer *et al.*, 1991), tomato (W. Gruissem, personal communication) and rubber (Chye *et al.*, 1991). Furthermore, rapid changes in HMG-CoA reductase activity have been seen in higher plants responding to stimuli such as light, sterols and hormones, and preliminary evidence has been presented that these changes in HMG-CoA reductase activity are brought about by reversible phosphorylation (Russell, Knight & Wilson, 1985).

The studies on the mammalian AMP-activated protein kinase were greatly helped when a novel and simple assay was developed for the enzyme, involving phosphorylation of a synthetic peptide based on the regulatory phosphorylation site (serine-79) on rat acetyl-CoA carboxylase. This assay is specific for this kinase in rat liver extracts (Davies, Sim & Hardie, 1990). Using the peptide assay we have identified in both monocotyledonous (wheat) and dicotyledonous (cauliflower inflorescence, avocado fruit, pea leaf, carrot cells, oilseed rape seeds, and potato tubers) plants a calcium-independent protein kinase that appears by a variety of functional criteria to be a homologue of mammalian AMP-activated protein kinase (MacKintosh *et al.*, 1992).

The plant and animal kinases fractionate in a similar manner through five purification steps, and comigrate on gel filtration with an apparent molecular mass of 65±5 kDa. The substrate specificities of the plant and animal enzymes are also identical. Apart from efficiently phosphorylating the synthetic peptide substrate, both enzymes phosphorylate and inactivate mammalian HMG-CoA reductase and acetyl-CoA carboxylase. Strikingly, when these proteins are phosphorylated with either the mammalian or plant kinase, digested and analysed by HPLC, the resulting phosphopeptide maps are indistinguishable. This strongly suggests that the plant kinase phosphorylates the same residues as its mammalian counterpart (MacKintosh *et al.*, 1992).

As with the mammalian AMP-activated protein kinase, the plant peptide kinase is itself regulated by reversible phosphorylation. Endogenous PP2A or PP2C can inactivate the peptide kinase in plant extracts (MacKintosh *et al.*, 1992) in a similar manner to the mammalian system, where PP2A and PP2C (and more slowly, PP1) inactivate the AMP-activated protein kinase in cell-free assays (Clarke, Moore & Hardie, 1991). PP2C is thought to perform this function *in vivo* in mammalian cells because okadaic acid (which acts on PP1 and PP2A) has no effect on the activity of the AMP-activated kinase in isolated hepatocytes (Moore *et al.*, 1991).

The inactivated plant peptide kinase can be reactivated in crude

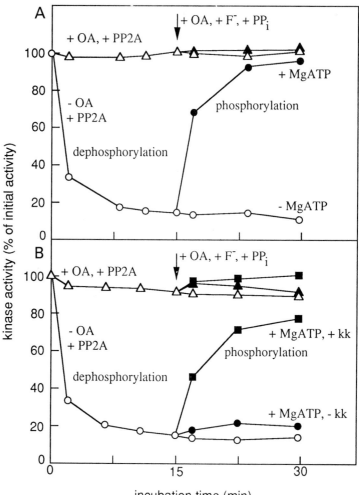

Fig. 5. Regulation of the plant kinase by phosphorylation. (A) Inactivation of crude plant protein kinase by mammalian PP2A and reactivation by endogenous kinase kinase; (B) the same experiment using highly purified cauliflower kinase, showing reactivation by mammalian AMP-activated protein kinase kinase. Extracts were prepared from cauliflower inflorescence in buffer lacking NaF and Na pyrophosphate. This extract was used directly in (A) but in (B) was purified, resulting in a 1000-fold increase in specific activity.

(A) The crude cauliflower kinase was incubated at 30 °C with the purified catalytic subunit of PP2A (30 U ml^{-1}) in the absence (circles) or presence (triangles) of okadaic acid (OA) (1 μM). At the point shown by extracts by addition of MgATP, after inhibiting protein phosphatase activities. However, highly purified peptide kinase could not be reactivated with MgATP (Figure 5). Very similar results are obtained with the animal system, and suggest that the plant peptide kinase is reactivated not by autophosphorylation but by a distinct kinase kinase that is lost during the purification procedure. We are still in search of the plant peptide kinase kinase. However, significantly we have found that the mammalian AMP-activated protein kinase kinase reactivates the dephosphorylated plant kinase (MacKintosh *et al.*, 1992).

Plant fatty acid biosynthesis occurs exclusively in chloroplasts or related non-photosynthetic plastids (Caughey & Kekwick, 1982). Consistent with this finding, plant acetyl-CoA carboxylase is only found in plastids (Mohan & Kekwick, 1980). In contrast, subcellular distribution studies show that HMG-CoA reductase is distributed among the microsomal, chloroplastic and mitochodrial plant cell fractions (Licum *et al.*, 1985). However, to date, all of the full-length cDNA clones for plant HMG-CoA reductase are likely to represent microsomal forms of the enzyme and none have been found with the transit sequences that would be required for import into chloroplasts or mitochondria (von Heijne, Steppuhn & Herrmann, 1989). Subcellular fractionation of leaves on sucrose density gradients shows that the plant kinase is soluble and not chloroplastic (MacKintosh *et al.*, 1992). Similarly, the protein phosphatases that form part of this cascade (PP1, PP2A and PP2C) are not associated with chloroplasts (MacKintosh *et al.*, 1990*a*). Consistent with a cytoplasmic location for the plant kinase cascade, we have found that the highly purified plant and animal kinases can both inactivate potato microsomal HMG-CoA reductase but not acetyl-CoA carboxylase from wheatgerm. The inactivation of HMG-CoA reductase was clearly a result of phosphorylation, since mammalian PP2A reactivated HMG-CoA reductase unless the specific inhibitor okadaic acid was added.

Caption to Fig. 5 (*cont.*)
the arrow, NaF (25 mM) and Na pyrophosphate (PP$_i$) (1 mM) were added, plus okadaic acid (1 μM) if not already present, either with (filled symbols) or without (open symbols) MgCl$_2$ (5 mM) and ATP (0.5 mM). At various times samples were removed for peptide kinase assay. (B) The experiment with purified peptide kinase in panel B is identical except that in some incubations mammalian AMP-activated protein kinase kinase (kk) (0.4 μg/ml, squares) was added at the point shown by the arrow. Kinase kinase had no effect in the absence of MgATP. These data were originally reported by MacKintosh *et al.* (1992).

The only significant difference we have found so far between the plant and animal protein kinases is the lack of AMP activation of the plant peptide kinase (MacKintosh *et al.*, 1992). In addition, the reactivation of the plant kinase by mammalian kinase kinase is not stimulated by AMP. This latter result is significant, because it suggests that in the homologous animal kinase cascade the stimulation by AMP on reactivation of dephosphorylated mammalian kinase (Moore *et al.*, 1991) is due to binding of AMP to the substrate (the kinase) and not the enzyme (the kinase kinase). The lack of any AMP activation of the plant peptide kinase is intriguing. It may be that the plant protein kinase is AMP activated but has a much higher binding affinity for AMP than the mammalian kinase, so the nucleotide remains bound even after 1000-fold purification. Alternatively, a different effector may be used for its regulation in plants. Because the plant kinase apparently lacks activation by AMP, we refer to it as HMG-CoA reductase kinase.

A variety of external stimuli result in rapid inactivation of plant HMG-CoA reductase, including red light via phytochrome (Brooker & Russell, 1979; Bach & Lichtenthaler, 1984), and the isoprenoids abscisic acid, cholesterol, or stigmasterol (Russell & Davidson, 1982). These effects were much more rapid than was seen with protein synthesis inhibitors. Sterols will stimulate phosphorylation of HMG-CoA reductase in animal cells (Beg *et al.*, 1984, 1986), and it may be that extracellular sterols have the same mechanism of action in both plants and animals. We do not yet know whether plant HMG-CoA reductase kinase plays a part in these responses to external stimuli.

Although further investigation is required for complete elucidation of the physiological role(s) of HMG-CoA reductase kinase in higher plants, it would appear that the basic system for controlling isoprenoid biosynthesis has been highly conserved since the estimated divergence of animals and plants around 1000 million years ago. This conservation is highlighted by the finding that heterologous systems of animal or plant kinase, kinase kinase and substrate are fully functional. Detailed structural information will determine how this conservation of catalytic properties is determined at the amino acid sequence level.

Concluding remarks

By making comparisons among mammalian, yeast and plant systems, it is possible to gain a perspective on the principles that govern signal transduction processes and to evaluate how these systems have changed during evolution. Because plants have evolved in response to different selective pressures than animals, it seems likely that more detailed investigation of plant signalling systems will enrich our understanding of the unifying control processes in all eukaryotic cells as well as those unique to plants.

Acknowledgements

The work described here was supported by grants from the AFRC, SERC, MRC, Wellcome Trust, and a Personal Fellowship from the Royal Society of Edinburgh to C. MacK. Many thanks to our colleagues in the Dundee University Biochemistry Department, MRC Protein Phosphorylation Unit and the Scottish Crops Research Institute. We would especially like to thank Grahame Hardie and Philip and Tricia Cohen for their advice and encouragement. Also thanks to Stephen Davies, Paul Clarke, John Weekes, John Gillespie and Barry Gibb for their contributions to the plant kinase cascade work.

References

Bach, T.J. & Lichtenthaler, H.K. (1984). Application of modified Lineweaver-Burk plots to studies of kinetics and regulation of radish 3-hydroxy-3-methylglutaryl Coenzyme A reductase. *Biochimica et Biophysica Acta* **794**, 152–61.

Beg, Z.H., Reznikov, D.C. & Avigan, J. (1986). Regulation of 3-hydroxy-3-methylglutaryl Coenzyme A reductase activity in human fibroblasts by reversible phosphorylation: modulation of enzymic activity by low density lipoprotein, sterols, and mevalonolactone. *Archives of Biochemistry and Biophysics* **244**, 310–22.

Beg, Z.H., Stonik, J.A. & Brewer, H.B. (1984). In vivo modulation of rat liver 3-hydroxy-3-methylglutaryl-coenzyme A reductase, reductase kinase, and reductase kinase kinase by mevalonolactone. *Proceedings of the National Academy of Sciences USA* **81**, 7293–7.

Brooker, J.D. & Russell, D.W. (1979). Regulation of 3-hydroxy-3-methylglutaryl coenzyme A reductase from pea seedlings: Rapid post-translational phytochrome-mediated decrease in activity and *in vivo* regulation by isoprenoid products. *Archives of Biochemistry and Biophysics* **198**, 323–4.

Campbell, W.H. (1989). Structure and regulation of nitrate reductase in higher plants. In *Molecular and Genetic Aspects of Nitrate Assimilation* (ed. J.L. Wray & J.R. Kinghorn), pp. 125–54. Oxford University Press.

Carling, D., Clarke, P.R., Zammit, V.A. & Hardie, D.G. (1989). Purification and characterization of the AMP-activated protein kinase. Copurification of acetyl-CoA carboxylase kinase and 3-hydroxy-3-methylglutaryl-CoA reductase kinase activities. *European Journal of Biochemistry* **186**, 129–36.

Caughey, I. & Kekwick, G.O. (1982). The characteristics of some components of the fatty acid synthetase system in the plastids from the mesocarp of avocado (*Persea americana*) fruit. *European Journal of Biochemistry* **123**, 553–61.

Chye, M.-L., Kush A., Tan, C.-T., & Chua, N.-H. (1991). Characterization of cDNA and genomic clones encoding 3-hydroxy-3-methylglutaryl-coenzyme A reductase from *Hevea brasiliensis*. *Plant Molecular Biology* **16**, 567–77.

Clarke, P.R. & Hardie, D.G. (1990). Regulation of HMG-CoA reductase: identification of the site phosphorylated by the AMP-activated protein kinase *in vitro* and in intact rat liver. *EMBO Journal* **9**, 2439–46.

Clarke, P.R. & Karsenti, E. (1991). Regulation of p34^{cdc2} protein kinase: new insights into protein phosphorylation and the cell cycle. *Journal of Cell Science* **100**, 409–14.

Clarke, P.R., Moore, F. & Hardie, D.G. (1991). The regulation of HMG-CoA reductase and the AMP-activated protein kinase by protein phosphatases. *Advances in Protein Phosphatases* **6**, 187–209.

Cohen, P. (1989). The structure and regulation of protein phosphatases. *Annual Review of Biochemistry* **58**, 453–508.

Cohen, P., Klumpp, S. & Schelling, D.L. (1989). An improved procedure for identifying and quantitating protein phosphatases in mammalian tissues. *FEBS Letters* **250**, 596–600.

Davies, S.P., Sim, A.T.R. & Hardie, D.G. (1990). Location and function of three sites phosphorylated on rat acetyl-CoA carboxylase by the AMP-activated protein kinase. *European Journal of Biochemistry* **187**, 183–90.

Hardie, D.G. (1992). Regulation of fatty acid and cholesterol metabolism by the AMP-activated protein kinase. *Biochimica et Biophysica Acta* **1123**, 231–8.

Hardie, D.G., Carling, D. & Sim, A.T.R. (1989). The AMP-activated protein kinase-a multisubstrate regulator of lipid metabolism. *Trends in Biochemical Sciences* **14**, 20–3.

von Heijne, G., Steppuhn, J. & Herrmann, R.G. (1989). Domain structure of mitochondrial and chloroplast targeting peptides. *European Journal of Biochemistry* **180**, 535–45.

Holmes, C.F.B., Luu, H.A., Carrier, F. & Schmitz, F.J. (1990). Inhibition of protein phosphatases-1 and -2A with acanthafolicin. Comparison with diarrhetic shellfish toxins and identification of a region on okadaic acid important for phosphatase inhibition. *FEBS Letters* **270**, 216–18.

Huber, J.L.A. & Huber, S.C. (1992). Site-specific seryl phosphorylation of spinach leaf sucrose-phosphate synthase. *Biochemical Journal.* **283**, 877–82.

Kaiser, W.M. & Brendle-Behnisch, E. (1991). Rapid modulation of spinach leaf nitrate reductase by photosynthesis I. Modulation *in vivo* by CO$_2$ availability. *Plant Physiology* **96**, 363–7.

Kaiser, W.M. & Spill, D. (1991). Rapid modulation of spinach leaf nitrate reductase by photosynthesis II. *In vitro* modulation by ATP and AMP. *Plant Physiology* **96**, 368–75.

Kaiser, W.M., Spill, D. & Brendle-Behnisch, E. (1992). Adenine nucleotides are apparently involved in the light-dark modulation of spinach-leaf nitrate reductase. *Planta* **186**, 236–40.

Learned, R.M. & Fink, G.R. (1989). 3-Hydroxy-3-methylglutaryl-coenzyme A reductase from *Arabidopsis thaliana* is structurally distinct from the yeast and animal enzymes. *Proceedings of the National Academy of Sciences USA* **86**, 2779–83.

Lillo, C. (1991). Diurnal variations of corn leaf nitrate reductase: an experimental distinction between transcriptional and post-transcriptional control. *Plant Science* **73**, 149–54.

Liscum, L., Cummings, R.D., Anderson, R.G.W., De Martino, G.N., Goldstein, J.L. & Brown, M.S. (1983). 3-Hydroxy-3-methylglutaryl-CoA reductase: A transmembrane glycoprotein of the endoplasmic reticulum with N-linked 'high-mannose' oligosaccharides. *Proceedings of the National Academy of Sciences USA* **80**, 7165–9.

MacKintosh, C. (1992). Regulation of nitrate reductase in spinach leaves by reversible phosphorylation. Dephosphorylation by protein phosphatase 2A. *Biochimica et Biophysica Acta* **1137**, 121–6.

MacKintosh, C., Beattie, K.A., Klumpp, S., Cohen, P. & Codd, G.A. (1990*a*). Cyanobacterial microcystin-LR is a potent and specific inhibitor of protein phosphatases 1 and 2A from both mammals and higher plants. *FEBS Letters* **264**, 187–92.

MacKintosh, C., Coggins, J.R. & Cohen, P. (1991). Plant protein phosphatases: Subcellular distribution, detection of protein phosphatase 2C and identification of protein phosphatase 2A as the major quinate dehydrogenase phosphatase. *Biochemical Journal* **273**, 733–8.

MacKintosh, C. & Cohen, P. (1989). Identification of high levels of type 1 and 2A protein phosphatase in higher plants. *Biochemical Journal* **262**, 335–9.

MacKintosh, R.W., Davies, S.P., Clarke, P.R., Weekes, J., Gillespie, J.G., Gibb, B.J. & Hardie, D.G. (1992). Identification of the first protein kinase cascade in higher plants: HMG-CoA reductase kinase. *European Journal of Biochemistry* (in press).

MacKintosh, R.W., Haycox, G., Hardie D.G. & Cohen, P.T.W. (1990*b*). Identification by molecular cloning of two cDNA sequences from the plant *Brassica napus* which are very similar to mammalian protein phosphatases 1 and 2A. *FEBS Letters* **276**, 155–60.

Mohan, S.B. & Kekwick, R.G.O. (1980). Acetyl-CoA carboxylase from avocado (*Persea americana*) plastids and spinach (*Spinacia oleracea*) chloroplasts. *Biochemical Journal* **187**, 667–76.

Moore, F., Weekes, J. & Hardie, D.G. (1991). AMP triggers phosphorylation as well as direct allosteric activation of rat liver AMP-activated protein kinase. A sensitive mechanism to protect the cell against ATP depletion. *European Journal of Biochemistry* **199**, 691–7.

Quy, L.-V., Foyer, C. & Champigny, M.-L. (1991). Effect of light and NO₃⁻ on wheat leaf phosphoenolpyruvate carboxylase activity. *Plant Physiology* **97**, 1476–82.

Remmler, J.L. & Campbell, W.H. (1986). Regulation of corn leaf nitrate reductase. I. Synthesis and turnover of the enzyme's activity and protein. *Plant Physiology* **80**, 442–7.

Russell, D.W. & Davidson, H. (1982). Regulation of cytosolic HMG-CoA reductase activity in pea seedlings: contrasting responses to different hormones, and hormone-product interaction, suggest hormonal modulation of activity. *Biochemical and Biophysical Research Communications* **104**, 1537–43.

Russell, D.W., Knight, J.S. & Wilson, T.M. (1985). Pea seedling HMG-CoA reductases: regulation of activity in vitro by phosphorylation and Ca²⁺, and posttranslational control in vivo by phytochrome and isoprenoid hormones. *Current Topics in Plant Biochemistry and Physiology* **4**, 191–206.

Siegl, G., MacKintosh, C. & Stitt, M. (1990). Sucrose-phosphate synthase is dephosphorylated by protein phosphatase 2A in spinach leaves: Evidence from the effects of okadaic acid and microcystin. *FEBS Letters* **270**, 198–202.

Stermer, B.A., Edwards, L.A., Edington, B.V. & Dixon, R.A. (1991). Analysis of elicitor-inducible transcripts encoding 3-hydroxy-3-methylglutaryl-coenzyme A reductase in potato. *Physiological and Molecular Plant Pathology* **39**, 135–45.

L. FAYE, A.-C. FITCHETTE-LAINE,
V. GOMORD, A. CHEKKAFI,
A.-M. DELAUNAY
and A. DRIOUICH

Detection, biosynthesis and some functions of glycans N-linked to plant secreted proteins

Introduction

When considering the first steps in plant glycobiology, summarised in this review, it may be wise not to forget the suggestion of a French philosopher about the period that precedes attempts at addressing a new issue:

> L'important est plus de poser les vraies questions que d'apporter les vraies réponses.
>
> Levi-Strauss

Laboratories interested in plant glycobiology are currently asking some very basic questions such as:

> What is a plant glycoprotein glycan made of?
> How is a peptide glycosylated?
> How is an N-linked glycan processed in a plant cell?
> Why are some plant proteins N-glycosylated?

The present review is an attempt to summarise the way these questions are being addressed and the preliminary answers that have already been obtained.

Structures of N-linked glycans

Plant glycoproteins have oligosaccharide sidechains attached to their protein backbone via an N-linkage (amide nitrogen of asparagine) or O-linkage (hydroxyl group of serine, threonine or hydroxyproline). The N-linked oligosaccharides, or glycans, found on plant glycoproteins fall into two general categories already described for other eukaryotes: high-mannose and complex glycans. The high-mannose type oligosaccharides

Society for Experimental Biology Seminar Series 53: *Post-translational modifications in plants*, ed. N.H. Battey, H.G. Dickinson & A.M. Hetherington. © Cambridge University Press 1993, pp. 213–242.

have the general structure $Man_{5-9}(GlcNAC)_2$. High-mannose type oligosaccharides are found associated with immature and mature glycoproteins of higher plants (Faye *et al.*, 1986). High-mannose glycans, first described for soybean lectin (Lis & Sharon, 1978), were observed in many other mature, vacuolar or extracellular plant lectins and enzymes (Paul & Stigbrand, 1970; Ericson & Chrispeels, 1973; Basha & Beevers, 1976; Sturm *et al.*, 1987*a*). The $GlcMan_9(GlcNAc)_2$ glycan of jack bean α-mannosidase (Sturm *et al.*, 1987*a*) has the high-mannose structure with the closest degree of identity with the lipid-linked precursor oligosaccharide $Glc_3Man_9(GlcNAc)_2$, which is co-translationally transferred to glycoproteins during the first step of N-glycosylation (see later). The amount of information available on plant oligosaccharide structures of the complex type has rapidly increased during recent years. Complex oligosaccharides from plant glycoproteins have a $Man_3(GlcNAc)_2$ core structure in common, to which is attached one or more of the following sugar residues: xylose, fucose, N-acetylglucosamine and galactose (see Fig. 1). The complex oligosaccharides in plants differ from those found in

Small biantennary complex oligosaccharide "PHA type"

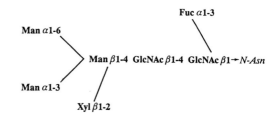

Large biantennary complex oligosaccharide "Laccase type"

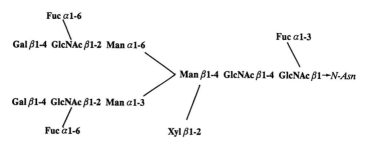

Fig. 1. Structures of the typical complex N-linked oligosaccharides from plant glycoproteins.

mammals in that they lack sialic acid and: (i) have a xylose residue linked β1–2 to the β-linked mannose residue of the glycan core; (ii) may have a fucose in an α1–3 linkage rather than α1–6 linked to the proximal N-acetylglucosamine of the chitobiose unit. The presence of a fucose and/or a xylose residue, in the position and linkage described above, first appeared to be specific to plant glycoproteins. However, the glycan of *Helix pomatia* α-haemocyanin contains a β1–2 xylose (Van Kuik *et al.*, 1985) and recently, an α1–3 linked fucose residue was described in some oligosaccharides N-linked to honey bee venom phospholipase A_2 (Prenner *et al.*, 1991).

Immunochemical studies using anti-carbohydrate antibodies specific for plant complex glycans containing either β1–2 xylose or α1–3 fucose have detected these residues in many glycoproteins from plant seeds, molluscs and insects. They have, however, never been observed in mammalian glycoproteins (Faye & Chrispeels, 1988; L. Faye, unpublished results).

Detection of N-linked glycans

A detailed characterisation of the oligosaccharide sidechains of a glycoprotein can be carried out by a combination of physical, chemical and enzymatic methods. In contrast to these conventional approaches, which require purified glycans and hence purified glycoproteins, we have developed alternative methods involving affino- or immunoblotting for a structural analysis of glycoprotein glycans. In the blotting approach, glycoproteins are electrophoretically separated and then immobilised onto a solid support prior to the interaction with lectins or glycan-specific antibodies. As compared to conventional techniques, the use of blotting for glycoprotein characterisation allows an easy and fast screening to differentiate classical glycan structures from unusual oligosaccharide sidechains.

Affinodetection of high-mannose type glycans with lectins

Lectins are useful analytical tools for the characterisation of glycoprotein or oligosaccharide populations in crude mixtures (for review, see Faye & Salier, 1989). The specificity of plant lectins for animal glycoprotein glycans has been extensively investigated but little is known about the specificity of most lectins for plant glycoproteins. High-mannose glycans have the same structure in plant and animal cells. Therefore lectins specific for these oligosaccharides, such as concanavalin A (ConA), are useful analytical tools for the characterisation and purification of plant glycoproteins and oligosaccharides. However, as described above, complex oligosaccharides N-linked to plant glycoproteins differ from those found in mammals. With the exception of a few lectins specific for α1–6 linked fucose (*Ulex europeus* agglutinin I) and terminal galactose (*Ricinus communis* agglutinin) residues in the complex type glycan of sycamore laccase, there is no other lectin available for a non-ambiguous detection of specific carbohydrate structures in plant complex glycans.

In the lectin-affinoblotting procedure, glycoproteins are separated by electrophoresis and then transferred onto a membrane. After incubation of the protein blot with a given lectin, several methods are available for the detection of glycoprotein–lectin complexes (for review see Faye & Salier, 1989). Glycoprotein detection with ConA is useful because ConA has several sugar-binding sites and displays a strong affinity for horseradish peroxidase oligosaccharide sidechains. The blot is first incubated with ConA, then with peroxidase that binds the residual free site(s) in ConA, and finally with a peroxidase substrate as illustrated in Fig. 2. This is a rapid (3–5 h) and efficient glycan detection procedure that provides a clean blot without any background staining, provided prior saturation of the blot with Tween 20 has taken place (Faye & Chrispeels, 1985). Owing to the use of a covalent lectin–peroxidase complex, this general procedure has been extended to other lectins lacking peroxidase affinity.

Immunodetection of complex type glycans

Plant glycoproteins generally cross-react antigenically and their shared epitopes are most often of a carbohydrate type (Lainé & Faye, 1988). Further studies on this cross-reactivity have shown that it originates from the common presence of highly immunogenic α1–3 linked fucose and/or β1–2 linked xylose residues associated with a $Man_3(GlcNAc)_2$ core unit, constitutive of plant complex N-linked oligosaccharide sidechains (Faye *et al.*, 1989). In a recent study, we have shown that most plant glycoproteins with complex N-linked glycans are probably good sources for the production of glycan-specific immunsera. Antibodies specific for given carbohydrate determinants are easily selected from crude polyclonal immunsera raised against plant glycoproteins. For instance, affinity chromatography on an immobilised honey bee venom phospholipase A_2 column allows for the separation of xylose-specific (unretained) from fucose-specific (retained) antibodies (L. Faye *et al.*, unpublished results).

The amount of information available in plant glycobiology is very limited. It is of major interest to develop new tools to study glycan structures through glycoprotein populations to select species of interest for further detailed studies. Fig. 3 is a good example of the use of fucose-

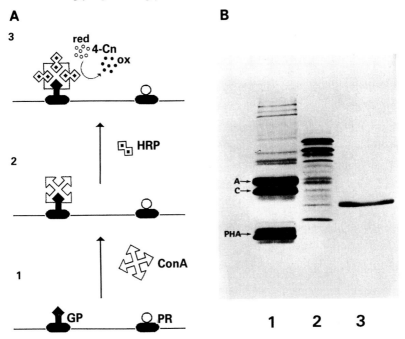

Fig. 2. Affinoblotting with concanavalin A/peroxidase for visualisation of glycoproteins with high mannose type glycans. Proteins (Pr) and glycoproteins (GP) were submitted to SDS–PAGE and blotted onto nitrocellulose. As illustrated in (A), the blot was saturated with 0.1% Tween 20 and successively incubated with ConA (1), then with horseradish peroxidase (HRP) (2). Glycoprotein–ConA–peroxidase complexes are visualised (3) in the presence of 4-chloro-1-naphthol (4-CN) and H_2O_2. Protein extracts analysed in (B) are, respectively: lane 1, extract of cotyledons from developing *Phaseolus vulgaris* seeds; lane 2, extract from sweet pepper fruit; lane 3, purified sweet pepper invertase. Arrowheads on lane 1 point to phytohaemagglutinin (PHA) and to phaseolin polypeptides A and C.

and xylose-specific antibodies in conjunction with immunoblotting for the characterisation of storage glycoprotein glycans in different seeds.

The comparison of blots stained for fucose- and xylose-containing glycoproteins (Fig. 3) confirms that the presence of both β1–2 linked xylose and α1–3 linked fucose in their complex glycan is a common feature for plant glycoproteins. Also consistent with the literature is the immunodetection (Fig. 3) of a limited number of plant glycoproteins with a xylose and no fucose residue in their oligosaccharide sidechains, while no glycoprotein with a fucose but without xylose could be identified.

CB   **Xyl**   **Fuc**

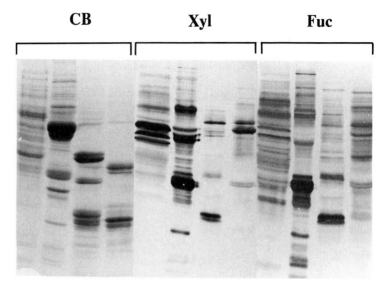

1 2 3 4   1 2 3 4   1 2 3 4

Fig. 3. Immunodetection of storage glycoproteins with fucose and/or xylose residues in their complex glycans. Storage proteins from seeds of wheat (lane 1), bean (lane 2), sunflower (lane 3) and flax (lane 4) were separated by SDS–PAGE and stained with Coomassie brilliant blue in the gel (CB) or blotted onto nitrocellulose. Blots were treated for glycoprotein immunodetection using purified antibodies specific for xylose-(Xyl) or fucose (Fuc)-containing plant complex glycans.

Despite a few drawbacks, affino- or immunoblotting is a sensitive and inexpensive technique that is presently in use in a growing number of laboratories for studies of plant glycoproteins. The blotting approach for glycan characterisation is a fast and easy procedure since it does not require a purification step prior to glycoprotein analysis. A major advantage of this technique is its ability to use denatured glycoproteins. The potential drawback of using a crude tissue extract, namely the presence of endogeneous glycosidase or lectin activities that may induce misleading results, is easily prevented by using denaturing conditions during the extraction procedure prior to the analysis on blots. The glycans may not be accessible to the lectin when the proteins are in their native configuration, but they may be so after denaturation. This is exemplified by the H subunit of jack bean α-mannosidase. This subunit is a ConA-reactive glycopeptide when analysed using affinoblotting, but native α-mannosidase does not bind to ConA on a ConA–Sepharose column (Faye & Salier, 1989).

Biosynthesis of N-linked glycans

N-linked glycosylation of proteins

Oligosaccharides N-linked to plant glycoproteins originate from a common precursor with a $Glc_3Man_9(GlcNAc)_2$ structure. As illustrated in Fig. 4, this oligosaccharide precursor is built up, sugar after sugar, on a

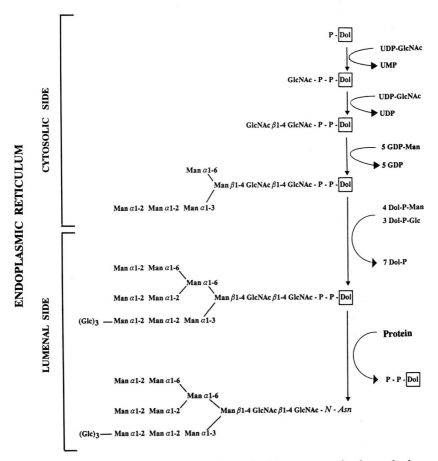

Fig. 4. Biosynthesis of the oligosaccharide precursor in the endoplasmic reticulum. Dol, dolichol; GlcNAc, N-acetylglucosamine; Man, mannose; Glc, glucose. Modified from Abeijon & Hirschberg (1992).

dolichol lipid carrier of the rough endoplasmic reticulum (RER) membrane. In animal cells and probably also in plant cells, the first two GlcNAc residues of the precursor are added to the dolichol carrier, probably on the cytosolic side of the RER membrane. The addition of the first five Man residues from GDP–Man to the $(GlcNAc)_2$–P–P–dolichol then takes place on the cytosolic face. The $Man_5(GlcNAc)_2$–P–P–dolichol next flips from the cytosolic to the lumenal face of the RER membrane. The final stages of elongation of the precursor are mediated through the transfer of four Man and three Glc residues from Man-P–dolichol and Glc-P–dolichol intermediates, which flip from the cytosolic to the lumenal face of the RER (see Hirschberg & Snider, 1987; Abeijon & Hirschberg, 1992).

The oligosaccharide precursor, with a structure $Glc_3Man_5(GlcNAc)_2$ is then co-translationally transferred from the lipid carrier to specific asparagine residues on the nascent polypeptide chain. Protein N-glycosylation occurs on asparagine residues of amino acid sequences Asn–X–Ser/Thr, where X is any amino acid except proline or aspartic acid. However, not all potential glycosylation sequences are glycosylated on a protein backbone, as illustrated by phaseolin polypeptides B and D (Sturm *et al.*, 1987c). In glycosylation sequences, the glycosylated Asn residues are located in peptide regions where the formation of β-turns is favoured (reviewed in Kornfeld & Kornfeld, 1985). Therefore, there are at least two structural requirements for the glycosylation of an Asn residue on a protein backbone: (i) it must be part of a correct tripeptide amino acid sequence; and (ii) it must be located in a β-turn region of the polypeptide.

Maturation of N-linked glycans

Mature N-linked oligosaccharides in plant glycoproteins fall into two general categories: high-mannose-type and complex type oligosaccharides. Complex oligosaccharides have fewer mannose residues than high-mannose glycans and they contain additional sugar residues such as fucose, xylose and galactose. These structural differences between glycan categories result from more or less extensive modifications of their common precursor $Glc_3Man_9(GlcNAc)_2$. High-mannose glycans are formed by limited processing of the precursor, which involves the removal of the three Glc residues and one or more Man residues. Complex oligosaccharides are built up after extensive trimming and addition of new sugar residues. Rapid progress has been made in the characterisation and subcellular localisation of the maturation events of N-linked oligosaccharide side chains in plant glycoproteins. Different glycosidases responsible for the trimming steps have recently been characterised and localised in the RER and Golgi apparatus of plant cells. The oligosaccharide precursor first undergoes a removal of three glucose residues by two membrane-bound α-glucosidases I and II (Szumilo, Kaushal & Elbein, 1986a; Szumilo *et al.*, 1986b). These glucosidases probably act quickly after glycosylation of the polypeptide. However, a transient reglucosylation of plant N-linked glycans may occur subsequently to the removal of these three glucoses in RER lumen (Trombetta, Bosch & Parodi, 1989). The GlcMan$_9$(GlcNAc)$_2$ glycan of the jack bean α-mannosidase (Sturm *et al.*, 1987a) originates from a precursor either exclusively trimmed by the glucosidase I, or completely deglucosylated and then reglucosylated by the RER glucosyltransferase.

En route to their final location, glycoproteins that are not resident in the RER enter the Golgi apparatus, where most other processing enzymes responsible for glycan maturation are located (Sturm *et al.*, 1987b). Removal of mannose residues occurs in the Golgi apparatus of plant cells by the successive action of two α-mannosidases. First, an α1–2 mannosidase (α-mannosidase I) (Szumilo *et al.*, 1986b) removes one to four mannose residues to yield Man$_5$(GlcNAc)$_2$. This Man$_5$(GlcNAc)$_2$ structure is the smallest high-mannose glycan found in plants, but a more limited trimming by α-mannosidase I results in other high-mannose glycan structures such as Man$_{9-5}$(GlcNAc)$_2$, already described for plant glycoproteins (Ericson & Chrispeels, 1973; Basha & Beevers, 1976; Lis & Sharon, 1978).

The sequence of events by which a Man$_5$(GlcNAc)$_2$ structure is converted into a biantennary complex glycan was described by Johnson & Chrispeels (1987). First, a GlcNAc residue is transferred on the α1–3 mannose branch of the Man$_5$(GlcNAc)$_2$ structure. Then two additional mannose residues are removed by α-mannosidase II, prior to the addition of another outer chain GlcNAc residue. As illustrated in Fig. 5, Johnson & Chrispeels (1987) and Kimura *et al.* (1987) proposed two different processing pathways for the biosynthesis of complex oligosaccharide side chains in plants. They are based either on glycosyltransferase substrate specificities in bean cotyledons (Johnson & Chrispeels, 1987) or on a detailed description of the structural diversity of mature sugar chains N-linked to *Ricinus communis* lectins (Kimura *et al.*, 1987). The major differences between these pathways is in the timing of fucosylation and xylosylation events.

We have recently described the subcompartmentation of plant-specific terminal glycosylation events using antibodies reacting with β1–2 xylose- or α1–3 fucose-containing complex glycans. This immunocytochemical approach allows for an indirect localisation of glycosyltransferases by

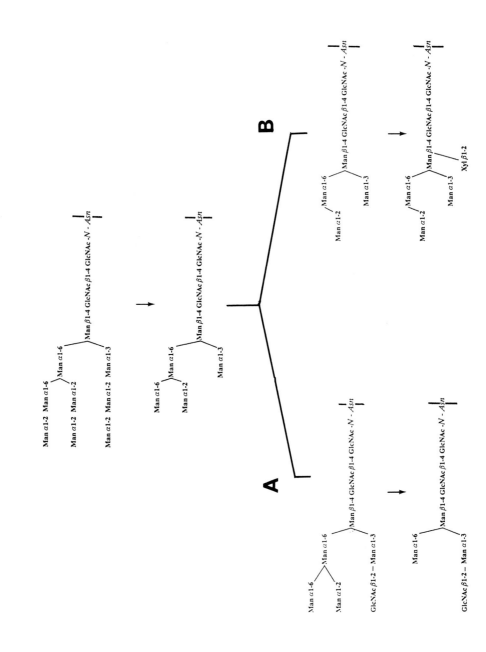

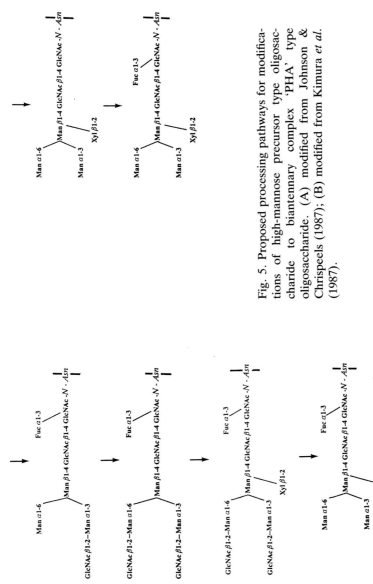

Fig. 5. Proposed processing pathways for modifications of high-mannose precursor type oligosaccharide to biantennary complex 'PHA' type oligosaccharide. (A) modified from Johnson & Chrispeels (1987); (B) modified from Kimura *et al.* (1987).

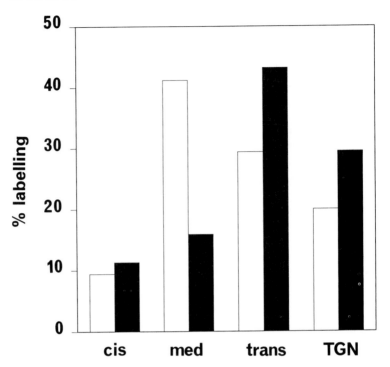

Fig. 6. Histogram illustrating the immunolabelling patterns of Golgi stacks of sycamore cells with fucose- (open bars) and xylose-specific (filled bars) antibodies. The Golgi stacks were subdivided into 4 types of cisternae: *cis* cisternae (*cis*), *medial* cisternae (*med*), *trans* cisternae (*trans*) and *trans*-Golgi network (TGN). The number of gold particles over each type of cisterna is expressed as a percentage of total amount of labelling in the dictyosomes.

mapping the corresponding oligosaccharide products (Lainé, Gomord & Faye, 1991; A.-C. Lainé *et al.*, unpublished results).

Consistent with results summarised in Fig. 6, a possible scheme for N-linked complex glycan biosynthesis in plant cells involves an addition of β1–2 xylose residues, initiated in *cis*-cisternae, but mostly occurring in the *medial* cisternae. Fucosylation of the complex glycans occurs mainly in the *trans*-Golgi cisternae. Thus in plant cells, the sequential nature of complex glycan biosynthesis, or at least of the fucosylation and xylosylation steps is likely controlled by substrate specificity and/or availability rather than by a strict compartmentation of these terminal glycosyltransferases.

The extent of glycan modifications is related to their accessibility to the Golgi processing enzymes

As illustrated above, glycosyltransferases responsible for the biosynthesis of complex glycans are localised in the Golgi apparatus of plants cells (Sturm & Kindl, 1983; Chrispeels, 1985; Sturm *et al.*, 1987c; Lainé *et al.*, 1991; A.-C. Lainé *et al.*, unpublished results). Consequently, the presence of complex glycans on a protein is circumstantial evidence that this protein passes through the Golgi during its intracellular transport (Chrispeels, 1983). The fact that several storage proteins and lectins, such as phaseolin polypeptides A and C (Sturm *et al.*, 1987c) or soybean agglutinin (Lis & Sharon, 1978), have glycans exclusively of the high-mannose type does not reinforce the hypothesis of a direct transport of proteins from the RER to the vacuole or protein bodies. This only shows that some oligosaccharide side chains are modified in the Golgi apparatus while others are not. Phytohaemagglutinin (PHA) glycan structure illustrates this situation. Mature PHA, stored in the protein bodies of *Phaseolus vulgaris* seed cotyledons, has one high-mannose and one complex glycan per polypeptide (Vitale, Warner & Chrispeels, 1984). Newly synthesised PHA in the RER has two high-mannose chains, one attached to Asn^{60}, the other to Asn^{12}. By determining the effect of digestion with various glycosidases, we have shown that native PHA obtained from the RER has only one chain readily accessible to α-mannosidase and endo-glycosidase H, in the same position on the polypeptide (Asn^{60}) as the complex glycan on mature PHA. The second oligosaccharide side chain, N-linked to Asn^{12}, is not very accessible to glycosidases on the native lectin and remains as a high-mannose chain on mature PHA (Faye *et al.*, 1986). Consistent with this is the observation that the oligosaccharide side chains of phaseolin, which are accessible to glycosidases *in vitro*, are also modified in the Golgi apparatus *in vivo* (Sturm *et al.*, 1987b). Furthermore, it was observed that the modified glycan is in a hydrophilic region of the phaseolin (Asn^{252}), whereas the glycan which remains in a high-mannose form is in a hydrophobic region of the polypeptide (Asn^{341}) (Fig. 7). Therefore, it is tempting to speculate that high-mannose chains buried in hydrophobic pockets of the protein backbone are not accessible to processing enzymes and thus remain unmodified. A clear-cut example of this situation is the high-mannose glycan of jack bean α-mannosidase, a $GlcMan_9(GlcNAc)_2$ structure (Sturm *et al.*, 1987a) which, as described above, is not accessible to the lectin concanavalin A (ConA) without complete denaturation of the glycoprotein (Bowles, Chaplin & Marcus, 1983; Faye & Chrispeels, 1985). Results obtained from studies on yeast (Trimble, Maley & Chu, 1983) and viral glycoproteins (Hsieh, Rosner &

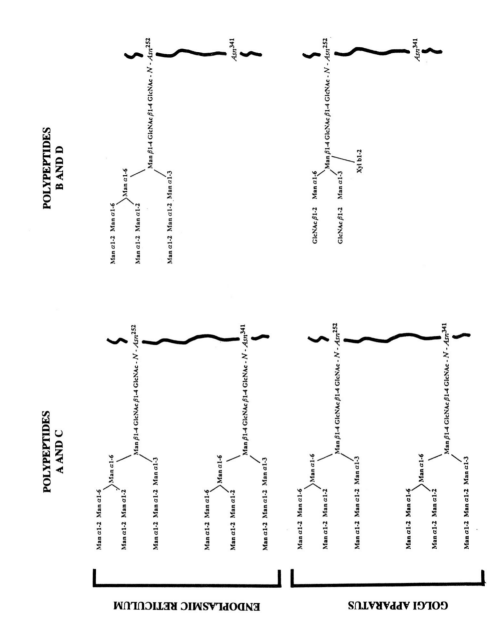

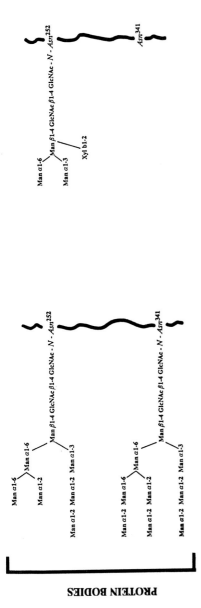

PROTEIN BODIES

Fig. 7. Proposed pathway for the processing of phaseolin oligosaccharides. The characterisation and localisation of the different phaseolin polypeptide oligosaccharide intermediaries were studied using labelling with [^{3}H]glucosamine for different periods of time. In the endoplasmic reticulum, all glycans detected have a $Man_9(GlcNAc)_2$ form. In the Golgi apparatus, unprocessed high-mannose and fully processed complex-type glycans are found. N-acetylglucosamine and some mannose terminal residues are removed in the protein bodies. The oligosaccharide side chains on Asn^{252}, localised in a hydrophilic region of the polypeptide, are processed either into a $Man_7(GlcNAc)_2$ structure for polypeptides A and C or into $Xyl-Man_3(GlcNAc)_2$ for polypeptides B and D. Asn^{341} is located in a hydrophobic pocket, which may indicate that either this glycosylation site is not used (polypeptides B and D), or when used (polypeptides A and C) the oligosaccharide chain N-linked to Asn^{341} remains unmodified. (Modified from Sturm et al., 1987b).

Robbins, 1983) have also shown that the extent of glycan modifications is related to their accessibility to the Golgi processing enzymes. The structure of the protein matrix probably provides the necessary signal(s) for glycan processing to take place. This was also illustrated by the study of PHA synthesised in various heterologous systems. Whether synthesised in tobacco seeds (Sturm et al., 1987d), yeast (Tague & Chrispeels, 1987), *Xenopus* oocytes (Vitale, Sturm & Bollini, 1986) or monkey COS cells (Voelker, Florkiewicz & Chrispeels, 1986), PHA always has a modified and a high-mannose type oligosaccharide chain, as is the case for mature PHA isolated from developing bean cotyledons. Therefore the decision whether or not a glycan will be processed depends on the protein backbone and is independent of the cell type in which protein synthesis occurs. However, the final structure of processed glycans is cell-type-specific.

The exposure of a glycan outside the protein matrix probably determines whether or not the processing is possible, but it does not explain the differences observed in the final structures of plant glycoprotein complex glycans. Some other regulatory factors, such as the concentration of nucleotide sugars or the abundance and specificity of processing enzymes, could determine structural differences observed in complex glycans between (i) different species, (ii) different tissues of the same species, or (iii) different cells of the same tissue.

A single component of the processing machinery, from a homogeneous cellular population or a cellular clone, produces different complex structures on a single glycosylation site of a glycoprotein, as recently illustrated for some extracellular glycoproteins secreted by suspension-cultured carrot cells (Sturm, 1991). To account for this observation, it is necessary to postulate the existence of additional regulatory factors acting at the cellular level. For instance, it is tempting to speculate that the protein matrix of a soluble glycoprotein could interact very heterogeneously with glycan processing enzymes, thus affecting the conformation and catalytic efficiency of the latter.

Some functions of N-linked glycans

Glycans have numerous roles depending on the particular protein to which they are attached. They may contain targeting information (for lysosomal enzymes), prevent the proteolytic degradation of proteins, be necessary for the correct folding and/or biological activity of proteins, allow proteins to interact with carbohydrate-binding sites on other proteins, or alter immunological and physiochemical properties of proteins.

Tools to study the roles of N-linked glycans in glycoprotein functions

There are different approaches to studying the function(s) of the carbo-hydrate moiety of a plant glycoprotein. Generally it is difficult fully to deglycosylate plant glycoproteins enzymatically. Furthermore, chemical procedures developed (Edge *et al.*, 1981) for their deglycosylation are efficient but often affect their native (i.e. biologically active) conforma-tion. Consequently, information on plant glycan functions have rarely been obtained from *in vitro* deglycosylated glycoproteins.

In contrast, glycosylation inhibitors and glycan-processing inhibitors are very useful tools for exploring the roles of N-linked glycans in gly-coprotein functions. Among glycosylation inhibitors whose action has been well studied in plant cells, tunicamycin is the most commonly used. Tunicamycin is a nucleoside antibiotic which inhibits the first enzyme of the dolichol cycle that is responsible for the formation of GlcNAc–P–P–Dol. Tunicamycin prevents N-glycosylation in yeast, animal and plant cells (Hori & Elbein, 1981). The stability and physical properties of some glycoproteins are often dependent on the presence of their oligosac-charide side chains, therefore the glycan-processing inhibitors are often preferred to glycosylation inhibitors for studies on the role of sugar residues in targeting or recognition phenomena. These inhibitors specifi-cally act on the glycosidases responsible for N-glycan processing (reviewed in Elbein, 1987). Castanospermine (Cast) inhibits glucosidase I and gives rise to $Glc_3Man_{7-9}(GlcNAc)_2$ in animal cells (Elbein, 1987) and to $Glc_3Man_{5-7}(GlcNAc)_2$ structures in plants (Hori *et al.*, 1984). Deoxy-mannojirimycin (DMM), which inhibits α-mannosidase I, causes an accumulation of glycoproteins with $Man_{8-9}(GlcNAc)_2$ glycans. Swain-sonine inhibits α-mannosidase II, which results in the production of hybrid glycans in animal and plant cells (Chrispeels & Vitale, 1985). Such hybrid glycan structures are not normally observed in plants. In a recent study, using these glycan-processing inhibitors, it has been shown that the inhibition of complex oligosaccharide side chain biosynthesis does not prevent the secretion of glycoproteins in suspension cultured sycamore cells. This was considered as evidence that a carbohydrate-type signal is not necessary for the targeting of secreted glycoproteins to the extracel-lular compartment in plant cells (Driouich *et al.*, 1989).

Site-directed mutagenesis of glycosylation sites on plant glycoproteins has been successfully used to study the role of N-linked glycans in the intracellular transport of plant vacuolar glycoproteins (Voelker, Herman & Chrispeels, 1989). When a cDNA mutated at glycosylation sites is expressed in transgenic plant cells, glycans are not attached to the recombinant polypeptide backbone. A major advantage of this approach over the use of tunicamycin is that the unglycosylated protein is expressed in a plant where glycosylation normally occurs, so the study of glycan role(s) is restricted to the protein of interest without any non-specific background related to modifications in the function of other cellular glycoproteins. However, as observed in the presence of tunicamycin, unglycosylated mutated glycoproteins may fail to fold properly and eventually become insoluble or be rapidly degraded by proteases. These non-specific roles of glycosylation in glycoprotein stability may be a limitation to the description of more specific glycan functions.

Mutants in the glycosylation pathway have been recently isolated in *Arabidopsis thaliana* (M.J. Chrispeels *et al.*, personal communication). The analysis of these mutants is complementary to the use of processing inhibitors. In particular, these glycosylation mutants are useful in addressing some very specific questions that cannot be resolved using glycan processing inhibitors, such as (i) is xylosylation required for the fucosylation of complex glycans in plants; or (ii) is $\alpha 1$–3 fucosylation of extracellular glycoproteins necessary for somatic embryogenesis?

Glycans affect glycoprotein conformation and stability

The transfer of one or several oligosaccharide side chains to a nascent polypeptide affects its initial folding in the RER. The glycan contribution to biologically active or stable conformations of glycoproteins has generally been supported by results obtained from the *in vivo* inhibition of glycosylation. These studies have shown that glycans are responsible for the thermal stability, solubility, and/or biological activity of some glycoproteins by promoting their folding in the most stable three-dimensonal structure (Tsaftari, Sorenson & Scandalios, 1980; Gibson, Kornfeld & Schlesinger, 1981; Hsieh *et al.*, 1983; Trimble *et al.*, 1983). A major difficulty in studying the roles of glycans *in vivo*, using glycosylation inhibitors such as tunicamycin, is related to the increased susceptibility of unglycosylated glycoproteins to proteolysis. Indeed, glycans are attached to protein backbones at the level of amino acid sequences which are highly exposed to proteases (e.g. β turns).

Glycans protect these regions from proteolytic degradation, probably by means of steric hindrance (Gottschalk & de Groth, 1960; Winkler & Segal, 1984). Surprisingly, a number of storage proteins in legume seeds are glycosylated (for review see Chrispeels, 1984) even though the fate of this protein family is typically a proteolysis in the vacuolar compartment of plant cells. Glycans affect the physiochemical properties of mature glycoproteins *in vivo* in so many ways that results obtained by using

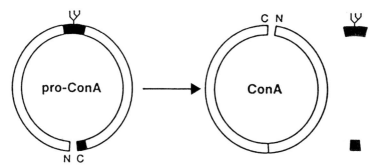

Fig. 8. Processing of pro-ConA into ConA. On this diagram, N and C indicate the N-terminal and C-terminal ends. The black sections are removed during the processing of pro-ConA. The branched symbol indicates the high-mannose oligosaccharide.

glycosylation inhibitors should be interpreted with caution. For instance, the observation that tunicamycin prevents the secretion of a glycoenzyme activity from plant cells can be interpreted in several ways, according to the different roles ascribed to glycans *in vivo*: (i) the unglycosylated enzyme has not folded properly into an enzymatically active conformation; (ii) the proper, glycan-dependent, folding of a soluble enzyme is necessary to permit its transport through the secretion pathway; (iii) oligosaccharide sidechains are necessary for the post-translational maturation of an inactive precursor of the enzyme; (iv) the glycan is necessary for proper targeting of the enzyme to the extracellular compartment; (v) the unglycosylated enzyme is degraded by proteases during the intracellular transport or after secretion.

The importance of N-glycosylation for the folding and assembly of plant proteins is illustrated throughout studies on the biosynthesis of ConA. This lectin is synthesised in the RER as an inactive, glycosylated precursor, pro-ConA. The processing of pro-ConA occurs in the protein bodies and involves the excision of a small glycopeptide from the centre of the precursor molecule and the ligation of the two polypeptides (see Fig. 8, and Bowles, this volume). Using pulse–chase experiments in the presence of tunicamycin, in conjunction with subcellular fractionation, it has been shown that there is very little transport of unglycosylated pro-ConA out of the RER. In addition, the absence of glycan also inhibits the processing of the small amounts of pro-ConA which do reach the protein bodies (Chrispeels *et al.*, 1986; Faye & Chrispeels, 1987). These results have been explained by changes in the physical properties of unglycosylated pro-ConA, which may have a low solubility in the absence of carbohydrates. The altered conformation of the unglycosylated precursor may also prevent its recognition by the processing machinery responsible for ConA maturation in the protein bodies.

Glycans and cellular interactions: self-incompatibility

Plant and animal cells have developed the capacity to discriminate between self and non-self. A well-known recognition system, responsible for self-incompatibility in flowering plants, mediates the interaction between pollen grains and pistil (Haring *et al.*, 1990; Nasrallah, Nishio & Nasrallah, 1991). This system prevents self-fertilisation by rapidly stopping pollen tube growth when the pollen grain has the same phenotype as the pistil. In most cases, self-incompatibility is controlled by a single, multiallelic locus, the S-locus. Products of the S-locus, the S-glycoproteins, have been identified and characterised in Cruciferae, Solanaceae and Papaveraceae (Haring *et al.*, 1990; Nasrallah *et al.*, 1991; Ioerger *et al.*, 1991; Dickinson, Crabbe & Gaude, 1992). The role that might be played by the glycosylated portion of the S-products in the self-incompatibility reaction still remains unclear. It was postulated that the glycan chains might act in the recognition between pistil and pollen. This interaction might be mediated through adhesion between a carbohydrate-reactive protein (i.e. lectin-like protein, glycosidase or glycosyltransferase) of pollen and the S-protein glycan. However, more data regarding the nature and specificity of the putative receptor involved in the interaction between pollen and pistil are needed. The common structure observed in the complex glycans of S-proteins from different strains of *Brassica campestris* (Takayama *et al.*, 1986) does not favour the hypothesis of carbohydrate type recognition signals in this mechanism. However, the number and position of potential N-glycosylation sites was shown to be characteristic for each allele (Nasrallah *et al.*, 1987; Dickinson *et al.*, 1992) and indicates that glycosylation may contribute to S-allele specificity. Besides, Sarker, Ellemen & Dickinson (1988) demonstrated by the use of metabolic inhibitors that glycosylation processing of stigma glycoproteins was essential for the successful operation of self-incompatibility in *Brassica oleracea*. The recent discovery of enzyme activities associated with the locus products in Solanaceae (RNase activity) (McClure *et al.*, 1989) and Cruciferae (protein kinase activity) (Stein *et al.*, 1991) underlines the complexity of these self/non-self recognition mechanisms in flowering plants.

There is a strong analogy between this putative recognition mechanism and the receptor–ligand interaction implicated in adhesion between mouse sperm and egg. The jelly coat of mouse eggs contains a glycoprotein bearing GlcNAc residues, which serve as recognition sites for sperm through interaction with a surface galactosyltransferase (Flornan & Wassarman, 1985; Lopez *et al.*, 1985). Compatible yeasts are also coated with glycoproteins able to bind together, resulting in the aggregation that facilitates yeast mating (Crandail & Brock, 1968).

Glycans and cellular interactions: multicellular organisation

Glycoproteins may be involved in recognition phenomena at the cell surface because they are part of the extracellular matrix. Thus, the significance of some extracellular glycoproteins probably '. . . lies in their being the morphogenic substrate of multicellular organisms . . .' (Lamport, 1980). In animal cells, glycoproteins containing hydroxyproline and O-linked glycans are always present in the extracellular matrix, which plays an important role in cell–cell interactions. The similarities between these proteins from animal cells and the hydroxyproline-rich glycoproteins and arabinogalactan proteins of plants may indicate that these proteins have similar roles, but as yet there is no evidence for such a role in plant cells. Numerous experiments using tunicamycin indicate that N-linked glycans at the cell surface or in extracellular proteins are involved in embryonic differentiation in animals. Tunicamycin blocks the developmental programme of embryos in sea urchins, amphibians and mammals at specific stages, generally when extensive movement of the cells begins (Schneider, Nguyen & Lennarz, 1978; Ivatt, 1985). When tunicamycin is added to the water used for radish seed imbibition, germination takes place and seedlings develop during 3 or 4 days, almost in the same manner as the controls (L. Faye, unpublished results). Tunicamycin does not block the development of 'predetermined' structures in plants. This was also observed in the somatic embryogenesis obtained from carrot cells cultured in the absence of the phytohormone 2,4-D. When added to carrot cell culture, tunicamycin does not prevent an unorganised cell proliferation. The antibiotic is capable of arresting somatic embryogenesis, in a reversible way, at any stage of the developmental programme (LoSchiavo, Quesada-Allue & Sung, 1986).

This inhibition could be alleviated by addition of properly glycosylated extracellular proteins (de Vries *et al.*, 1988). The activity of some of these extracellular glycoproteins is dependent on the correct fucosylation of their oligosaccharide sidechains (LoSchiavo *et al.*, 1990). These results point to an as yet unknown role for extracellular glycoproteins in plant embryogenesis. That oligosaccharides may act as extracellular signals in plants has been clearly illustrated through studies of plant defence responses. In these systems, well-characterised oligo- or polysaccharides from plant or fungus cell walls act as primary signals in the defence gene activation (Nothnagel *et al.*, 1983; Sharp, McNeil & Albersheim, 1984).

Glycosylation and protein targeting

The presence of numerous subcellular compartments is one of the distinctive features of eukaryotic cells. The biogenesis and maintenance of these compartments depend upon the import of specific proteins.

In animal cells, a role for N-linked oligosaccharide side chains in the intracellular targeting of glycoproteins has been shown for lysosomal enzymes (Sly & Fischer, 1982). The latter carry phosphorylated mannose residues that are recognised by mannose-6-P receptors and subsequently transported to lysosomes. In the past few years, investigations into the role of N-linked glycans in glycoprotein targeting to the plant vacuole have included (i) the use of drugs that inhibit N-glycosylation and (ii) the use of transgenic systems for expression of cloned vacuolar glycoprotein (first mutated *in vitro* to lack specific glycosylation sites). In transgenic tobacco plants containing a bean lectin PHA gene, PHA is normally glycosylated and transported to tobacco vacuoles. When both PHA glycosylation sites are mutated, unglycosylated PHA is also transported to the storage vacuoles of transgenic tobacco (Voelker *et al.*, 1989). Many other vacuolar glycoproteins have been investigated using these complementary approaches; all results lead to the conclusion that glycans have no specific role in the targeting of plant glycoproteins to the vacuole. This means that there is no carbohydrate receptor involved in glycoprotein transport to the plant vacuole.

To study the roles of glycosylation in plant glycoprotein secretion into the extracellular compartment, suspension-cultured sycamore and carrot cells have been used. The majority of proteins secreted by these cells have N-linked glycans. When N-glycosylation is inhibited by tunicamycin, there is a dramatic decrease (70–90%) in the accumulation of newly synthesised secreted proteins in the extracellular compartment of these suspension-cultured cells (Driouich *et al.*, 1989). Therefore N-glycosylation is required for protein accumulation in the extracellular compartment. In suspension-cultured carrot cells, interest has also focused on cell-wall-bound β-fructosidase. This enzyme has been used as a model to study the *devenir* of unglycosylated secreted proteins. Using pulse-labelling experiments, radioactive β-fructosidase has been shown to accumulate in the cell wall of control cells but not in the wall of the tunicamycin-treated cells (Faye & Chrispeels, 1989). This apparent inhibition of β-fructosidase secretion does not result from the enzyme biosynthesis inhibition, nor does it result from a lower ability to be transported out of the ER as described for the unglycosylated pro-ConA (Faye & Chrispeels, 1987).

To investigate whether this apparent inhibition of glycoprotein secretion was related to a specific or unspecific role of glycans, we looked for targeting signals in the oligosaccharide sidechains of extracellular glycoproteins secreted by sycamore cells. We used specific glycan-processing inhibitors, such as Cast and DMM, which prevent the formation of complex-type glycans. In the presence of DMM, high-mannose type glycans are produced. Glucosylated high-mannose oligosaccharides are made by Cast-treated sycamore cells. When these inhibitors of glycan processing are added to sycamore cells, there is no quantitative difference in the secretion of glycoproteins as compared to control cells. However, in these conditions, secreted glycoproteins have almost exclusively unprocessed glycans of the high-mannose type (Driouich *et al.*, 1989; L. Faye *et al.*, unpublished results).

Together these results, obtained using either carrot or sycamore cells, show that N-glycosylation but not oligosaccharide processing is required for secretion of plant glycoproteins (Faye & Chrispeels, 1989; Driouich *et al.*, 1989). It has recently been observed that unglycosylated secreted glycoproteins probably never reach the plasma membrane and are degraded during their intracellular transport.

Questions for the future of plant glycobiology research

Recent progress in carbohydrate technology provides analytical tools for the isolation and recognition of complex glycans. Currently, a very limited number of plant glycan structures has been determined. This is probably a poor reflection of the actual structural diversity of carbohydrate moieties of plant glycoproteins. The use of recombinant DNA technologies offers new approaches for future research in plant glycobiology. For instance, it is now possible to express high amounts of a given recombinant glycoprotein in different plants, different tissues of the same plant, or in cellular clones grown in different conditions. This approach will help in studying plant- or cell-specific glycosylation and also will provide insights on how glycosylation is influenced by the extracellular environment.

The possibility to express foreign genes in transgenic plant systems also provides the opportunity to study the structure and function of glycans in relation to glycoprotein intracellular targeting. For instance, site-directed mutagenesis of glycosylation sites in recombinant glycoproteins allows for a comparison of the stability of extracellular and vacuolar glycoproteins in the presence or absence of their glycans. Peptide sequences necessary and sufficient to target extracellular proteins to the vacuole have recently been described (Bednareck & Raikhel, 1991; Matsuoka & Nakamura, 1991; Neuhaus *et al.*, 1991). These vacuolar targeting signals are the necessary tools to investigate the use of a glycosylation site and the maturation of oligosaccharide side chains in connection with the intracellular transport of a secretory glycoprotein.

The production of recombinant glycoproteins in transgenic plants for human therapy is completely dependent on a better understanding of complex glycan biosynthesis in plant cells. As plant complex glycans are highly immunogenic, the manipulation of the processing pathway is one of the conditions necessary to use plants for the production of recombinant glycoproteins for *in vivo* administration in mammals.

Acknowledgements

We are grateful to Dr T. Gaude and J.P. Salier for critical reading of the manuscript and helpful suggestions. This work was supported by grants from Centre National de la Recherche Scientifique, the University of Rouen and Region Haute-Normandie. Financial support for collaborative research with Professor M.J. Chrispeels was provided by NATO, CNRS, NSF and the University of California, San Diego.

References

Abeijon, C. & Hirschberg, C.B. (1992). Topography of glycosylation reactions in endoplasmic reticulum. *Trends in Biochemical Sciences* **17**, 32–6.

Basha, S.M. & Beevers, L. (1976). Glycoprotein metabolism in the cotyledon of *Pisum sativum* during development and germination. *Plant Physiology* **57**, 93–7.

Bednarek, S.Y. & Raikhel, N.V. (1991). The barley lectin carboxyl-terminal propeptide is a vacuolar protein sorting determinant in plants. *Plant Cell* **3**, 1195–206.

Bowles, D.J., Chaplin, M.F. & Marcus, S.E. (1983). Interaction of Concanavalin A with native and denatured forms of jack-bean α-D-mannosidase. *European Journal of Biochemistry* **130**, 613–18.

Chrispeels, M.J. (1983). Incorporation of fucose into the carbohydrate moiety of phytohemagglutinin in developing *Phaseolus vulgaris* cotyledons. *Planta* **157**, 454–61.

Chrispeels, M.J. (1984). Biosynthesis, processing and transport of storage proteins and lectins in cotyledons of developing legume seeds. *Philosophical Transactions of the Royal Society London* B**304**, 309–22.

Chrispeels, M.J. (1985). UDP-GlcNAc: glycoprotein GlcNAc transferase is located in the Golgi apparatus of developing bean cotyledons. *Plant Physiology* **78**, 835–8.

Chrispeels, M.J., Hartl, P.M., Sturm A. & Faye, L. (1986). Characterization of the endoplasmic reticulum-associated precursor of concanavalin A. *Journal of Biological Chemistry* **261**, 10021–4.

Chrispeels, M.J. & Vitale, A. (1985). Abnormal processing of the modified oligosaccharide side chains of phytohemagglutinin in the presence of swainsonine and deoxynojirimycin. *Plant Physiology* **78**, 704–9.

Crandail, M. & Brock, T. (1968). Molecular basis of mating in the yeast *Hanselula wingi*. *Bacteriology Review* **32**, 139–42.

deVries, S.C., Booij, H., Janssens, R., Vogels, R., Saris, L., LoSchiavo, F., Terzi, M. & van Kammen, A. (1988). Carrot somatic embryogenesis depends on the phytohormone-controlled presence of correctly glycosylated extracellular proteins. *Genes and Development* **2**, 462–76.

Dickinson, H.G., Crabbe, J. & Gaude, T. (1992). Sporophytic self-incompatibility systems: S-gene products. *International Review of Cytology*, in press.

Driouich, A., Gonnet, P., Makkie, M., Lainé, A.-C. & Faye, L. (1989). The role of high-mannose and complex asparagine-linked glycans in the secretion and stability of glycoproteins. *Planta* **180**, 96–104.

Edge, A.S.B., Faltyneck, C.R., Hof, L., Le Reichet, J.R. & Weber, P. (1981). Deglycosylation of glycoproteins by trifluoromethane sulfonic acid. *Analytical Biochemistry* **118**, 131–7.

Elbein, E.A. (1987). Inhibitors of the biosynthesis and processing of N-linked oligosaccharide chains. *Annual Review of Biochemistry* **56**, 497–537.

Ericson, M.C. & Chrispeels, M.J. (1973). Isolation and characterization of glucosamine-containing glycoproteins from cotyledons of *Phaseolus aureus*. *Plant Physiology* **52**, 98–104.

Faye, L. & Chrispeels, M.J. (1985). Characterization of N-linked oligosaccharides by affinoblotting with concanavalin A-peroxidase and treatment of the blots with glycosidases. *Analytical Biochemistry* **149**, 218–24.

Faye, L. & Chrispeels, M.J. (1987). Transport and processing of the glycosylated precursor of concanavalin A in jack-bean. *Planta* **170**, 217–24.

Faye, L. & Chrispeels, M.J. (1988). Common antigenic determinants in the glycoproteins of plants, molluscs and insects. *Glycoconjugate Journal* **5**, 245–56.

Faye, L. & Chrispeels, M.J. (1989). Apparent inhibition of β-fructosidase secretion by tunicamycin may be explained by breakdown of the unglycosylated protein during secretion. *Plant Physiology* **89**, 845–51.

Faye, L., Johnson, K.D., Sturm, A. & Chrispeels, M.J. (1989). Structure, biosynthesis and function of asparagine-linked glycans of plant glycoproteins. *Physiologia Plantarum* **75**, 309–14.

Faye, L. & Salier, J.P. (1989). Crossed affino-immunoelectrophoresis or affinoblotting with lectins: Advantages and limitations for glycoprotein studies. *Electrophoresis* **10**, 841–7.

Faye, L., Sturm, A., Bollini, R., Vitale, A. & Chrispeels, M.J. (1986). The position of the oligosaccharide side chains of phytohemagglutinin and their accessibility to glycosidases determined their subsequent processing in the Golgi. *European Journal of Biochemistry* **158**, 655–61.

Flornan, H.M. & Wassarman, P.M. (1985). O-linked oligosaccharides of mouse egg ZP_3 account for its sperm receptor activity. *Cell* **41**, 313–24.

Gibson, R., Kornfeld, S. & Schlesinger, S. (1981). The effect of oligosaccharide chains of different sizes on the maturation and physical properties of vesicular stomatitis virus. *Journal of Biological Chemistry* **256**, 456–62.

Gottschalk, A. & de Groth, S.F. (1960). Studies on mucoproteins III. The accessibility to trypsin of the susceptible bonds in ovin submaxillary gland mucoproteins. *Biochimica et Biophysica Acta* **43**, 513–19.

Haring, V., Gray, J.E., McClure, B.A., Anderson, M.A. & Clarke, A.E. (1990). Self-incompatibility: A self-recognition system in plants. *Science* **250**, 937–41.

Hirschberg, G. & Snider, M.D. (1987). Topography of glycosylation in the rough endoplasmic reticulum and Golgi apparatus. *Annual Review of Biochemistry* **56**, 63–87.

Hori, H. & Elbein, A.D. (1981). Tunicamycin inhibits protein glycosylation in suspension cultured soybean cells. *Plant Physiology* **67**, 882–6.

Hori, H., Pan, Y.T., Molineaux, R.J. & Elbein, A.D. (1984). Inhibition of processing of plant N-linked oligosaccharides by castanospermine. *Archives of Biochemistry and Biophysics* **228**, 525–33.

Hsieh, P., Rosner, M.R. & Robbins, P.W. (1983). Selective cleavage by endo β-N-acetylglucosaminidase H at individual glycosylation sites of Sindbis virion envelope glycoproteins. *Journal of Biological Chemistry* **258**, 2555–61.

Ioerger, T.R., Gohlke, J.R., Xu, B. & Kao, T.H. (1991). Primary structural features of the self-incompatibility protein in Solanaceae. *Sexual Plant Reproduction* **4**, 81–7.

Ivatt, R.J. (1985). Role of glycoproteins during early mammalian embryogenesis. In *The Biology of Glycoproteins* (ed. R.J. Ivatt), pp. 81–95. New York: Plenum Press.

Johnson, K.D. & Chrispeels, M.J. (1987). Substrate specificities of N-acetylglucosaminyl-, fucosyl-, and xylosyltransferases that modify glycoproteins in the Golgi apparatus of bean cotyledons. *Plant Physiology* **84**, 1301–8.

Kimura, Y., Hase, S., Kobayashi, Y., Kyogoku, Y., Funatsu, G. & Ikenata, Y. (1987). Possible pathway for the processing of sugar chains containing xylose in plant glycoproteins deduced on structural analysis of sugar chains from *Ricinus communis* lectins. *Journal of Biochemistry* **101**, 1051–4.

Kornfeld, R. & Kornfeld, S. (1985). Assembly of asparagine-linked glycoproteins. *Annual Review of Biochemistry* **54**, 631–64.

Lainé, A.-C. & Faye, L. (1988). Significant immunological cross-reactivity of plant glycoproteins. *Electrophoresis* **9**, 841–4.

Lainé, A.-C., Gomord, V. & Faye, L. (1991). Xylose-specific antibodies as markers of subcompartmentation of terminal glycosylation in the Golgi apparatus of sycamore cells. *FEBS Letters* **295**, 179–84.

Lamport, D.T.A. (1980). Structure and function of plant glycoproteins. In *The Biochemistry of Plants* (ed. J. Preiss), pp. 501–41. New York: Academic Press.

Lis, H. & Sharon, N. (1978). Soybean agglutinin-A plant glycoprotein. *Journal of Biological Chemistry* **253**, 3468–76.

Lopez, L.C., Bayna, E.M., Litoff, E.M., Shaper, N.L., Schaper, J.H. & Shur, D.B. (1985). Receptor function of mouse sperm surface galactosyltransferase during fertilization. *Journal of Cell Biology* **101**, 1501–10.

LoSchiavo, F., Giuliano, G., de Vries, S.C., Genga, A., Bollini, R., Pitto, L., Cozzani, F., Nuti-Ronchi, V. & Terzi, M. (1990). A carrot cell variant temperature sensitive for somatic embryogenesis reveals a defect in the glycosylation of extracellular proteins. *Molecular and General Genetics* **223**, 385–93.

LoSchiavo, F., Quesada-Allue, L.A. & Sung, Z.R. (1986). Tunicamycin affects somatic embryogenesis but not cell proliferation of carrot. *Plant Science* **44**, 65–71.

Matsuoka, K. & Nakamura, K. (1991). Propeptide of a precursor to a plant vacuolar protein required for vacuolar targeting. *Proceedings of the National Academy of Sciences USA* **88**, 834–8.

McClure, B.A., Haring, V., Ebert, P.R., Anderson, M.A., Simpson, R.J., Sakiyama, F. & Clarke, A.E. (1989). Style self-incompatibility gene products in *Nicotiana alata* are ribonucleases. *Nature* **342**, 955–7.

Nasrallah, J.B., Doney, R.C. & Nasrallah, M.E. (1985). Biosynthesis of glycoproteins involved in the pollen-stigma interaction of incompatibility in developing flowers of *Brassica oleracea* L. *Planta* **165**, 100–7.

Nasrallah, J.B., Kao, T.-H., Chen, C.-H., Goldberg, M.L. & Nasrallah, M.E. (1987). Amino-acid sequence of glycoproteins encoded by three alleles of the S-locus of *Brassica oleracea*. *Nature (London)* **326**, 617–19.

Nasrallah, J.B., Nishio, T. & Nasrallah, M.E. (1991). The self-incompatibility genes of Brassica: expression and use in genetic ablation of floral tissues. *Annual Review of Plant Physiology and Plant Molecular Biology* **42**, 393–422.

Neuhaus, J.M., Sticher, L., Meins, F.J. & Boller, T. (1991). A short C-terminal sequence is necessary and sufficient for the targeting of chitinases to the plant vacuole. *Proceedings of the National Academy of Sciences USA* **88**, 10362–6.

Nothnagel, E.A., McNeil, M., Albersheim, P. & Dell, A. (1983). Host pathogen interactions: XXII. A galacturonic acid oligosaccharide from plant cell wall elicits phytoalexins. *Plant Physiology* **71**, 916–26.

Paul, K.G. & Stigbrand, T. (1970). Four isoperoxidases from horseradish roots. *Acta Chimica Scandinavica* **24**, 3607–17.

Prenner, C., Mach, L., Glössl, J. & März, L. (1991). The specificity of an antibody against the N-glucan of an insect glycoprotein, phospholipase A$_2$. *Glycoconjugate Journal* **8**, 241.

Sarker, R.H., Elleman, C.J. & Dickinson, H.G. (1988). The control of pollen hydration in *Brassica* requires continued protein synthesis whilst glycosylation is necessary for intraspecific incompatibility. *Proceedings of the National Academy of Sciences USA* **88**, 4340–4.

Schneider, E.G., Nguyen, H.T. & Lennarz, W.J. (1978). The effect of tunicamycin, an inhibitor of protein glycosylation, on embryonic development in the sea urchin. *Journal of Biological Chemistry* **253**, 2348–55.

Sharp, J.K., McNeil, M. & Albersheim, P. (1984). The primary structures of one elicitor-active and seven elicitor-inactive hexa(β-D-glucopyranosyl)-D-glucitols isolated from the mycelial cell walls of *Phytophthora megasperma* f. sp. *glycinea. Journal of Biological Chemistry* **259**, 11321–36.

Sly, W.S. & Fischer, M.D. (1982). The phosphomannosyl recognition system for intracellular and intercellular transport of lysosomal enzymes. *Journal of Cell Biochemistry* **18**, 67–85.

Stein, J.C., Howlett, B., Boyes, D.C., Nasrallah, M.E. & Nasrallah, J.B. (1991). Molecular cloning of a putative receptor protein kinase gene encoded at the self-incompatibility locus of *Brassica oleraceae. Proceedings of the National Academy of Sciences USA* **88**, 8816–20.

Sturm, A. (1991). Heterogeneity of the complex N-linked oligosaccharides at specific glycosylation sites of two secreted carrot glycoproteins. *European Journal of Biochemistry* **199**, 169–79.

Sturm, A., Chrispeels, M.J., Wieruszesli, J.M., Strecker, G. & Montreuil, J. (1987*a*). Structural analysis of the N-linked oligosaccharides from jack bean α-mannosidase. In *Glycoconjugates, Proceedings of the 9th International Symposium*, Lille-France, Poster A107.

Sturm, A., Johnson, K.D., Szumilo, T., Elbein, A.D. & Chrispeels, M.J. (1987*b*). Subcellular localization of glycosidases and glycosyltransferases involved in the processing of N-linked oligosaccharides. *Plant Physiology* **85**, 741–5.

Sturm, A. & Kindl, H. (1983). Fucosyl transferase activity and fucose incorporation *in vivo* as markers for subfractionating cucumber microsomes. *FEBS Letters* **160**, 165–8.

Sturm, A., van Kuik, J.A., Vliegenthart, J.F.G. & Chrispeels, M.J. (1987*c*). Structure, position and biosynthesis of the high-mannose and the complex oligosaccharide side chains of the bean storage protein phaseolin. *Journal of Biological Chemistry* **262**, 13992–4003.

Sturm, A., Voelker, T.A., Tague, B.W., Faye, L. & Chrispeels, M.J. (1987*d*). Glycan modification and targeting during the biosynthesis of phytohemagglutinin in beans and in 3 different transgenic systems (Tobacco, Yeast, COS Cells). In *Glycoconjugates, Proceedings of the 9th International Symposium*, Lille-France, Poster B28.

Szumilo, T., Kaushal, G.P. & Elbein, A.D. (1986*a*). Purification and properties of glucosidase I from mung bean seedlings. *Archives of Biochemistry and Biophysics* **247**, 261–71.

Szumilo, T., Kaushal, G.P., Hori, H. & Elbein, A.D. (1986*b*). Purification and properties of a glycoprotein processing α-mannosidase from mung bean seedling. *Plant Physiology* **81**, 383–9.

Tague, B.W. & Chrispeels, M.J. (1987). The plant vacuolar protein, phytohemagglutinin, is transported to the vacuole of transgenic yeast. *Journal of Cell Biology* **105**, 1971–9.

Takayama, S., Isogai, A., Tsukamoto, C., Ueda, Y., Hinata, K., Okazaki, K. & Suzuki, A. (1986). Structure of carbohydrate chains of S-glycoproteins in *Brassica campestris* associated with self-incompatbility. *Agricultural and Biological Chemistry* **50**, 1673–6.

Trimble, R.B., Maley, F. & Chu, F.K. (1983). Glycoprotein biosynthesis in yeast. Protein conformation affects processing of high-mannose oligosaccharides on carboxypeptidase and invertase. *Journal of Biological Chemistry* **258**, 2562–7.

Trombetta, S.E., Bosch, M. & Parodi, A.J. (1989). Glucosylation of glycoproteins by mammalian, plant, fungal and trypanosomatid protozoa microsomal membranes. *Biochemistry* **28**, 8108–16.

Tsaftari, A.S., Sorenson, J.C. & Scandalios, J.C. (1980). Glycosylation of catalase inhibitor is necessary for activity. *Biochemical and Biophysical Research Communications* **92**, 889–95.

Van Kuik, J.A., van Halbeek, H., Kamerling, J.P. & Vliegenthart, J.F.G. (1985). Primary structure of the low-molecular-weight carbohydrate chains of *Helix pomatia* α-hemocyanin xylose as a constituent of N-linked oligosaccharides in an animal glycoprotein. *Journal of Biological Chemistry* **260**, 13984–8.

Vitale, A., Sturm, A. & Bollini, R. (1986). Regulation of processing of a plant glycoprotein in the Golgi complex: a comparative study using *Xenopus* oocytes. *Planta* **169**, 108–16.

Vitale, A., Warner, T.G. & Chrispeels, M.J. (1984). *Phaseolus vulgaris* phytohemagglutinin contains high-mannose and modified oligosaccharide chains. *Planta* **160**, 256–63.

Voelker, T.A., Florkiewicz, R.Z. & Chrispeels, M.J. (1986). Secretion of phytohemagglutinin by monkey COS cells. *European Journal of Cell Biology* **42**, 218–23.

Voelker, T.A., Herman, E.M. & Chrispeels, M.J. (1989). In vitro mutated phytohemagglutinin genes expressed in tobacco seeds: role of glycans in protein targeting and stability. *Plant Cell* **1**, 95–104.

Winkler, J.R. & Segal, H.C. (1984). Swainsonine inhibits glycoprotein degradation by isolated rat liver lysosomes. *Journal of Biological Chemistry* **259**, 15369–72.

Note added in proof

Recent results obtained either from expression of a non-glycosylated pro-ConA in *Escherichia coli* or from *in vitro* deglycosylation of pro-ConA with N-glycanase have shown that the non-glycosylated lectin precursor: (i) has a very low solubility; and (ii) exhibits the carbohydrate-binding activity of the mature lectin (Min, Dunn & Jones, 1992; Sheldon & Bowles, 1992). Such a low solubility confirms our suggestion that unglycosylated pro-ConA partly precipitates in the RER, giving both physical and biological explanations of this aggregation. Interestingly, these results also put the emphasis on a new activation mechanism of a precursor by the removal of the glycan.

References

Min, W., Dunn, A.J. & Jones, D.H. (1992). Non-glycosylated recombinant pro-concanavalin A is active without polypeptide cleavage. *EMBO Journal* **11**, 1303–7.

Sheldon, P.S. & Bowles, D.J. (1992). The glycosylated precursor of concanavalin A is converted to an active lectin by deglycosylation. *EMBO Journal* **11**, 1297–301.

J.M. LORD and L.M. ROBERTS

Biosynthesis, intracellular transport and processing of ricin

Introduction

Many plant tissues produce ribosome-inactivating proteins (RIPs) which act as N-glycosidases removing a specific adenine residue from a highly conserved surface loop present in 23S, 26S and 28S ribosomal RNA (Lord, Hartley & Roberts, 1991). The adenine residue in question (adenine 4324 in rat liver 28S rRNA) (Endo et al., 1987), plays a necessary role in the binding of elongation factors, and ribosomes that have been depurinated by RIPs can no longer function in protein synthesis. Typically, a single molecule of RIP can depurinate 1500–2000 susceptible ribosomes per minute. Plant RIPs usually occur as monomeric proteins with molecular masses of around 30 kDa and are frequently but not always N-glycosylated. Although these RIPs can potently and irreversibly inactivate mammalian ribosomes they are not cytotoxic to mammalian cells since they are unable to enter such cells and reach the cytosol where their ribosome substrates are located. It has recently been found that these single-chain RIPs are also active against prokaryotic ribosomes (Hartley et al., 1991). In some instances, however, the RIP is joined via a disulphide bond to a second polypeptide which, in all cases described to date, is a galactose-binding lectin whose molecular mass is also around 30 kDa. These heterodimeric plant toxins are able to bind opportunistically to eukaryotic cells by interacting with galactose residues present on cell-surface glycoproteins and glycolipids. Such cytotoxic lectins are amongst the most potent cytotoxins in Nature. Although the RIP portion of these heterodimeric molecules are able to depurinate eukaryotic ribosomes in common with single-chain RIPs, they are unable to depurinate prokaryotic ribosomes in contrast to single-chain RIPs (Hartley et al., 1991). This group of cytotoxic lectins includes ricin (from *Ricinus communis*) and abrin (from *Abrus precatorius*). Ricin was the first lectin to be des-

Society for Experimental Biology Seminar Series 53: *Post-translational modifications in plants*, ed. N.H. Battey, H.G. Dickinson & A.M. Hetherington. © Cambridge University Press 1993, pp. 243–255.

cribed, over a century ago (Stillmark, 1888), and it remains the most thoroughly characterised plant RIP.

Occurrence and structure of ricin

There are several forms of ricin including ricin D, ricin E and the closely related *Ricinus communis* agglutinin (RCA), which are encoded by a small multigene family composed of approximately eight members (Tregear & Roberts, 1992). At least three members of this lectin gene family are non-functional (Tregear & Roberts, 1992). Expression of the genes is both developmentally regulated and tissue-specific. Ricin is synthesised in the endosperm cells of maturing *Ricinus* seeds; it is stored within the protein bodies of mature seeds and, in common with the other protein body storage proteins, it is hydrolysed during the first few days of post-germinative growth (Tulley & Beevers, 1976; Youle & Huang, 1976; Roberts & Lord, 1981*a*).

Ricin is a heterodimer consisting of two distinct polypeptides held together by a single disulphide bond (Olsnes & Pihl, 1981) (Fig. 1). One of these polypeptides (molecular mass 32 kDa, designated the A chain) is the RIP, while the other (34 kDa, designated the B chain) is a galactose or N-acetylgalactosamine-binding lectin (Nicolson, Blanstein & Etzler, 1974). RCA is a tetramer consisting of two ricin-like heterodimers, each of which contains an A chain (32 kDa) and a galactose-binding B chain (36 kDa) (Olsnes, Saltvedt & Pihl, 1974). Ricin is a potent cytotoxin but is a weak haemagglutinin, whereas RCA is only weakly toxic to intact cells but is a strong haemagglutinin (Olsnes *et al.*, 1974). Although RCA is only weakly toxic to intact cells, its isolated A chain has rRNA N-glycosidase activity (O'Hare, Roberts & Lord, 1992) and is of comparable toxicity to ricin A chain when added to a cell-free protein synthesis system (Cawley, Hedblom & Houston, 1978).

Ricin and RCA are closely related proteins. It has been known for some time that antisera raised against individual ricin A or B chains cross-react with the corresponding RCA chains and *vice versa* (Saltvedt, 1976; Olsnes & Saltvedt, 1975). More recently the complete sequences of the A and B chains of ricin and RCA have been deduced from the nucleotide sequences of cDNA and genomic clones (Lamb, Roberts & Lord, 1985; Halling *et al.*, 1985; Roberts *et al.*, 1985). The RCA A chain is one residue shorter than that of ricin, corresponding to the omission of the alanine residue at position 130 in ricin A chain. Overall, the A chains differ in 18 out of 267 residues and are thus 93% homologous at the amino acid level, while the B chains differ in 41 out of 262 residues, giving 84% homology (Roberts *et al.*, 1985).

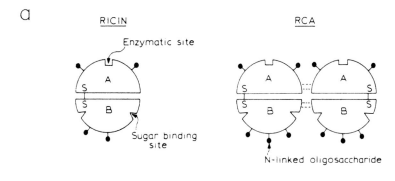

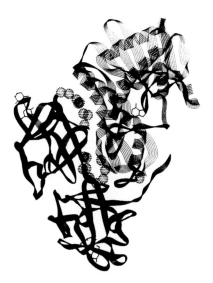

Fig. 1. Structure of ricin and *Ricinus communis* agglutinin. (a) Schematic illustration of quaternary structure. (b) The three-dimensional structure of ricin. The α-carbon backbone of the protein is displayed as ribbons. The A chain is shown in the upper right as a multi-stranded ribbon; the B chain is the solid ribbon. The spheres between the chains mark the positions of solvent water molecules trapped in the interface. Taken from Robertus (1991), with permission.

Both the A and B chains of ricin (and their RCA counterparts) are N-glycosylated (Olsnes & Pihl, 1982). The A and B chains each contain two (Asn–X–Ser/Thr) potential N-glycosylation sites. Purified ricin A chain contains two components of molecular mass 32 kDa and 34 kDa. The lighter component has been designated the A1-chain and the heavier component the A2-chain. These two components represent glycosylation variants of a single A chain polypeptide. The A1-chain contains a single oligosaccharide side chain whereas the A2 chain contains two (Foxwell *et al.*, 1985). Ricin B chain contains two N-linked oligosaccharide side chains (Foxwell *et al.*, 1985).

Structure of the sites in ricin that confer biological activity

The extensive analysis of the primary sequence and X-ray structure of ricin by Robertus and his colleagues (Rutenber *et al.*, 1991) has identified the catalytic site of ricin A chain (Katzin, Collins & Robertus, 1991) and the sugar binding sites of ricin B chain (Rutenber & Robertus, 1991).

Ricin A chain has a considerable amount of secondary structure. It contains seven α helices, which account for about 30% of the total protein, and contains about 15% β sheet structure. The X-ray structure reveals a reasonably prominent active site cleft in ricin A chain which includes the residues Glu177 and Arg180. During ricin A chain modification, the ribosome is recognised and bound by a number of contacts, which produce a distortion of the susceptible adenine ring into the high energy *syn* conformation. Bond breakage is enhanced by the juxtaposition of Glu177 to the developing oxycarbonium ion on the ribose sugar. Arg180 stabilises substrate binding by forming an ion pair with a phosphate group of the rRNA.

Ricin B chain is a bilobal protein, which forms two globular domains with almost identical folding properties. Each globular domain binds one molecule of galactose. Both of the sugar binding sites are structurally very similar; the sugar lies in a pocket formed in part by a kink in the polypeptide chain. In each site the kink is formed by the homologous tripeptide Asp–Val–Arg. In the first domain binding site, galactose hydrogen bonds to ricin B chain residues Lys40 and Asn46. Asn46 is itself stabilised by hydrogen bonding to Asp22 of the first site Asp–Val–Arg tripeptide. In the second domain binding site, galactose hydrogen bonds to Asn255, which is in turn stabilised by hydrogen bonding to Asp234 of the second site Asp–Val–Arg tripeptide (Rutenber, Ready & Robertus, 1987).

Biosynthesis of ricin

Ricin and RCA are coordinately synthesised in equivalent amounts in the endosperm cells of ripening *Ricinus* seeds (Gifford, Greenwood & Bewley, 1982). In common with the other major storage components of the protein bodies, synthesis occurs during and after testa formation when these storage organelles are being rapidly formed (Roberts & Lord, 1981*a*). The gene encoding ricin has been cloned (Lamb *et al.*, 1985; Halling *et al.*, 1985) showing that both the A and B chains of ricin are synthesised together as components of a single preprotein precursor. The schematic structure of preproricin is illustrated in Fig. 2. During ricin biosynthesis a series of co-translational and post-translational steps have been defined (Roberts, Lamb & Lord, 1987) and these are described briefly below.

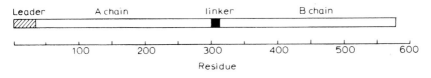

Fig. 2. Schematic structure of preproricin.

Co-translational modifications

The first indication that ricin A and B chains were synthesised together in a precursor polypeptide (proricin) came from cell-free translation studies (Roberts & Lord, 1981*b*; Butterworth & Lord, 1983). Antibodies were raised in rabbits against the individual purified A and B chains. Although A chain antibodies did not cross-react with B chain and *vice versa*, each of the antisera recognised a single polypeptide of 59 kDa when total mRNA from ripening *Ricinus* endosperm tissue was translated *in vitro* in the rabbit reticulocyte system (Butterworth & Lord, 1983). When dog pancreas microsomes were included in the translation system, the ricin precursor was co-translationally translocated into the lumen of the microsomal vesicles (Roberts & Lord, 1981*b*). In this case the translocated precursor, precipitated with anti-A chain or anti-B chain antibodies, appeared as a group of polypeptides of molecular mass 64–68 kDa. Subsequent studies showed that the translocation of proricin into microsomal vesicles *in vitro* was an accurate reflection of the biosynthetic events occurring during ricin biosynthesis. Protein translocation into heterologous microsomes *in vitro* (Roberts & Lord, 1981*b*) or into homologous (Lord, 1985*a*) or heterologous (Richardson *et al.*, 1989) endoplasmic reticulum (ER) *in vivo* is accompanied by a series of co-translational modifications. Firstly, membrane traversal is accompanied by the proteolytic removal of the N-terminal signal peptide which directs ribosomes engaging preprocin mRNA to the membrane of rough ER for co-translational segregation into the ER lumen (Roberts & Lord, 1981*b*). The mechanism of translocation appears, and is assumed, to be identical to that established in the case of mammalian secretory proteins (Walter, Gilmore & Blobel, 1984). Secondly, proricin entering the ER lumen is N-glycosylated by an ER glycosyltransferase that transfers *en bloc* an oligosaccharide moiety from a lipid carrier to appropriate Asn residues in the nascent protein (Lord, 1985*b*). The appearance of a group of translocated ricin precursors of molecular mass 64–68 kDa (Roberts & Lord, 1981*b*; Lord, 1985*b*) was due to heterogeneity in glycosylation, since treating this group with endo-N-acetylglucosaminidase H removed the oligosaccharide side chains converting the precursors into a single polypeptide of 57.5 kDa (Roberts & Lord, 1981*b*). The difference in size between the non-segregated precursor synthesised *in vitro* in the absence of microsomes (59 kDa) and the deglycosylated precursor (57.5 kDa) is accounted for by the N-terminal signal peptide, which is removed by ER signal peptidase during the membrane translocation step.

A third modification to the ricin precursor in the ER lumen, occurring during or immediately after synthesis, is disulphide bond formation (Richardson *et al.*, 1989). In mature ricin, the B chain contains four intrachain disulphide bonds and is joined to the A chain by a single disulphide bond. These disulphide bonds are formed enzymically by ER protein disulphide isomerase (Freedman, 1984) which introduces five intrachain disulphide bonds into proricin; the bond destined to become the interchain disulphide bond joining the A and B chains of mature ricin is formed between cysteine residues present in the A and B chain sequences in the precursor.

Post-translational modifications

The biosynthesis and co-translational modifications of ricin result in the deposition of core-glycosylated, disulphide-stabilised proricin in the ER lumen. Subsequently, proricin is transported, via the Golgi complex, to the protein bodies where native heterodimeric ricin is proteolytically generated and stored. Transport through the Golgi complex is accompanied by the enzymatic modification of the oligosaccharide side chains of proricin. Although the details of the sugar modifications are not known, they probably involve typical Golgi oligosaccharide trimming and monosaccharide additions to the side chains (Schekman, 1985; Farquhar,

1985). The latter include addition of fucose and xylose to oligosaccharides on ricin A chain (Lord, 1985*b*; Lord & Harley, 1985). The sugar modifications confer partial endo-N-acetylglucosaminidase H resistance on the proricin oligosaccharide side chains. Proricin is then transported from the Golgi complex within Golgi-derived vesicles to the protein bodies where, after membrane fusion, they discharge their contents into the protein body matrix. Within the protein bodies proricin is processed by at least one acid endopeptidase which liberates free A and B chains, still covalently joined by a disulphide bond. This is because cleavage occurs within a disulphide loop between the A and B chain sequences (Harley & Lord, 1985). This processing step involves the removal of a twelve amino acid peptide, which links the A and B chain sequences in proricin. The carboxy-terminal residue of the linker peptide is Asn. Endoproteolytic cleavage after an Asn residue is a common feature in the processing of plant proproteins (Lord & Robinson, 1986); the enzyme responsible has recently been purified (Hara-Nishimura *et al.*, 1991). The same protein body endopeptidase may also be responsible for the N-terminal trimming of proricin. The preproricin precursor includes a 35 amino acid N-terminal extension (Fig. 2) which includes but does not entirely consist of the signal peptide. After co-translational removal of the signal peptide the segregated precursor is still believed to possess an N-terminal presequence (Roberts *et al.*, 1987). The final amino acid of this presequence is Asn, and proteolytic removal to generate the native N-terminal of ricin A chain probably also takes place in the protein bodies.

There is evidence that oligosaccharide modification during ricin biosynthesis is not limited to processing the proricin side chains during intracellular transport through the ER and Golgi complex. *In vivo* [^{35}S]methionine-labelling of *Ricinus* endosperm tissue for a short time period (45 min) followed by immunoprecipitation with anti-RCA antibodies identified three polypeptides of 39, 38 and 33 kDa (Roberts & Lord, 1981*b*). These represented RCA B chain (39 kDa), ricin B chain (38 kDa) and a mixture of RCA and ricin A chains (33 kDa). When the 45 min radioactive pulse was followed by a 17 h chase with unlabelled methionine, the three immunoreactive polypeptides had molecular masses of 36, 34 and 31 kDa (the mature sizes of RCA B, ricin B and both RCA and ricin A chains, respectively). A kinetic analysis showed that newly synthesised lectin subunits undergo a reproducible trimming in size with time (Roberts & Lord, 1981*b*). This processing occurs within the protein bodies since proteolytic cleavage of proricin and proRCA had already taken place to liberate the individual A and B chains. The observed processing represented oligosaccharide trimming rather than proteolysis since non-glycosylated lectin subunits synthesised in tunicamycin-treated endosperm tissue under identical conditions did not show a corresponding size reduction (J.M. Lord & S.M. Harley, unpublished). A schematic account of ricin biosynthesis is shown in Fig. 3.

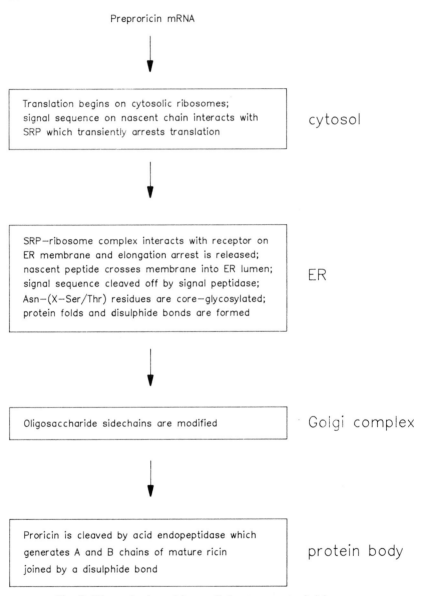

Preproricin mRNA

| Translation begins on cytosolic ribosomes; signal sequence on nascent chain interacts with SRP which transiently arrests translation | cytosol |

SRP—ribosome complex interacts with receptor on ER membrane and elongation arrest is released; nascent peptide crosses membrane into ER lumen; signal sequence cleaved off by signal peptidase; Asn—(X—Ser/Thr) residues are core—glycosylated; protein folds and disulphide bonds are formed   ER

Oligosaccharide sidechains are modified   Golgi complex

Proricin is cleaved by acid endopeptidase which generates A and B chains of mature ricin joined by a disulphide bond   protein body

Fig. 3. Biosynthesis and intracellular transport of ricin.

Significance of the post-translocational modifications

Ricinus endosperm ribosomes are sensitive to the RNA N-glycosidase activity of ricin A chain. However, in contrast to free ricin A chain, the ricin A chain portion of proricin is completely devoid of rRNA N-glycosidase activity, even though the ricin B chain portion has lectin activity (Richardson *et al.*, 1989). Thus proricin moves from the ER, via the Golgi, to the protein bodies in a catalytically inactive form. In this way, any miscompartmentation of proricin into the cytosol during its biosynthesis would not expose the sensitive *Ricinus* ribosomes to a catalytically active RIP. Catalytically active ricin A chain is generated in the protein bodies where it is stored in the mature seed and rapidly degraded during germination (Lord, Lamb & Roberts, 1984).

N-glycosylation is not required for intracellular transport since non-glycosylated proricin produced *in vivo* in tunicamycin-treated endosperm tissue moves from the ER to the protein bodies at the same rate as the normal N-glycosylated precursor (Lord, 1985*b*). Likewise, non-glycosylated proricin is effectively processed by the protein body endopeptidase into the A and B chains (Lord, 1985*b*). Glycosylation is not required for the stability or catalytic activity of ricin A chain; recombinant non-glycosylated ricin A chain produced in *Escherichia coli* is indistinguishable from native ricin A chain in all its properties (O'Hare *et al.*, 1987; Piatak *et al.*, 1988). N-glycosylation does, however, affect the stability and lectin activity of ricin B chain. Non-glycosylated recombinant ricin B chain produced in *E. coli* (Hussain *et al.*, 1989), in tunicamycin-treated *Xenopus* oocytes (Richardson *et al.*, 1991) or in oocytes after changing the Asn residues of the two B chain N-glycosylation sites to Gln (Wales *et al.*, 1991) rapidly loses lectin activity and aggregates in contrast to glycosylated recombinant ricin B chain (Richardson *et al.*, 1988). Recent work indicates that N-glycosylation may be less important for stability in the case of truncated, biologically active ricin B chain fragments (Wales *et al.*, 1992).

We have explored the possibility that the portion of the presequence which remains after signal sequence cleavage, and is removed in the protein bodies, might include a protein body targeting signal. No evidence for this was found since recombinant fusion proteins in which neomycin phosphotransferase was preceded by the entire proricin leader (35 amino acids) or just the first 22 amino acids of the proricin leader (the signal peptide) were both secreted when transiently expressed in tobacco protoplasts (Westby, 1992). If the full presequence had possessed a targeting signal we should have observed retention of the neomycin phosphotransferase reporter.

Finally, it is possible that the intracellular transport of proricin through the *Ricinus* endomembrane system may also protect the host plant against the potentially harmful effects of ricin A chain. During ricin intoxication of mammalian cells, the holotoxin enters the cells by endocytosis, usually via coated pits and coated vesicles. Several immunoelectron microscopy studies have shown that the endocytosed ricin is first delivered to the endosomes and that a proportion of the toxin subsequently appears within the Golgi complex, in particular the *trans*-Golgi network (van Deurs *et al.*, 1988). It appears that ricin has to reach the Golgi complex in order to exert its cytotoxic effects. Several treatments that induce morphological changes in the Golgi complex sensitise cells to ricin (Sandvig, Tonnesseu & Olsnes, 1986) and disruption of the Golgi by brefeldin A prevents ricin intoxication (Yoshida *et al.*, 1991). Indeed, incoming ricin may move from the *trans*-Golgi network to *trans*-, *medial* or *cis*-Golgi compartments, perhaps even to the ER, before translocation of the A chain into the cytosol occurs. Since newly synthesised proricin may pass through the corresponding cell compartment during intracellular transport to the protein bodies, the proprotein precursor form may ensure that membrane translocation of the ricin A chain does not inadvertently occur.

References

Butterworth, A.G. & Lord, J.M. (1983). Ricin and *Ricin communis* agglutinin subunits are all derived from a single-sized polypeptide precursor. *European Journal of Biochemistry* **137**, 57–65.

Cawley, D.B., Hedblom, M.W. & Houston, L.L. (1978). Homology between ricin and *Ricinus communis* agglutinin. *Archives of Biochemistry and Biophysics* **190**, 744–55.

Endo, Y., Mitsiu, K., Motizuki, M. & Tsurugi, K. (1987). The mechanism of ricin and related toxins on eukaryotic ribosomes. *Journal of Biological Chemistry* **262**, 5908–12.

Farquhar, M.G. (1985). Progress in unraveling pathways of Golgi traffic. *Annual Review of Cell Biology* **1**, 447–88.

Foxwell, B.M.J., Donovan, T.A., Thorpe, P.E. & Wilson, G. (1985). The removal of carbohydrates from ricin with endoglycosidases H, F and D and α-mannosidase. *Biochimica et Biophysica Acta* **840**, 193–203.

Freedman, R.B. (1984). Protein disulphide isomerase. *Trends in Biochemical Sciences* **9**, 438–41.

Gifford, D.J., Greenwood, J.S. & Bewley, J.D. (1982). Deposition of matrix and crystalloid storage proteins during protein body development in the endosperm of *Ricinus communis* L. cv. Hale seeds. *Plant Physiology* **69**, 1471–8.

Halling, K.C., Halling, A.C., Murray, E.E., Ladin, B.F., Houston, K.L. & Weaver, R.F. (1985). Genomic cloning and characterization of a ricin gene from *Ricinus communis*. *Nucleic Acids Research* **13**, 8019–33.

Hara-Nishimura, I., Inoue, K. & Nisimura, M. (1991). A unique vacuolar processing enzyme responsible for conversion of several pro-protein precursors into the mature forms. *FEBS Letters* **294**, 89–93.

Harley, S.M. & Lord, J.M. (1985). *In vitro* endoproteolytic cleavage of castor bean lectin precursors. *Plant Science* **41**, 111–16.

Hartley, M.R., Legname, G., Osborn, R., Chen, Z. & Lord, J.M. (1991). Single-chain ribosome inactivating proteins from plants depurinate *E. coli* 23S ribosomal RNA. *FEBS Letters* **290**, 65–8.

Hussain, K., Bowler, C., Roberts, L.M. & Lord, J.M. (1989). Expression of ricin B chain in *Escherichia coli*. *FEBS Letters* **244**, 383–7.

Katzin, B.J., Collins, E.J. & Robertus, J.D. (1991). Structure of ricin A chain at 2.5 Å. *Proteins* **10**, 251–9.

Lamb, F.I., Roberts, L.M. & Lord, J.M. (1985). Nucleotide sequence of cloned cDNA coding for preproricin. *European Journal of Biochemistry* **148**, 265–70.

Lord, J.M. (1985*a*). Synthesis and intracellular transport of lectin and storage protein precursors in endosperm from castor bean. *European Journal of Biochemistry* **146**, 403–9.

Lord, J.M. (1985*b*). Precursors of ricin and *Ricinus communis* agglutinin. Glycosylation and processing during synthesis and intra-cellular transport. *European Journal of Biochemistry* **146**, 411–16.

Lord, J.M. & Harley, S.M. (1985). *Ricinus communis* agglutinin B chain contains a fucosylated oligosaccharide side chain not present on ricin B chain. *FEBS Letters* **189**, 72–6.

Lord, J.M., Hartley, M.R. & Roberts, L.M. (1991). Ribosome inac-tivating proteins of plants. *Seminars in Cell Biology* **2**, 15–22.

Lord, J.M., Lamb, F.I. & Roberts, L.M. (1984). Ricin: structure, bio-logical activity and synthesis. *Oxford Surveys of Plant Molecular and Cell Biology* **1**, 85–101.

Lord, J.M. & Robinson, C. (1986). Role of proteolytic enzymes in the post-translational modification of proteins. In *Plant Proteolytic Enzymes* (ed. M.J. Dalling), pp. 69–80. Boca Raton, Florida: CRC Press.

Nicolson, G.L., Blanstein, J. & Etzler, M. (1974). Characterization of two plant lectins from *Ricinus communis* and their quantitative inter-action with a murine lymphoma. *Biochemistry* **13**, 196–204.

O'Hare, M., Roberts, L.M. & Lord, J.M. (1992). Biological activity of recombinant *Ricinus communis* agglutinin A chain produced in *Escherichia coli*. *FEBS Letters* **299**, 209–12.

O'Hare, M., Roberts, L.M., Thorpe, P.E., Watson, G.J., Prior, B. & Lord, J.M. (1987). Expression of ricin A chain in *Escherichia coli*. *FEBS Letters* **216**, 73–8.

Olsnes, S. & Pihl, A. (1982). Toxic lectins and related proteins. In *Molecular Action of Toxins and Viruses* (ed. P. Cohen & S. van Heyningen), pp. 51–105. New York: Elsevier.

Olsnes, S. & Saltvedt, E. (1975). Conformation-dependent antigenic determinants in the toxic lectin ricin. *Journal of Immunology* **114**, 1743–8.

Olsnes, S., Saltvedt, E. & Pihl, A. (1974). Isolation and comparison of galactose-binding lectins from *Abrus precatorius* and *Ricinus communis. Journal of Biological Chemistry* **249**, 803–10.

Piatak, M., Lane, J.A., Laird, W., Bjorn, M.J., Wang, A. & Williams, M. (1988). Expression of soluble and fully functional ricin A chain in *Escherichia coli* is temperature-sensitive. *Journal of Biological Chemistry* **263**, 4837–43.

Richardson, P.T., Gilmartin, P., Colman, A., Roberts, L.M. & Lord, J.M. (1988). Expression of functional ricin B chain in *Xenopus* oocytes. *Bio/Technology* **6**, 565–70.

Richardson, P.T., Westby, M., Roberts, L.M., Gould, J.H., Colman, A. & Lord, J.M. (1989). Recombinant proricin binds galactose but does not depurinate 28S ribosomal RNA. *FEBS Letters* **255**, 15–20.

Richardson, P.T., Hussain, K., Woodland, H.L., Lord, J.M. & Roberts, L.M. (1991). The effect of N-glycosylation on the lectin activity of recombinant ricin B chain. *Carbohydrate Research* **213**, 19–25.

Roberts, L.M., Lamb, F.I. & Lord, J.M. (1987). Biosynthesis and molecular cloning of ricin and *Ricinus communis* agglutinin. In *Membrane-mediated cytotoxicity* (ed. B. Bonavida & R.J. Collier), pp. 73–82. New York: Alan R. Liss.

Roberts, L.M., Lamb, F.I., Pappin, D.J.C. & Lord, J.M. (1985). The primary sequence of *Ricinus communis* agglutinin: Comparison with ricin. *Journal of Biological Chemistry* **260**, 15682–6.

Roberts, L.M. & Lord, J.M. (1981*a*). Protein biosynthetic capacity in the endosperm tissue of ripening castor bean seeds. *Planta* **152**, 420–7.

Roberts, L.M. & Lord, J.M. (1981*b*). The synthesis of *Ricinus communis* agglutinin. Co-translational and post-translational modification of agglutinin polypeptides. *European Journal of Biochemistry* **119**, 31–41.

Robertus, J.D. (1991). The structure and action of ricin, a cytotoxic N-glycosidase. *Seminars in Cell Biology*, **2**, 23–30.

Rutenber, E., Katzin, B.J., Ernst, S., Collins, E.J., Mesna, D., Ready, M.P. & Robertus, J.D. (1991). Crystallographic refinement of ricin to 2.5 Å. *Proteins* **10**, 240–50.

Rutenber, E., Ready, M. & Robertus, J.D. (1987). Structure and evolution of ricin B chain. *Nature (London)* **236**, 624–6.

Rutenber, E. & Robertus, J.D. (1991). Structure of ricin B chain at 2.5 Å resolution. *Proteins* **10**, 260–9.

Saltvedt, E. (1976). Structure and toxicity of pure *Ricinus* agglutinin. *Biochimica et Biophysica Acta* **451**, 536–46.

Sandvig, K., Tonnessen, T.I. & Olsnes, S. (1986). Ability of inhibitors of glycosylation and protein synthesis to sensitize cells to abrin, ricin, *Shigella* toxin and *Pseudomonas* toxin. Cancer Research **46**, 6418–22.

Schekman, R. (1985). Protein localization and membrane traffic in yeast. *Annual Review of Cell Biology* **1**, 115–43.

Stillmark, H. (1888). Uber Ricin, eines gifiges Ferment aus den Samen von *Ricinus communis* L. und anderen Euphorbiaceen. Inaugural Dissertation, University of Dorpat, Estonia.

Tregear, J.W. & Roberts, L.M. (1992). The lectin gene family of *Ricinus communis*: cloning of a functional ricin gene and three lectin pseudogenes. *Plant Molecular Biology* **18**, 515–25.

Tulley, R.E. & Beevers, H. (1976). Protein bodies of castor bean endosperm. Isolation, fractionation and characterization of protein components. *Plant Physiology* **58**, 710–16.

van Deurs, B., Sandvig, K., Peterson, O.W., Olsnes, S., Simons, K. & Griffiths, G. (1988). Estimation of the amount of internalized ricin that reaches the trans-Golgi network. *Journal of Cell Biology* **106**, 253–67.

Wales, R., Richardson, P.T., Roberts, L.M., Woodland, H.R. & Lord, J.M. (1991). Mutational analysis of the galactose binding ability of recombinant ricin B chain. *Journal of Biological Chemistry* **266**, 19172–9.

Wales, R., Richardson, P.T., Roberts, L.M., Woodland, H.R. & Lord, J.M. (1992). Recombinant ricin B chain fragments containing a single galactose binding site retain lectin activity. *Archives of Biochemistry and Biophysics* **294**, 291–6.

Walter, P., Gilmore, R. & Blobel, G. (1984). Protein translocation across the endoplasmic reticulum. *Cell* **38**, 5–7.

Westby, M. (1992). Studies on the targeting and processing of proricin. Ph.D. thesis, University of Warwick.

Yoshida, T., Chen, C., Zhang, M. & Wu, H.C. (1991). Disruption of the Golgi apparatus by brefeldin A inhibits the cytotoxicity of ricin, modeccin and *Pseudomonas* toxin. *Experimental Cell Research* **192**, 389–95.

Youle, R.J. & Huang, A.H.C. (1976). Protein bodies from the endosperm of castor bean. Subfractionation, protein components, lectins and changes during germination. *Plant Physiology* **58**, 703–9.

D. J. BOWLES

Post-translational processing of concanavalin A

Concanavalin A (Con A) is a plant lectin, making up some 30% of the protein in jackbean seeds. At seed maturity, the protein is located in the protein bodies of the storage parenchyma cells within the cotyledons of the seed. The biosynthesis and processing of Con A have been studied in a number of laboratories; the resulting data indicate a complex and novel series of events.

Glycoconjugates in jackbean seeds are recognised by the endogenous lectin

At Leeds, our studies on concanavalin A (Con A) began more than ten years ago, when we were interested in determining the range of endogenous glycoconjugates in jackbeans that could interact with the lectin. Our aim was to gain insight into the nature of lectin–receptor interactions within the seed, and thereby to understand the potential function of the lectin through the types of glycoproteins and/or glycolipids available for complexing.

Using the technique of Con A overlays of jackbean polypeptides separated by SDS–PAGE, we found that only one polypeptide bound the lectin, the heavy subunit (66 kDa) of the enzyme α-mannosidase (Bowles, Andralojc & Marcus, 1982). However, the interaction between the lectin and the hydrolase could only be demonstrated if the enzyme was first denatured; in the native form of the enzyme, the N-glycan was inaccessible for binding to Con A (Bowles, Chaplin & Marcus, 1983). This suggested that in native tetrameric form, the N-glycan is sterically hindered from binding through positioning of the heavy and light subunits within the oligomer.

The studies were carried out using extracts prepared from dry, mature jackbean seeds. The apparent lack of a glycoconjugate receptor to Con A

Society for Experimental Biology Seminar Series 53: *Post-translational modifications in plants*, ed. N.H. Battey, H.G. Dickinson & A.M. Hetherington. © Cambridge University Press 1993, pp. 257–266.

in the extracts led to the idea that a receptor might be developmentally regulated. To investigate this possibility we prepared a developmental time course of jackbean seeds from early stages through to maturity and desiccation. We investigated the timing of the appearance of Con A using antibodies to the lectin, as well as assays based on haemagglutination activity (Marcus *et al.*, 1984; Maycox *et al.*, 1988). In particular, we followed the accumulation of Con A by immunoblotting of extracts following SDS–PAGE, and compared the profile of the immunoblots with Con A blots to correlate the appearance of the lectin with the appearance and disappearance of putative glycoprotein receptors (Marcus *et al.*, 1984).

These data led to a change in direction of the research programme at Leeds, since they showed that a polypeptide immunogenically related to Con A (i.e. labelled with anti-Con A) also bound to the lectin in the overlays, implying that it was N-glycosylated. The data also showed that as the seed matured and Con A accumulated, progressively fewer Con A-binding polypeptides could be detected, until eventually at the dry seed stage, the only one remaining was the heavy subunit of α-mannosidase. The progressive decline in N-glycosylation was confirmed by using metabolic labelling with glycan precursors and analysis of the radioactive oligosaccharides following their release from the proteins (S.E. Marcus & D.J. Bowles, unpublished data).

As yet, the significance of the data showing the progressive decline in N-glycans during seed development is unclear; owing to the very obvious significance of a glycosylated Con A species, the research followed the direction of the biosynthesis and assembly of the lectin. At the same time, the existence of a Con A glycoprotein precursor was shown in Chrispeels' laboratory by using a different approach, and the presence of a single N-glycan attached to the polypeptide was confirmed (Herman, Shannon & Chrispeels, 1985; Chrispeels *et al.*, 1986; Faye & Chrispeels, 1987).

Biogenesis of protein bodies in jackbean seeds

In order to place the molecular data concerning jackbean polypeptides within a cellular context, we also analysed events in the developing seed by electron microscopy (Maycox *et al.*, 1988). The transport route and traffic of newly synthesised proteins destined for a membrane-bound compartment within the cell is well characterised and has been reviewed a number of times (Chrispeels 1984, 1991; Bowles, 1990*a,b*). The biogenesis of protein bodies is particularly interesting, since the compartments are derived from vacuoles (Craig, Goodchild & Miller, 1980). The storage parenchyma of the cotyledons initially contain a single large vacuole at early stages in seed maturation. Activation of genes encoding the storage proteins and lectins, and the accumulation of the seed proteins, occurs concurrently with changes to the vacuole. In some way, protein loading by vesicular traffic and fusion of the vesicles with the tonoplast occurs at the same time as (or drives?) fragmentation of the vacuole into the progressively smaller compartments that ultimately become the protein bodies.

Data from Leeds (Maycox *et al.*, 1988; Maycox 1986) confirmed earlier studies (Herman & Shannon, 1984) that showed that the intracellular traffic of Con A followed a typical transport route. The newly synthesised protein was co-translationally segregated into the rough endoplasmic reticulum, transported in vesicles to the Golgi apparatus and from there to the protein bodies. We also performed immunocytochemical double-labelling to compare the distribution of Con A, α-mannosidase and canavalin simultaneously. The data showed that the transport vesicles and the protein bodies contained mixtures of the three jackbean proteins, implying that they followed an identical intracellular route during their traffic from the endoplasmic reticulum to the vacuole (Maycox *et al.*, 1988).

Curiously, the typical stacked cisternae of the Golgi apparatus were not observed as abundant organelles in the EM sections of the storage parenchyma, an observation noted by others (Craig *et al.*, 1980; Herman & Shannon, 1984). Instead, the bulk of the cytoplasm was taken up with sheets of rough endoplasmic reticulum and the vacuolar compartments. This may well reinforce an opinion, first formulated by de Duve (1969), that the Golgi apparatus essentially represents a key funnelling system analogous to a canal lock. Again, as shown in Grimstone's seminal paper (Grimstone, 1959) and discussed by Bowles (1973), the observations may well endorse the notion that the Golgi stack is a measure of membrane inflow versus outflow at a particular locus in time and space within the cell. Rapid traffic through the locus, for example as in the protein loading period of seed formation, may be sufficient to lead to an absence of stacked cisternae and their replacement by morphologically indistinct transit vesicles.

Biosynthesis of Con A

The biosynthesis of Con A was studied at Leeds using an approach involving metabolic labelling and pulse–chase analysis. These data are described by Bowles *et al.* (1986), and will not be discussed in detail here. To summarise, evidence was found for three novel events: (1) removal of an N-glycan from the 33.5 kDa lectin precursor first labelled in the pulse–

chase leading to a small decrease in molecular mass consistent with the loss of a core-oligosaccharide; (2) construction of a post-translational peptide bond such that peptide fragments observed in fluorographs of immunoprecipitates from extracts during the pulse–chase appeared to re-ligate into a polypeptide of 30 kDa (the protein re-ligation event was confirmed by N-terminal sequencing of the different molecular species and indicated that the N-terminal sequence of the first-labelled lectin precursor was found in the middle of the sequence of mature Con A); (3) monensin treatment of the jackbean cotyledons and ionophore-induced disruption of traffic through the *trans*-Golgi network led to loss of newly synthesised Con A out of the cells; the form of Con A that accumulated outside of the plasma membrane was found to consist of only the first, glycosylated precursor.

Thus, in total, we provided evidence of a complex series of post-translational events during the processing of Con A, involving a putative deglycosylation, a number of proteolytic events and a peptide ligation. In terms of viewing these events as changes to linear sequence or as changes in patterns of polypeptides in SDS–PAGE analyses, the processing events were very difficult to understand. However, when viewed within the three-dimensional shape of the Con A precursor and its conversion to the mature lectin, the processing events became extraordinarily simple and straightforward. Following a completely different strategy, work at Cambridge had led to the cloning of the gene for Con A (Carrington, Auffret & Hanke, 1985). The lectin cDNA was sequenced, and the predicted polypeptide sequence encoded by the mRNA transcript was compared with the known polypeptide sequence of mature Con A. The two did not match since it seemed that the predicted translation product would start with a sequence found in the middle of the mature protein. This circular permutation had been noted earlier in relation to the sequence of Con A and another closely related legume lectin, favin (Cunningham *et al.*, 1979). Our data confirmed the suggestion of protein re-ligation during Con A assembly but showed that essentially no structural rearrangement was necessary. All of the processing events we described involved changes to the surface of a protein molecule whose shape did not alter from that established during synthesis and folding.

The analyses described in Bowles *et al.* (1986) and Marcus & Bowles (1988) also provided a way forward to studying the enzymes involved in the processing events. Following monensin treatment and loss of the precursors down the secretory default pathway, the species of Con A that accumulated was the glycoprotein precursor. This implied that the processing enzymes that removed the N-glycan, proteolytically processed the lectin and caused the re-ligation were all located at a site on the transport route after the point at which monensin exerted its effect. A vacuolar site for processing endopeptidases of storage proteins and lectins has been shown for a number of plant species, including legumes and other dicots, as well as cereals (Chrispeels, 1991).

The work on Con A processing in Leeds was carried out prior to information now known on vacuolar targeting signals in plants (reviewed by Bowles, 1990*a*). At the time, we raised the suggestion that perhaps, given the immensity and rate of protein loading, the transport route to the vacuole might constitute the 'default' pathway, i.e. the pathway that would occur in the absence of any active targeting signal (Bowles & Pappin, 1988). This suggestion has now been clearly shown to be incorrect, since it seems that targeting signals are required for correct transport to the vacuoles (Denecke, Botterman & Deblaere, 1990; Dorel *et al.*, 1989). Further, there is good evidence to indicate that the targeting signal–receptor system is conserved irrespective of cell type, tissue or even plant species (Sonnewald *et al.*, 1989). There is also good evidence to indicate the existence of clathrin-coated vesicles in cells of developing seeds and that the vesicles contain storage protein precursors (Robinson, Babusek & Freundt, 1989; Hurley & Beevers, 1989). Thus, there is every probability that the strategy for sorting and targeting in the *trans*-Golgi network in plants will be analogous to the receptor-based systems known for other eukaryotic cells. Given the extent of rough endoplasmic reticulum in the seed storage parenchyma cells and the rate of vacuolar loading, the receptor-mediated shuttle, receptor and clathrin recycling, and their re-use, must be extraordinarily rapid.

Processing enzymes of Con A: reconstitution of events *in vitro*

The post-translational processing of Con A is interesting because of the deglycosylation event as well as the peptide bond construction. To follow these processing events and dissect the mechanisms involved, it became essential to purify the glycoprotein precursor. This has recently been achieved and provides the basis for reconstitution of the processing events *in vitro* (Sheldon & Bowles, 1992). It is clear that the enzymes responsible for the processing exist in jackbean seeds and we have evidence that both the N-glycanase and the endopeptidase are developmentally regulated (P. Sheldon & D.J. Bowles, unpublished data). The N-glycanase can be isolated from dry jackbean seeds (Sugiyamas *et al.*, 1983; Yet & Wold, 1988) but the endopeptidase activity of dry seeds is very different from the endopeptidase activity involved in the proteolytic processing of the Con A precursor (Dalkin, Marcus & Bowles, 1983).

Since we had shown that the removal of the N-glycan from the glycoprotein precursor *in vivo* appeared to involve the complete removal of the oligosaccharide, i.e. cleavage between the internal GlcNAc and the linkage Asn, we decided to analyse the consequences of this complete deglycosylation. To date this event is unique in eukaryotic cells. Deglycosylation as a regulatory mechanism would possess an on-off simplicity, and its potential elegance as a means of regulating protein activity led us to investigate its role in Con A assembly.

N-glycanases are available commercially, and whilst purification of the jackbean enzyme was in progress, we used a commercial preparation to assay the effects of deglycosylation on the lectin precursor. We had previously shown that the glycoprotein precursor did not bind to carbohydrate, confirming that the precursor was inactive as a lectin (Bowles *et al.*, 1986). At that time we could not distinguish between deglycosylation and the first proteolytic event as the decisive step for activation, since *in vivo* proteolysis very rapidly followed removal of the N-glycan and prevented isolation of deglycosylated precursors with intact polypeptide chains.

We found that N-glycanase rapidly removed the N-glycan from the native form of the Con A precursor (Sheldon & Bowles, 1992). This contrasts sharply with the effect of endoglycanases on α-mannosidase, when cleavage of the N-glycan was only possible after the tetrameric form of the enzyme had been denatured (Bowles *et al.*, 1983). These data suggest that the GlcNAc–Asn bond of the N-glycan on the Con A precursor is sterically accessible to an endo-glycanase, whereas the internal GlcNAc–GlcNAc bond of the N-glycan on α-mannosidase is not. The removal of the oligosaccharide from the Con A precursor was carried out experimentally *in vitro*, but can be thought of as mimicking the known processing events *in vivo*.

Removal of the N-glycan was found to activate the lectin function of the precursor (Sheldon & Bowles, 1992). The glycoprotein was inactive as a lectin, whereas the deglycosylated protein became active. This change in activity was assayed by the ability of the Con A precursors to bind to an ovalbumin–Sepharose affinity matrix. The glycoprotein precursor was routinely recovered in the effluent. In contrast, a substantial proportion of the deglycosylated precursor (*ca.* 40%) bound to the matrix and was recovered in the α-methylmannoside eluate. We believe that only a proportion of the deglycosylated precursor bound to the matrix, because removal of the N-glycan destabilised/denatured the polypeptide. This was reflected in a progressive tendency, over time, for the deglycosylated proteins to precipitate. It had been noted earlier that tunicamycin treatment of jackbeans led to non-glycosylated Con A accumulating in the microsomes and this may have reflected an instability or precipitation *in vivo* (Faye & Chrispeels, 1987). Similarly, when recombinant non-glycosylated Con A precursor was synthesised in *Escherichia coli*, the species was again found to have a tendency to precipitate (Min, Dunn & Jones, 1992). Significantly, the recombinant Con A precursor also showed an ability to bind to an affinity matrix (Min *et al.*, 1992).

Thus, whether the native glycoprotein precursor was deglycosylated *in vitro*, or whether a recombinant non-glycosylated precursor was synthesised, both species exhibited lectin activity. These data confirm that the N-glycan alone determines the lectin activity of the Con A precursor; proteolysis is an irrelevance for this activation of function. To date, this is the first example in any eukaryotic cell in which an N-glycan has been found to regulate protein function. Naturally, it would be reassuring were this mechanism of regulation to be found to extend beyond the regulation of lectin activity. Time will tell; an increasing awareness of the range and distribution of N-glycanases will help.

In the meantime, we are keen to understand the molecular mechanism by which the N-glycan regulates the activity. Is the oligosaccharide in the binding pocket of the lectin? Neutralisation was first discussed in 1979 (Bowles, 1979) as a means of either regulating the binding capacity of a lectin, or regulating indirectly an additional activity of a lectin and/or receptor, such as allosteric regulation of enzyme activity, gating of ion channels, etc. Interestingly, some 12 years later, the neutralisation model may well be proved to be correct for Con A processing. Intra- or intermolecular self-neutralisation with the N-glycan in the binding pockets of self monomers or adjacent monomers may indeed be the means by which regulation of function is achieved. Certainly the glycoprotein precursor of Con A cannot be purified on a lectin–affinity matrix (Marcus, 1988). This suggests that the N-glycan is not available for complexing 'outwardly'. This observation, coupled with the accessibility of the GlcNAc–Asn bond to hydrolysis, at least suggests that the bulk of the oligosaccharide may well be masked within the pocket, whereas the hinge-attachment region of the N-glycan to the polypeptide remains external and available for enzymic cleavage.

The relevance of this lectin activation is as yet also unclear. Other lectin precursors, in particular those of the structurally related pea lectin, are made in active form and affinity precipitation has been used to purify the precursors (Higgins *et al.*, 1983). It is tempting to suggest that Con A must be made in inactive form since otherwise it would bind to lumenal high-mannose N-linked membrane glycoproteins within the rough endoplasmic reticulum, and complexing to the membrane of precursor would prevent traffic. Although the ribophorins are known to be Con A-binding polypeptides (Kreibich *et al.*, 1982) and the data from tunicamycin-treated jackbean cells (Faye & Chrispeels, 1987) certainly support such a story, one is left with the apparently conflicting data from the other lectins. It is possible that the fine differences in specificity for N-linked glycoconjugates between the different plant lectins provide an answer. This possibility also provides the basis for designing yet more experiments to improve our understanding of the biological significance of the post-translational processing of Con A.

Acknowledgements

The research described at Leeds has been supported by SERC grants to D.J.B.

References

Bowles, D.J. (1973). A function of the Golgi apparatus in root tissue of maize. Ph.D. thesis, University of Cambridge.

Bowles, D.J. (1979). Lectins as membrane components: implications of lectin-receptor interactions. *FEBS Letters* **102**, 1–3.

Bowles, D.J. (1990*a*). Endomembrane traffic and targeting in plant cells. *Current Opinion in Cell Biology* **2**, 673–80.

Bowles, D.J. (1990*b*). Defence-related proteins in higher plants. *Annual Review of Biochemistry* **59**, 873–907.

Bowles, D.J., Andralojc, J. & Marcus, S.E. (1982). Identification of an endogenous ConA-binding polypeptide as the heavy subunit of α-mannosidase. *FEBS Letters* **140**, 234–6.

Bowles, D.J., Chaplin, M.F. & Marcus, S.E. (1983). Interaction of Concanavalin A with native and denatured forms of jackbean α-mannosidase. *European Journal of Biochemistry* **130**, 613–18.

Bowles, D.J., Marcus, S.E., Pappin, D.J., Findlay, J.B.C., Eliopoulas, E., Maycox, P.R. & Burgess, J. (1986). Post-translational processing of Concanavalin A precursors in jackbean cotyledons. *Journal of Cell Biology* **102**, 1284–97.

Bowles, D.J. & Pappin, D.J. (1988). Traffic and assembly of Con-canavalin A. *Trends in Biological Sciences* **13**, 60–4.

Carrington, D.H., Auffret, A. & Hanke, D.G. (1985). Polypeptide ligation occurs during post-translational modification of ConA. *Nature (London)* **313**, 64–7.

Chrispeels, M.J. (1984). Biosynthesis, processing and transport of storage proteins and lectins in cotyledons of developing legume seeds. *Philosophical Transactions of the Royal Society London* **B304**, 309–22.

Chrispeels, M.J. (1991). Sorting of proteins in the secretory system. *Annual Review of Plant Physiology and Plant Molecular Biology* **42**, 21–53.

Chrispeels, M.J., Hartl, P.M., Sturm, A. & Faye, L. (1986). Characterisation of the endoplasmic reticulum associated precursor of Concanavalin A. *Journal of Biological Chemistry* **261**, 10021–4.

Craig, S., Goodchild, D.J. & Miller, C. (1980). Structural aspects of protein accumulation in developing pea cotyledons 2. 3-D reconstructions of vacuoles and protein bodies from several sections. *Australian Journal of Plant Physiology* **7**, 329–37.

Cunningham, B.A., Hemperley, J.J., Hopp, T.P. & Edelman, G.M. (1979). Favin versus Concanavalin A: circularly-permutated amino acid sequences. *Proceedings of the National Academy of Sciences USA* **76**, 3218–22.

Dalkin, K., Marcus, S.E. & Bowles, D.J. (1983). Endopeptidase activity in jackbeans and its effect on Concanavalin A. *Planta* **157**, 531–5.

deDuve, C. (1969) In *Lysosomes in Biology and Pathology*, Vol. 1 (ed. J.T. Dingle & H.B. Fell), p. 340. North Holland Publishing Company.

Denecke, J., Botterman, J. & Deblaere, R. (1990). Protein secretion in plant cells can occur via a default pathway. *Plant Cell* **2**, 51–9.

Dorel, C., Voelker, T.A., Herman, E.M. & Chrispeels, M.J. (1989). Transport of proteins to the plant vacuole is not by bulk flow through the secretory system and requires positive sorting information. *Journal of Cell Biology* **108**, 327–37.

Faye, L. & Chrispeels, M.L. (1987). Transport and processing of the glycosylated precursor of Concanavalin A in jackbeans. *Planta* **170**, 217–24.

Grimstone, A.V. (1959). Cytoplasmic membranes and the nuclear membrane in the flagellate *Trichonympha*. *Journal of Biophysical and Biochemical Cytology* **6**, 369.

Herman, E.M. & Shannon, L.M. (1984). Immunocytochemical localisation of Concanavalin A in developing jackbean cotyledons. *Planta* **161**, 97–104.

Herman, E.M., Shannon, L.M. & Chrispeels, M.J. (1985). Concanavalin A is synthesized as a glycoprotein precursor. *Planta* **165**, 23–9.

Hurley, S.M. & Beevers, L. (1989). Coated vesicles are involved in the transport of storage protein during seed development in *Pisum sativum*. *Plant Physiology* **91**, 674–8.

Higgins, T.J.V., Chandler, P.M., Zwawski, G., Burton, S.C. & Spencer, D. (1983). The biosynthesis and primary structure of pea seed lectin. *Journal of Biological Chemistry* **258**, 9544–9.

Kreibich, G., Ojakian, E., Rodriguez-Boulan, E. & Sabatini, D.D. (1982). Recovery of ribophorins. *Journal of Cell Biology* **93**, 111–21.

Marcus, S.E. (1988). Studies on jackbean proteins. M.Phil. thesis, University of Leeds.

Marcus, S.E. & Bowles, D.J. (1988). Deglycosylation of a lectin intermediate during assembly of Con A. *Protoplasma* **147**, 113–16.

Marcus, S.E., Burgess, J., Maycox, P.R. & Bowles, D.J. (1984). Maturation events in jackbeans: *Canavalia ensiformis. Biochemical Journal* **222**, 265–8.

Maycox, P.R. (1986). Immunocytochemical studies on three jackbean seed proteins. Ph.D. thesis, University of Leeds.

Maycox, P.R., Burgess, J., Marcus, S.E. & Bowles, D.J. (1988). Studies on α-mannosidase and Concanavalin A during jackbean development and germination. *Protoplasma* **144**, 34–45.

Min, W., Dunn, A.J. & Jones, D.H. (1992). Non-glycosylated recombinant pro-concanavalin A is active without polypeptide cleavage. *EMBO Journal* **11**, 1303–7.

Robinson, D.G., Babusek, K. & Freundt, H. (1989). Legumin antibodies recognize polypeptides in coated vesicles isolated from developing pea cotyledons. *Protoplasma* **150**, 79–82.

Sheldon, P.S. & Bowles, D.J. (1992). The glycoprotein precursor of Concanavalin A is converted to an active lectin by deglycosylation. *EMBO Journal* **11**, 1297–301.

Sonnewald, U., Sturm, A., Chrispeels, M.J. & Willmitzer, L. (1989). Targeting and glycosylation of patatin, the major potato tuber protein in leaves of transgenic tobacco. *Planta* **179**, 171–80.

Sugiyama, K., Ishihara, H., Tejima, S. & Takahashi, N. (1983). Demonstration of a new glycopeptidase, from jack-bean meal, acting on aspartylglucosylamine linkages. *Biochemical and Biophysical Research Communications* **112**, 155–60.

Yet, M.-G. & Wold, F. (1988). Purification and characterization of two glycopeptide hydrolases from jackbeans. *Journal of Biological Chemistry* **263**, 118–22.

J.P. KNOX

The role of cell surface glycoproteins in differentiation and morphogenesis

Introduction

The differentiation of a plant cell subsequent to its origin in a meristem involves the alteration of its surface in terms of both cell shape and molecular composition. It is the anatomical complexity resulting from the highly ordered arrangement of differing cell morphologies that we can perceive as cell and tissue patterns. A useful starting point in understanding the mechanisms that lie behind the development of such complexity is to identify changes in the cell wall associated with these diverging cell morphologies and fates. Although the molecular architecture of the plant cell wall principally involves the organisation of polysaccharides there is an increasing awareness that glycoconjugates, although less abundant than polysaccharides, are likely to be extremely important for the integration of wall functions and cellular processes.

When considering cell surface glycoproteins it is useful to consider those associated with the outer face of the plasma membrane in addition to those clearly occurring in the wall. It is molecules at this location that will be involved in wall–cytoplasm interactions and are thus likely to mediate the assembly of specific wall architectures. In many cases we do not yet fully understand the functions of cell surface proteins in a bio-chemical or a developmental sense; the most abundant classes of surface proteins are still currently named in relation to aspects of their protein or carbohydrate structure. At this stage it is possible only to categorise their roles broadly as structural, enzymic or signalling.

The post-translational hydroxylation of proline within proteins destined for the cell surface is a major feature of cell metabolism in both plants and animals. The presence of hydroxyproline can contribute to distinctive conformations of collagens and hydroxyproline-rich plant pro-teins, and it has generally been surmised that hydroxyproline-containing molecules perform structural roles in the extracellular matrix (Cassab &

Society for Experimental Biology Seminar Series 53: *Post-translational modifications in plants*, ed. N.H. Battey, H.G. Dickinson & A.M. Hetherington. © Cambridge University Press 1993, pp. 267–283.

Varner, 1988; Guzman, Fuller & Dixon, 1990). Further modification occurs in plant cells where a large proportion of the hydroxyproline is glycosylated with arabinose- and/or galactose-containing oligosaccharides prior to reaching the surface. The three major classes of plant hydroxyproline-containing macromolecules are the cell wall hydroxyproline-rich glycoproteins (HRGPs, extensins), the Solanaceous lectins and the arabinogalactan-proteins (AGPs). Recent work has indicated several new classes of hydroxyproline- and proline-rich proteins in plant cells, some of which are highly developmentally regulated. Cell wall proteins have recently been reviewed by Cassab & Varner (1988), Showalter & Varner (1989) and Showalter & Rumeau (1990) and AGPs by Fincher, Stone & Clarke (1983).

A consideration of AGPs forms the basis of this chapter, which is largely concerned with the way in which immunolocalisation studies are widening our view of cell surface glycoproteins and of how their specific roles may be integrated with developmental mechanisms.

Monoclonal antibodies and the plant cell surface

The recognition specificity provided by mammalian immune systems continues to provide the best means for the identification and localisation of molecules of biological interest within complex mixtures. Although the advantages and attractiveness of hybridoma technology do not require reiteration here, one of its most important aspects is the fact that highly specific antibodies can be selected subsequent to immunisations with complex immunogens. It is at the hybridoma screening stage that antibodies with desired recognition specificities can be selected out from the full range of the immune response that is generally observed within an antiserum. This molecular dissection can be carried out to derive probes for distinct antigens occurring at the same cell surface or for different epitopes occurring on a single complex macromolecule. Such procedures are of particular importance and usefulness when dealing with carbohydrate structures and glycoconjugates (Kannagi & Hakomori, 1986; Feizi & Childs, 1987). Once derived, monoclonal antibodies are very powerful and versatile probes for immunolocalisation studies, microanalysis, and the determination of structure–function relations of polysaccharides, glycoproteins and glycolipids.

Immunisations with protoplasts or with preparations of plant cell surface or extracellular material have recently proved extremely effective for the generation of novel probes for glycoconjugates and polysaccharides of plant origin. In many cases these probes indicate extensive developmental regulation of surface glycoproteins (Knox, 1992b; Perotto et al.,

1991). With the immense array of carbohydrate structures occurring at the plant cell surface as part of both polysaccharides and glycoconjugates, this approach would appear to be far from exhausted. Although the derivation of monoclonal antibodies by methods involving the immunisation with whole cells or complex mixtures is not always predictable, the fortuitous aspects of such an approach should not be eschewed, nor the prospect of serendipity ignored.

AGP epitopes at the plasma membrane

Arabinogalactan-proteins generally contain a high proportion of carbohydrate (generally 50–90%) which is characterised by a $\beta1$–3-linked galactan backbone with varied amounts of $\beta1$–6-linked galactosyl branches, substituted with arabinose and other less abundant monosaccharides (Fincher *et al.*, 1983). Owing to the often extensive polymeric carbohydrate components, such molecules can be viewed as polysaccharides covalently attached to protein and have been regarded as proteoglycans (Clarke, Anderson & Stone, 1979; Fincher *et al.*, 1983). No sequence data are yet available for AGP polypeptide structure, but they are generally rich in hydroxyproline, alanine, serine and glycine (Gleeson *et al.*, 1989).

AGPs have a large water-holding capacity and are common soluble components of lubricants and plant secretions such as root cap slimes and exudate gums (Fincher *et al.*, 1983; Stephen, Churms & Vogt, 1990; Qi, Fong & Lamport, 1991). AGPs display a specific reactivity with synthetic multivalent phenyl-β-glycosides known as artificial carbohydrate antigens, first prepared by Yariv (Fincher *et al.*, 1983), which may correspond with a capacity for highly specialised interactions with endogenous molecules. The complex branching patterns of AGP carbohydrate structures, the features described above, and their presence in the extracellular and intercellular space of tissues has led to frequent speculation upon their possible functions other than as lubricants. These include the possible utilisation of their information potential in some aspect of plant or cell identity, recognition or signalling (see Fincher *et al.*, 1983).

It is now firmly established that arabinogalactan-rich glycoproteins are associated with the plasma membrane of all higher plant cells. The evidence for this has come from various sources. Yariv antigens are reactive with the outer face of protoplasts (Larkin, 1978; Nothnagel & Lyon, 1986). The preparation of anti-plasma membrane antibodies has indicated that glycoproteins with the structural features of AGPs are tightly associated with the plasma membrane of tobacco cells (Norman *et al.*, 1990). Other anti-plasma-membrane antibodies have been derived that cross-react with both soluble AGPs obtained from the media of cultured cells and AGP components of exudate gums such as gum arabic (Pennell *et al.*, 1989; Knox, Day & Roberts, 1989; Knox *et al.*, 1991). In these cases the epitopes common to both the soluble and membrane bound AGPs are carbohydrate structures. The binding of the antibodies is sensitive to periodate treatment indicating reaction with terminal sugars (Knox *et al.*, 1991). In some cases, the binding of the antibodies can be inhibited by monosaccharide haptens. For example, the binding of the monoclonal antibody MAC207 is most readily inhibited by L-arabinose (Pennell *et al.*, 1989) and that of JIM15 by D-glucuronic acid (J.P. Knox, unpublished observation). These antibodies therefore indicate that carbohydrate structures are common to soluble and membrane-bound AGPs. We know less about how the proteins of the soluble and bound AGPs compare, although the membrane-associated proteins are similar to the soluble AGPs in terms of the profile of amino acid composition (Norman *et al.*, 1990).

An extended range of monoclonal antibodies reacting with both soluble and membrane-bound AGPs has now been prepared, subsequent to immunisation with protoplasts or preparations of extracellular material (Knox *et al.*, 1991). These antibodies recognise different, but as yet structurally undetermined, carbohydrate epitopes. These epitopes are currently envisaged as distinct oligosaccharide structures, most likely involving the substituted side chains of the galactan backbones. The use of these antibodies in immunochemical and immunolocalisation studies has indicated complex biochemical relationships linking the epitopes and, in addition, that extensive glycosylic modulations of the membrane-bound glycoproteins correlate with the formation of plant cell and tissue patterns.

Patterns of AGP epitope expression and carrot cell development

Currently, the developmental systems that have been most characterised in terms of patterns of AGP epitope expression are those of the carrot root apex and carrot cells in culture (Knox *et al.*, 1991). Carrot has been utilised in these studies because of the ease in which rapidly proliferating and embryogenic cell cultures can be induced from hypocotyl or secondary phloem tissues. Suspension cell cultures also provide large amounts of conditioned media, allowing an easy access to extracellular material.

Several of the anti-AGP/plasma membrane antibodies so far characterised recognise one or other, or both, of two soluble extracellular AGPs, (M_r 70000 – 100000) obtained from the conditioned media of two

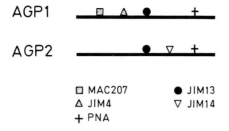

Fig. 1. Highly schematic diagram indicating the occurrence of four carbohydrate epitopes on two polymorphic extracellular AGPs obtained from the conditioned media of two distinct carrot suspension-cultured cell lines. The symbols indicate the presence of epitopes recognised by the monoclonal antibodies and also by peanut agglutinin (PNA).

distinct carrot cell lines (Knox *et al.*, 1991). The differing reactivities of four monoclonal antibodies with carrot AGP1 and AGP2 are shown schematically in Fig. 1. The glycan component forms over 90% of these AGPs. Their identical electrophoretic characteristics and variable binding of the antibodies to them indicates that AGP1 and AGP2 are polymorphic forms. The presence or absence of epitopes are likely to relate to differences in the peripheral substituents on the polysaccharide side chains, as discussed above. An important feature of these molecules is the fact that distinct epitopes, such as those recognised by JIM4 and MAC207, can occur on the same galactan backbone. This has been confirmed by immunoaffinity purification of the AGPs (J.P. Knox, unpublished observations). The galactan backbone itself, which is presumably identical or similar in both AGP1 and AGP2, is recognised by anti-galactose lectins such as peanut agglutinin (Knox & Roberts, 1989), which recognises both AGPs equally.

If we now consider the binding of these antibodies to the plasma membranes of the developing carrot root, we observe that epitopes common, for example, to AGP1 (such as those of JIM4 and JIM13) not only can occur on separate molecules but are also expressed by separate groups of cells. Glycosylic modulations of sets of membrane glycoproteins result in the presence or absence of the epitopes, presumably involving the modification of the terminal substituents, equivalent to the basis of the AGP1/AGP2 polymorphism.

The pattern of primary tissues at the carrot root

At a root apex files of cells can be followed spatially, from a region immediately proximal to the meristem to the region of differentiation, to

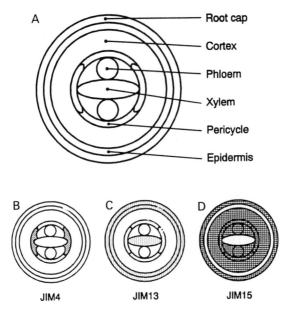

Fig. 2. A highly schematic diagram of the carrot root primary tissue
pattern seen in transverse section indicating the major tissue distinctions
(A). Cell populations recognised by the three monoclonal antibodies
(JIM4, JIM13 and JIM15) reflect this pattern and are shown by shaded
regions in (B)–(D).

reveal a temporal developmental sequence. The anatomical complication
of lateral meristems only arises after the major feats of cell development
have passed.

The primary tissues of the carrot root can be viewed in transverse
section as a concentric pattern of root cap, epidermis, cortex with its
inner layer of the endodermis and central stele containing a band of
xylem flanked by two regions of phloem (see Fig. 2A). The pericycle can
be regarded as the outer layer of the stele, and in the carrot root consists
of two distinctive cell populations (Esau, 1940; Knox *et al.*, 1989; Lloret *et
al.*, 1989). Adjacent to the region where the protoxylem will develop, the
pericycle cells are larger with distinct oblique divisions. The occurrence of
these distinctive cell shapes is one of the earliest signs of cell differentiation in the carrot root (Esau, 1940).

The patterns of occurrence of three carbohydrate AGP epitopes,
recognised by JIM4, JIM13 and JIM15, mark the major distinctions of
this tissue pattern (Fig. 2B–D; Knox *et al.*, 1991). If we consider a group
of cells as they mature and become separated from the meristem initials,

the first modulation of membrane-bound AGPs is indicated by the appearance of the JIM4 epitope (J4e) at the surface of the distinctly shaped pericycle cells and cells in adjacent files associated by position (Fig. 2B) (Knox *et al.*, 1989). J4e is initially expressed by just one or two cells of the future pericycle marking each end of the future xylem band, before any of the distinctive cell divisions occur. The expression of J4e then increases as cell number increases in these regions in conjunction with the development of the distinctive cell morphologies (Knox *et al.*, 1989). The expression of J4e by these two groups of cells can be taken as a marker of the stele–cortical division and also of the positioning of the xylem–phloem division within the stele. Later in development the same epitope is expressed by the surface of epidermal cells and in some cases by the cortical cells, although weakly.

Subsequently, a group of cells positioned between the two JIM4-reactive regions of the pericycle begins to express the J13e determinant (Fig. 2C). This group of cells is centred upon a region that will include the future xylem. The whole group appears to become reactive together well before visible signs of xylem differentiation. There is no indication of J13e expression correlating with the sequential development of proto-xylem and metaxylem. The J13e determinant is not expressed by meristem cells but does occur abundantly at the surface of all root cap cells and the epidermal cells.

The third modulation of glycosylation during development is the loss of the J15e determinant (it being expressed in the region of the meristem) from the cells forming the epidermis and the region of the future xylem band (Fig. 2D). To some extent this pattern is the opposite of J13e expression, but it does not correlate precisely and some cells can express both epitopes together (Knox *et al.*, 1991).

The modulations involving these three epitopes reflect the development of tissue complexity as a population of cells becomes separated from the meristem, and occur within a developmental distance of probably fewer than 10 cells from the meristem initials. Over this same developmental sequence cell fate becomes predictable in terms of cell shape.

The binding of these antibodies presents molecular markers of the emergence of tissue and cell distinctions in terms of a variety of the combination of AGP epitopes expressed at the cell surface. Different combinations of epitopes define most of the major developing tissue distinctions (Fig. 3). However, these individual modulations are not related to specific cell fate, nor are they specific differentiation antigens. They reflect cell positions and emerging boundaries between tissues. This is perhaps seen most clearly for J4e, the earliest epitope modulation to be observed in the developing stele. The patterns of expression reflect the

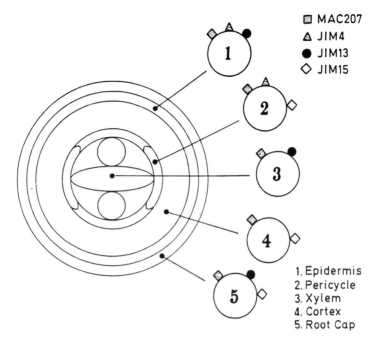

Fig. 3. A schematic diagram of the main tissues of the carrot root seen in transverse section, with five examples of cell types with the specific combination of AGP epitopes that they express at the plasma membrane.

partitioning of a cell population into a pattern or framework in which cell differentiation can subsequently take place.

The patterns of AGP epitope expression are generally maintained into maturity, although the combinations of AGP epitopes expressed by a cell in the region of the root apex are dynamic. In addition to the AGP epitope modulations that can be observed in terms of the development of spatial distinctions and complexity involving cells of similar developmental age, several modulations can also occur within a single file of cells over time. For example, as a file of cells develops into the epidermis J13e is expressed, J15e is lost and J4e expressed in a temporal sequence. These dynamic changes at the surface of a single cell as it matures are shown schematically for a file of cells in Fig. 4.

The biochemical basis of the differing combinations of AGP epitopes at the plasma membrane is far from clear. Each epitope occurs on a large range of membrane glycoproteins but how many occur on common antigens is uncertain.

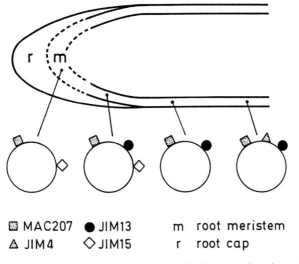

☒ MAC207   ● JIM13   m root meristem
△ JIM4   ◇ JIM15   r root cap

Fig. 4. Schematic diagram of a longitudinal section through a carrot root apex indicating the varying cell surface expression of specific AGP epitopes by the epidermal cells as they differentiate.

Developmental significance of plasma membrane AGPs

At this stage it is not clear how these markers integrate with other aspects of the developmental process. The development of a structure such as a root by a group of meristematic cells is clearly immensely complex, involving the regulation of processes at many levels of organisation. Lineage analyses, generally of shoot meristems, have demonstrated that the fate of meristematic cells is dependent upon cell position and not absolutely upon lineage (Irish, 1991). Such observations indicate that cell–cell interactions must play an important role in morphogenesis. The modulations indicated by the anti-AGP antibodies may thus reflect some aspect of these signalling mechanisms. They may be part of an intercellular signalling system across the apoplast or alternatively an aspect of the cellular response to signals and are associated with the assembly of specific cell wall architectures.

An important aspect of these restricted patterns of epitope expression is their inexactness in terms of pattern. Cells adjacent to cells with a high level of epitope expression are often weakly reactive with the antibody. For example, endodermal cells can express J4e (Knox *et al.*, 1989). In certain cases cells, seemingly at random, express the epitope. This is particularly observed in regions of the carrot root meristem as shown in Fig. 5. JIM4-reactive cells occur adjacent to unreactive cells and do not

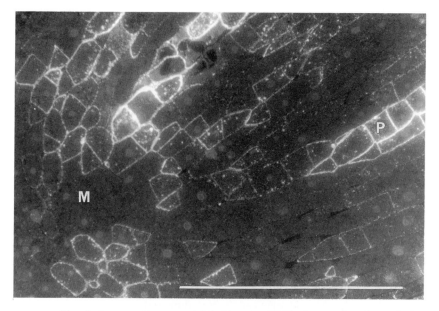

Fig. 5. Immunofluorescent labelling by JIM4 of a section through the carrot root meristem (M) showing the apparently random distribution of cells weakly expressing J4e. The future pericycle cells (P) express J4e strongly. Scale bar, 100 μm.

correlate with any apparent emerging complexity. This irregular expression is weak relative to that of the future pericycle cells. The extent of such developmentally inexact expression varies between root apices. Such observations occur most frequently in regions where cell interactions are probably at their most dynamic and may reflect the fact that regulatory interactions of signals and responses between cells are being established.

Taxonomic significance of AGP epitopes

The patterns of AGP expression, by the virtue of the fact that they reflect the emerging tissue pattern of the carrot root, are carrot-root-specific. The expression of an individual epitope is not specific to the root or to carrot. The epitopes are expressed in highly specific patterns in other developmental systems – see, for example, J4e expression in carrot cell clusters and developing somatic embryos (Stacey, Roberts & Knox, 1990) – and by other species (Rae *et al.*, 1991; Pennell & Roberts, 1990; Pennell *et al.*, 1991).

Any speculation as to the function of these cell surface AGPs must accommodate observations on the patterns of expression of the developmentally regulated epitopes in different species. Such studies are at a very preliminary stage but already offer intriguing puzzles. The initial observation of the reaction of JIM4 with the carrot root seemed to be confined to the Umbelliferae, with no cross-reaction occurring with roots of pea, maize, onion or *Arabidopsis* (Knox *et al.*, 1989).

However, JIM4 has subsequently been found to be reactive with cells of the mung bean root apex (Fig. 6A). In this case, the reaction was restricted to the root cap cells. There was no binding to cells in the stele. The intensity of immunofluorescent labelling of the root cap cell surface increased as the cells matured; in some cases cells close to the meristem, identifiable as root cap cells owing to the presence of amyloplasts, were unreactive with JIM4. This observation may indicate one of two things. In the case of mung bean J4e may be an indicator of root cap cell identity, a surface determinant associated with one specific cell type. Alternatively, it may be associated with root cap cell function. AGPs are known to be common components of root cap slimes, which contain a diverse range of glycosyl residues within one sample and also vary greatly in composition between species (Bacic, Moody & Clarke, 1986; Moody, Clarke & Bacic, 1988). In certain species root cap cells exude a large amount of material that passes through the cell wall to accumulate as slime droplets. In other cases material accumulates between the cell wall and the protoplast in the periplasmic space (Rougier, 1981). This may be the case with the material localised with JIM4 at the mung bean root apex. It is also worth noting that all of the anti-carrot AGP antibodies derived, other than JIM4, are abundantly reactive with carrot root cap cells.

Although J4e is not expressed by the root tissues of pea, JIM4 does react with a specific highly differentiated cell type: pollen (Fig. 6C). J4e is only detectable in mature pollen, approximately one day before anthesis. In the case shown in Fig. 6, the epitope was located throughout the cytoplasm of the vegetative cell and not at the cell surface. The regions of the vegetative nucleus and/or generative cell, identified by DAPI staining of the same section (Fig. 6D) appear as unlabelled regions.

It has been noted previously that AGPs occur in pollen (Clarke *et al.*, 1979) where they may function as lubricants for the germination of the pollen tube. There is currently no evidence for any role for AGPs in the molecular recognition processes occurring between pollen and stigma. The occurrence of J4e at this location in the pea flower does not appear to be related to the establishment of tissue or cell distinctions although the related MAC207 epitope has been implicated in such phenomena (Pennell & Roberts, 1990).

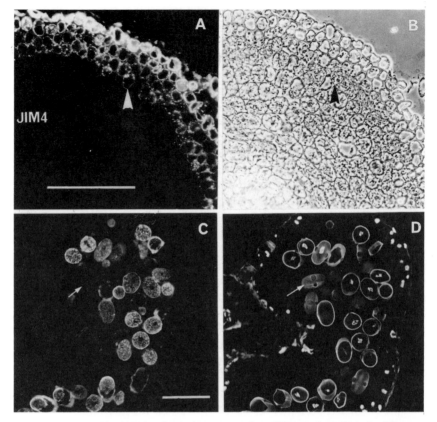

Fig. 6. The JIM4 epitope is expressed at different locations in different species. Immunofluorescent labelling by JIM4 of a transverse section of the mung bean root apex is restricted to the root cap cells (A). Epitope expression increases as the cells mature. Arrowhead indicates a reactive cell adjacent to an unreactive cell. A phase-contrast image of the same section is shown in B. A section through a mature pea anther indicates that JIM4 binds only to the cytoplasm of the vegetative cell of the pollen and to no other cells (C). The same section stained with DNA-reactive DAPI indicates the location of nuclei (D). The arrow indicates a pollen grain wall, which is not JIM4-reactive. Scale bars, 100 μm.

These preliminary, and somewhat tangential, observations indicate that in different species the JIM4 epitope (or a series of highly related epitopes) can be expressed in entirely different developmental locations. This can be termed heterotopic expression of the AGP epitopes between species.

The occurrence of J4e on developmentally regulated plasma membrane glycoproteins, pollen AGPs and also the gum exudates, gum arabic, gum karaya and gum ghatti (J.P. Knox, unpublished observations) indicate that related structures can occur as components of molecules of diverse origin, size, nature and possibly function.

AGPs are widespread throughout the flowering plants but have not been explored systematically in lower plants (Fincher *et al.*, 1983). There is some indication that terminal disaccharides of *Acacia* AGPs may have taxonomic significance (Anderson & Dea, 1969). A wider study of AGPs and their possible heterotopic expression may reveal further insights into function and specific monoclonal antibodies will be powerful tools for such studies. For example, a limited survey using MAC207 indicated that this epitope was restricted to the flowering plants (Pennell *et al.*, 1989).

Prospects

These recent observations have brought plasma membrane AGPs into focus as possible components of cell signalling and recognition mechanisms that are required for the establishment of tissue boundaries. The most pressing question concerns the precise function of these glycoproteins. Do they have a morphoregulatory role? Intriguing observations in leafy liverworts implicate AGPs as candidates, among hydroxyproline-containing proteins, in having a morphoregulatory function concerned with cell proliferation and organogenesis (Basile & Basile, 1987; Basile, 1990).

Many questions now arise with regard to these observations.

> What is the biochemical basis of epitope modulations?
> How many developmentally regulated AGP epitopes of the carrot root and of other systems remain undetected?
> Can we predict other likely patterns of expression?
> Are the glycosylic changes the result of the synthesis of new proteins or due to enzyme action at the surface?
> How rapidly are AGPs turned over?
> How diverse are the protein cores that carry the epitopes?
> What is the biosynthetic relationship of soluble and membrane-bound AGPs?
> What is the significance of heterotopic expression of the epitopes?

It is highly probable that AGPs have several molecular roles. They may have become important in multifarious roles associated with many of the complex processes that are required for the development of multicellular organisms. This appears to be the case for the large and complex families of animal cell adhesion molecules that have diverse morphoregulatory functions. These molecules are involved in dynamic regulatory loops intimately tied to various processes such as tissue boundary formation, cell migration, cell communication, embryonic induction and tissue stabilisation (Edelman & Crossin, 1991). The diversity of structure and location of AGPs seems akin to these cell adhesion molecules. Although there are fundamental differences in the molecular and cellular mechanisms underlying tissue cohesion in plants and animals (Knox, 1992a), the development of tissue boundaries and complexity may involve common regulatory processes, if not machinery.

As we learn more about the biochemistry and developmental regulation of AGPs and other plant cell surface glycoproteins clear functional subgroups may become apparent. Studies on the cell surface will, as suggested by Dansky & Bernfield (1991) when referring to animal cells, continue to provide surprises and puzzles in addition to explanations.

Acknowledgements

I am grateful to many colleagues, particularly Keith Roberts, at the John Innes Institute where this work began. I thank Anna Durbin and Philip Gilmartin for help with figure preparation.

References

Anderson, D.M.W. & Dea, I.C.M. (1969). Chemotaxonomic aspects of the chemistry of *Acacia* gum exudates. *Phytochemistry* **8**, 167–76.

Bacic, A., Moody, S.F. & Clarke, A.E. (1986). Structural analysis of the secreted root slime from maize (*Zea mays* L.). *Plant Physiology* **80**, 771–7.

Basile, D.V. (1990). Morphoregulatory role of hydroxyproline-containing proteins in liverworts. In *Bryophyte development: physiology and biochemistry* (ed. R.N. Chopra & S.C. Bhatla), pp. 225–43. Boca Raton, Florida: CRC Press.

Basile, D.V. & Basile, M.R. (1987). The occurrence of cell-wall associated arabinogalactan proteins in the Hepaticae. *The Bryologist* **90**, 401–4.

Cassab, G.I. & Varner, J.E. (1988). Cell wall proteins. *Annual Review of Plant Physiology and Plant Molecular Biology* **39**, 321–53.

Clarke, A.E., Anderson, R.L. & Stone, B.A. (1979). Form and function of arabinogalactans and arabinogalactan-proteins. *Phytochemistry* **18**, 521–40.

Dansky, C.H. & Bernfield, M. (1991). Cell-to-cell contact and extracellular matrix. *Current Opinion in Cell Biology* **3**, 777–8.

Edelman, G.M. & Crossin, K.L. (1991). Cell adhesion molecules: implications for a molecular histology. *Annual Review of Biochemistry* **60**, 155–90.

Esau, K. (1940). Developmental anatomy of the fleshy storage organ of *Daucus carota. Hilgardia* **13**, 175–226.

Feizi, T. & Childs, R.A. (1987). Carbohydrates as antigenic determinants of glycoproteins. *Biochemical Journal* **245**, 1–11.

Fincher, G.B., Stone, B.A. & Clarke, A.E. (1983). Arabinogalactan-proteins: structure, biosynthesis and function. *Annual Review of Plant Physiology* **34**, 47–70.

Gleeson, P.A., McNamara, M., Wettenhall, R.E.H., Stone, B.A. & Fincher, G.B. (1989). Characterization of the hydroxyproline-rich core of an arabinogalactan-protein secreted from suspension cultured *Lolium multiflorum* (Italian ryegrass) endosperm cells. *Biochemical Journal* **264**, 857–62.

Guzman, N.A., Fuller, G.C. & Dixon, J.E. (1990). Hydroxyproline-containing proteins and their hydroxylations by genetically distinct prolyl-4-hydroxylases. In *Organization and assembly of plant and animal extracellular matrix* (ed. W.S. Adair & R.P. Mecham), pp. 301–56. San Diego: Academic Press.

Irish, V.F. (1991). Cell lineage in plant development. *Current Opinion in Cell Biology* **3**, 983–7.

Kannagi, R. & Hakomori, S. (1986). Monoclonal antibodies directed to carbohydrate antigens. In *Handbook of Experimental Immunology*, vol. 4: *Applications of immunological methods in biomedical sciences* (ed. D.M. Weir), pp. 117.1–177.20. Oxford: Blackwell.

Knox, J.P. (1992a). Cell adhesion, cell separation and plant morphogenesis. *Plant Journal* **2**, 137–41.

Knox, J.P. (1992b). Molecular probes for the plant cell surface. *Protoplasma*, **167**, 1–9.

Knox, J.P., Day, S. & Roberts, K. (1989). A set of cell surface glycoproteins forms an early marker of cell position, but not cell type, in the root apical meristem of *Daucus carota* L. *Development* **106**, 47–56.

Knox, J.P., Linstead, P.J., Peart, J., Cooper, C. & Roberts, K. (1991). Developmentally regulated epitopes of cell surface arabinogalactan proteins and their relation to root tissue pattern formation. *Plant Journal* **1**, 317–26.

Knox, J.P. & Roberts, K. (1989). Carbohydrate antigens and lectin receptors of the plasma membrane of carrot cells. *Protoplasma* **152**, 123–9.

Larkin, P.J. (1978). Plant protoplast agglutination by artificial carbohydrate antigens. *Journal of Cell Science* **30**, 283–92.

Lloret, P.G., Casero, P.J., Pulgarín, A. & Navascués, J. (1989). The behaviour of two cell populations in the pericycle of *Allium cepa, Pisum sativum*, and *Daucus carota* during early lateral root development. *Annals of Botany* **63**, 465–75.

Moody, S.F., Clarke, A.E. & Bacic, A. (1988). Structural analysis of secreted slime from wheat and cowpea roots. *Phytochemistry* **27**, 2857–61.

Norman, P.M., Kjellbom, P., Bradley, D.J., Hahn, M.G. & Lamb, C.J. (1990). Immunoaffinity purification and biochemical characterization of plasma membrane arabinogalactan-rich glycoproteins of *Nicotiana glutinosa*. *Planta* **181**, 365–73.

Nothnagel, E.A. & Lyon, J.L. (1986). Structural requirements for the binding of phenylglycosides to the surface of protoplasts. *Plant Physiology* **80**, 91–8.

Pennell, R.I., Knox, J.P., Scofield, G.N., Selvendran, R.R. & Roberts, K. (1989). A family of abundant plasma membrane-associated glycoproteins related to the arabinogalactan proteins is unique to flowering plants. *Journal of Cell Biology* **108**, 1967–77.

Pennell, R.I., Janniche, L., Kjellbom, P., Scofield, G.N., Peart, J.M. & Roberts, K. (1991). Developmental regulation of a plasma membrane arabinogalactan protein epitope in oilseed rape flowers. *Plant Cell* **3**, 1317–26.

Pennell, R.I. & Roberts, K. (1990). Sexual development in the pea is presaged by altered expression of arabinogalactan protein. *Nature (London)* **344**, 547–9.

Perotto, S., VandenBosch, K.A., Butcher, G.W. & Brewin, N.J. (1991). Molecular composition and development of the plant glycocalyx associated with the peribacteroid membrane of pea root nodules. *Development* **112**, 763–73.

Qi, W., Fong, C. & Lamport, D.T.A. (1991). Gum arabic glycoprotein is a twisted hairy rope. A new model based on O-galactosylhydroxyproline as the polysaccharide attachment site. *Plant Physiology* **96**, 848–55.

Rae, A.L., Perotto, S., Knox, J.P., Kannenberg, E.L. & Brewin, N.J. (1991). Expression of extracellular glycoproteins in the uninfected cells of developing pea nodule tissue. *Molecular Plant-Microbe Interactions* **4**, 563–70.

Rougier, M. (1981). Secretory activity in the root cap. In *Encyclopaedia of Plant Physiology, New Series*, vol. 13B: *Plant Carbohydrates II* (ed. W. Tanner & F.A. Loewus), p. 542. Berlin: Springer.

Showalter, A.M. & Rumeau, D. (1990). Molecular biology of plant cell wall hydroxyproline-rich glycoproteins. In *Organization and Assembly of Plant and Animal Extracellular Matrix* (ed. W.S. Adair & R.P. Mecham), pp. 247–81. San Diego: Academic Press.

Showalter, A.M. & Varner, J.E. (1989). Plant hydroxyproline-rich glycoproteins. In *The Biochemistry of Plants* (ed. P.K. Stumpf & E.E. Conn), vol. 15, pp. 485–519. San Diego: Academic Press.

Stacey, N.J., Roberts, K. & Knox, J.P. (1990). Patterns of expression of the JIM4 arabinogalactan protein epitope in cell cultures and during somatic embryogenesis in *Daucus carota* L. *Planta* **180**, 285–92.

Stephen, A.M., Churms, S.C. & Vogt, D.C. (1990). Exudate gums. In *Methods in Plant Biochemistry*, vol. 2: *Carbohydrates* (ed. P.M. Dey), pp. 483–522. London: Academic Press.

S.E. COURTNEY, C.C. RIDER
and A.D. STEAD

Ubiquitination of proteins during floral development and senescence

Ubiquitin

Introduction

Ubiquitin is a small, 76 amino acid polypeptide (molecular mass 8.6 kDa) which is present in all eukaryotic cells. Its amino acid sequence shows an extraordinary degree of conservation, being invariant in all animal species examined, and differing by only three residues in yeast and higher plants and by only two residues in lower plants (Busch, 1984; Vierstra, Langan & Schaller, 1986; Callis *et al.*, 1989). This remarkable conservation of sequence suggests that ubiquitin has a fundamental cellular function common to all organisms.

Ubiquitin was first isolated by Goldstein *et al.* (1975) during the purification of peptide hormones from thymus, although it was subsequently demonstrated that the carboxy-terminal glycyl dipeptide had been cleaved off by protease(s) during the purification procedure (Haas, Murphy & Bright, 1985). Two years later, structural analysis of a purified chromosomal protein, A24, revealed that it consisted of ubiquitin conjugated to histone H2A via an isopeptide linkage (Goldknopf & Busch, 1977; Bonner, Hatch & Wu, 1988). Subsequently, ubiquitin was isolated independently by Ciechanover, Wilkinson and their co-workers (Ciechanover *et al.*, 1980; Wilkinson, Urban & Haas, 1980) as a cofactor (previously known as APF-1) required for ATP-dependent proteolysis in reticulocytes.

ATP/ubiquitin-dependent proteolysis

Over the past decade, numerous studies have indicated that the post-translational covalent conjugation of ubiquitin to cytoplasmic proteins of eukaryotic cells serves as a signal for the energy-dependent degradation

Society for Experimental Biology Seminar Series 53: *Post-translational modifications in plants*, ed. N.H. Battey, H.G. Dickinson & A.M. Hetherington. © Cambridge University Press 1993, pp. 285–303.

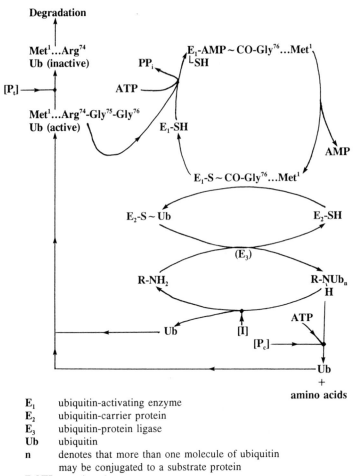

E_1	ubiquitin-activating enzyme
E_2	ubiquitin-carrier protein
E_3	ubiquitin-protein ligase
Ub	ubiquitin
n	denotes that more than one molecule of ubiquitin may be conjugated to a substrate protein
$R\text{-}NH_2$	protein substrate
[I]	isopeptidase
$[P_c]$	protease complex
$[P_t]$	trypsin-like protease

Fig. 1. Proposed pathway of ATP/ubiquitin-dependent protein degradation. (Scheme modified according to Pollmann & Wettern, 1989.)

of many such proteins, including those with short half-lives and/or structural abnormalities (for reviews see Hershko & Ciechanover, 1982; Rechsteiner, 1987; Finley & Chau, 1991). Ligation of ubiquitin to cellular proteins is ATP-dependent and is catalysed by a multi-enzyme system initially observed in rabbit reticulocytes (Hershko *et al.*, 1983) and more recently characterised in wheat germ extracts (Hatfield & Vierstra, 1989) and yeast (Seufert, McGrath & Jentsch, 1990). In this series of reactions (see Fig. 1), an unusual isopeptide bond is formed between the C-terminal glycine carboxyl group of ubiquitin and ε-amino groups of lysyl residues on the target protein. In the initial step of ligation, ubiquitin-activating enzyme (E1) adenylates the carboxyl terminus of ubiquitin using ATP. The activated molecule is then attached to the side chain of a specific cysteine of the same enzyme, forming an ubiquitin-E1 thiolester, with the concomitant release of AMP (Haas *et al.*, 1982). The ubiquitin moiety is then transferred by a transesterification reaction to ubiquitin carrier proteins (E2s), which comprise a family of related proteins (Pickart & Rose, 1985). Finally, ubiquitin is covalently attached to target proteins with or without the participation of a family of ubiquitin protein ligases (E3s) (Hershko *et al.*, 1983; Lee *et al.*, 1986).

Whereas monoubiquitination of proteins is implicated in diverse physiological processes (Ciechanover & Schwartz, 1989), it is insufficient to target proteins for degradation by the ATP/ubiquitin-dependent proteolytic system (Ball *et al.*, 1987). Instead, multiply ubiquitinated conjugates are formed by the subsequent addition of up to twenty ubiquitin peptides to the ubiquitin moiety of the monoconjugate (Chau *et al.*, 1989). This involves the generation of an ordered chain of branched ubiquitin–ubiquitin conjugates by a mechanism in which the carboxy-terminal Gly^{76} of one ubiquitin moiety is joined in isopeptide linkage to the internal Lys^{48} of an adjacent ubiquitin moiety. Target proteins are rapidly degraded by a high-molecular-mass (26*S*) ATP-dependent protease complex specific for multiply ubiquitinated protein conjugates, with the release of free, functional ubiquitin (Hough, Pratt & Rechsteiner, 1986; Driscoll & Goldberg, 1990).

Determinants for ubiquitination of protein substrates

In general, ubiquitin–protein conjugation of target proteins for degradation probably requires two structural features on protein substrates. The first requirement is a determinant for interaction with components of the ubiquitin–protein ligase system, that is, a free α-amino terminus (Hershko *et al.*, 1984). Bachmair, Finley & Varshavsky (1986) have proposed the 'N-end' rule, which is based on observations of the different half-lives of β-galactosidase with different N-termini expressed in bacteria and on a survey of N-termini of long-lived eukaryotic and prokaryotic intracellular proteins from the literature. The N-end rule indicates that proteins with certain N-terminal amino acids (which are termed 'destabilising') are rapidly degraded *in vivo* (half-lives less than 2 h). Conversely, other amino acids are stabilising. The second requirement is that the protein substrate has an exposed lysine residue, which serves as the acceptor site for ubiquitination, in close steric proximity to the ubiquitin–protein ligase system (Bachmair & Varshavsky, 1989). E2 systems which possess C-terminal (or N-terminal) extensions may suffice for both recognition and conjugation, whereas some conjugation reactions require a two-component system: a complex is initially formed with an E2 enzyme; subsequently, protein determinants are probably recognised by an N-end-recognising E3 enzyme. The E2 enzyme then catalyses the covalent attachment of ubiquitin to the acceptor lysine (Bachmair *et al.*, 1986).

Ubiquitin in higher plants

Until recently, research on the structure and function of ubiquitin was mainly restricted to animal sources (Hershko, 1988); however, in recent years, species from the plant kingdom have been included. In 1985, Vierstra and colleagues (Vierstra, Langan & Haas, 1985) used anti-human ubiquitin antibodies and Western blotting to identify ubiquitin in plant extracts from etiolated leaves, green leaves and dehydrated seeds of *Avena sativa* L. (oat). Ubiquitin is as conserved in the plant kingdom as it is in the animal kingdom: to date, the amino acid sequence has been determined in oat (Vierstra *et al.*, 1986), *Hordeum vulgare* (barley) (Gausing & Barkardottir, 1986), *Glycine max* (soybean) (Fortin, Purohit & Verma, 1988), *Helianthus annuus* (sunflower) (Binet, Steinmetz & Tessier, 1989), *Pisum sativum* (pea) (Watts & Moore, 1989), *Arabidopsis thaliana* (Callis, Raasch & Vierstra, 1990), *Zea mays* (maize) (Chen & Rubenstein, 1991) and *Linum usitatissimum* (flax) (Agarwal & Cullis, 1991) and shows 100% homology between species.

Studies on the ubiquitination of plant proteins

Studies on phytochrome turnover provided the first evidence for a natural endogenous substrate of the ATP/ubiquitin-dependent proteolytic pathway. Vierstra and colleagues demonstrated that in etiolated oat seedlings, after photoconversion of phytochrome from the stable red light-absorbing form (P_r) to the rapidly degraded far-red light-absorbing form (P_{fr}), ubiquitin–phytochrome conjugates appear and then disappear (Shanklin, Jabben & Vierstra, 1987). Furthermore, more than half of the phytochrome molecules were degraded before the ubiquitin-phytochrome conjugates accumulated to any great extent (Jabben, Shanklin & Vierstra, 1989), providing clear evidence that degradation of this plant photoreceptor is mediated via the ubiquitin pathway.

Subsequently, it has been demonstrated that calmodulin, which is one of the most important signal transducing molecules for metabolic regulation in plants and animals, is polyubiquitinated in a Ca^{2+}-dependent manner in plants and fungi (Ziegenhagen & Jennissen, 1990). Polyubiquitination of calmodulin has also been demonstrated in both rabbit tissue extracts (Laub & Jennissen, 1991) and yeast (Jennissen *et al.*, 1992), thereby providing evidence that ubiquitin has functions in higher plants similar to those it fulfils in animal tissues and yeast.

Several studies have demonstrated the presence of ubiquitin conjugates and ATP-dependent ubiquitin-conjugating activities in higher plants. Vierstra (1987) demonstrated the presence of ATP-dependent ubiquitin-conjugating activities in crude extracts prepared from etiolated seedlings of oat, barley, wheat, corn, rye, zucchini squash, soybean, sunflower and pea, and Hatfield & Vierstra (1989) showed that the same enzymes were present in wheat germ. Ferguson, Guikema and Paulsen (1990) demonstrated the presence of ubiquitin conjugates in wheat roots and studied the effect of high temperature stress on the ubiquitin conjugate protein pattern; Veierskov & Ferguson (1991) compared the ubiquitin conjugate protein pattern in green and etiolated oat leaves and oat roots. Neither of these studies, however, focused on individual ubiquitinated protein bands.

Bachmair *et al.* (1990) expressed a ubiquitin variant gene in *Nicotiana tabacum*. This mutation, in which Lys^{48} is replaced by Arg, thereby preventing polyubiquitination of proteins (since Arg cannot serve as an acceptor of ubiquitin) has previously been shown to act as an inhibitor of ubiquitin-dependent protein degradation *in vitro* (Chau *et al.*, 1989). Transformed plants showed leaf curling, vascular tissue alterations and necrotic lesions, demonstrating the importance of this system in the differentiation of plant tissues.

Although ubiquitination of plant proteins has clearly been demonstrated in a number of species, protein ubiquitination during the development and senescence of plant material has not so far been described.

Floral development and senescence

Floral senescence in daylilies

The control of flower senescence in many species has been extensively documented. In many species ethylene production has been shown to increase during senescence; exposure to ethylene accelerates senescence. In other species, however, ethylene does not appear to be involved in the control of flower senescence. The flowers of *Hemerocallis fulva* (daylilies), for example, produce very little ethylene either as buds or later as the flower collapses. Further evidence that ethylene does not influence flower opening or senescence of daylilies comes from the use of the ethylene synthesis inhibitor, amino-oxyacetic acid, and an antagonist of ethylene action, silver thiosulphate, neither of which influences flower longevity in daylilies (Lukaszewski & Reid, 1989).

Daylily flowers, as their name implies, persist for just one day. The flowers commence opening at about 2300 h and are fully open at about 0400 h. The first visible signs of senescence are seen soon after midday when either the petals wilt or the petal margins appear translucent.

Inhibition of daylily senescence by cycloheximide treatment

Flower buds were removed from the daylily plant at 1600 h, the day before they were due to open (day -1) and placed in small vials containing distilled water. These flowers showed the same increase and decrease in fresh mass and flower diameter during flower opening and senescence as was seen when attached to the plant. At 2300 h, that is, as the flowers commenced opening, the buds were transferred to vials containing 1 mg ml^{-1} cycloheximide, a protein synthesis inhibitor, or 25 µg ml^{-1} actinomycin D, an inhibitor of RNA biosynthesis, for 1 h and were subsequently returned to vials containing distilled water. The flowers were maintained throughout in a growth cabinet at 23 °C (day), 18 °C (night) with a 14 h photoperiod and maximum irradiance 450 µE m^{-2} s^{-1}.

Treatment with cycloheximide was found to slightly inhibit flower opening of daylilies and to delay the onset of visible signs of senescence by 5–6 days (Fig. 2). Treatment with actinomycin D, however, showed no such inhibition of senescence. Although the fresh and dry mass of the daylily flowers increased as the flowers opened, the protein content declined (results not shown). However, the inhibition of flower senescence by cycloheximide resulted in the maintenance of the protein population of petals (Lay-Yee, Reid & Stead, 1992).

The observation that cycloheximide inhibited petal senescence and protein degradation, while inhibitors of RNA biosynthesis were ineffective, implied that control of petal senescence in daylilies is dependent upon 80S ribosomal protein synthesis. The loss of protein starts before the petals senesce, although this is exaggerated by the increase in fresh mass during flower opening. Such rapid loss of protein implies rapid degradation of proteins. To date, ubiquitin has been demonstrated to be involved in the degradation of short-lived and abnormal proteins in mammalian cell lysates (Hershko & Ciechanover, 1986) and has also been shown to be involved in the selective catabolism of a number of proteins (reviewed by Rechsteiner, 1991). This study examines the involvement of

A

B

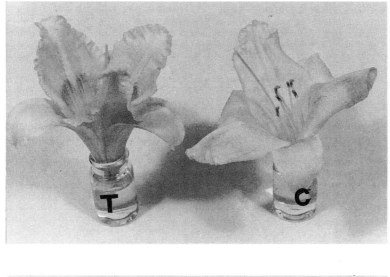

Fig. 2. Inhibition of senescence in daylilies by treatment with cyclohex-imide. Daylily buds were picked at 1600 h and held in water. Control flowers (C) were kept in water throughout, whereas treated flowers (T) received a 1 h pulse of cycloheximide (1 mg ml^{-1}) between 2300 h and midnight. Flowers pulsed with cycloheximide did not open quite as fully as those maintained in water. (A) The appearance of daylily flowers at 1000 h on the day of opening. (B) The same flowers photographed 24 h later.

ubiquitin in floral development and senescence, where there is almost total, as opposed to selective, protein degradation occurring.

Ubiquitination of petal proteins during floral development and senescence

Extraction of petal proteins

Daylilies (*Hemerocallis fulva* cv. Cradle Song) were grown in a glass-house with supplementary heat and light (21–25 °C day, 17 °C night, 16 h light photoperiod). Flowers were harvested at recognisable morphological stages: one, two and three days prior to opening (days −1, −2 and −3, respectively), on the day of opening, (day 0) or one day after opening (day +1). Petals were removed from daylily flowers at defined times before or after flower opening. The lower one quarter of the petals is heavily vascularised and was therefore discarded. Proteins were extracted from the petals of daylily flowers by freezing the petals in liquid nitrogen, followed by lyophilisation. A pre-cooled mortar and pestle were used to grind the petals to a fine powder in liquid nitrogen; the addition of fine sand enhanced this procedure. It was necessary to carry out the extractions in the presence of protease inhibitors, in particular to prevent isopeptidase(s) from disassembling the ubiquitin–protein conjugates. Pre-cooled extraction buffer consisting of 0.4 M Tris, pH 8.0, containing 2 mM dithiothreitol, 1 mM EDTA, 0.2% (w/v) polyvinylpyrrolidone K40, 40 mM sodium bisulphite and 2 mM *p*-chloromercuribenzoic acid was added to the ground petals to make a smooth paste, which was then centrifuged at 10000 *g* at 4 °C for 20 min.

Analysis of petal proteins by SDS–PAGE

Aliquots of the supernatant were diluted with an equal volume of two-times concentrated Laemmli (1970) sample buffer and the extracted proteins were resolved by one-dimensional electrophoresis on 12.5% discontinuous polyacrylamide gels (Laemmli, 1970). Gel loadings were equivalent to 0.5 mg dry mass of petals per track in order to obviate differences in protein concentration arising from varying degrees of hydration of the corolla at progressive stages of flower development; the proteins were visualised by staining with 0.1% (w/v) Coomassie blue R-250.

Analysis by SDS–PAGE of the extracted proteins from petals of daylily flowers of differing ages demonstrated the abrupt decrease in total protein content between the open flower (day 0) and senesced flowers (day +1) (Fig. 3A). Although there was an overall loss of protein, it was

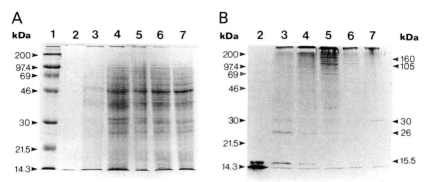

Fig. 3. Daylily flower: developmental time course. Petal extracts were resolved by SDS–(12.5%)PAGE and (*A*) stained with Coomassie blue R-250, or (*B*) Western blotted with affinity-purified antibody to ubiquitin. Lane 1, rainbow markers (molecular masses are indicated in kDa on the left of the gel and blot); lane 2, bovine ubiquitin, 0.5 μg; lanes 3–7, daylily petal extracts equivalent to 0.5 mg dry mass; lane 3, senesced flower, day +1; lane 4, open flower, day 0; lane 5, bud, day −1; lane 6, bud, day −2; lane 7, bud, day −3.

difficult to detect specific changes in the pattern of proteins extracted from either buds or fully open flowers; only a few faint protein bands were detected in extracts from senescing petals. This loss of total protein was not, however, uniform: the most abundant protein present in the developing bud and open flower of apparent molecular mass 46 kDa was barely discernable at day +1, yet less prominent bands between 38–40 kDa and 49–52 kDa were still clearly visible at day +1. Proteins of molecular mass greater than approximately 100 kDa were not detected during either floral development or senescence, except for material at the interface between the stacking and separating gels, which was most abundant in the open daylily flower (day 0).

Proteins were extracted from the petals of cycloheximide-treated flowers and controls as described above and were resolved by 12.5% SDS–PAGE. Cycloheximide treatment of daylily buds with as little as a 1 h pulse of cycloheximide prevented the abrupt decline in petal protein content observed in untreated flowers between day 0 and day +1 (Fig. 4A). The pattern of bands was indistinguishable from that found in the buds at the time that they were treated with cycloheximide (that is, at 2300 h).

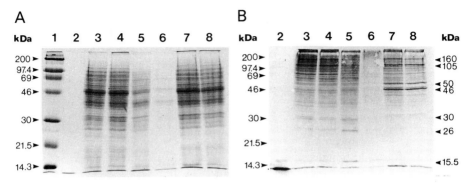

Fig. 4. Daylily flowers treated with cycloheximide. Petal extracts were resolved by SDS–(12.5%)PAGE and (A) stained with Coomassie blue R-250, or (B) Western blotted with affinity-purified antibody to ubiquitin. Lane 1, rainbow markers (molecular masses are indicated in kDa on the left of the gel and blot); lane 2, bovine ubiquitin, 0.5 µg; lanes 3–8, daylily petal extracts equivalent to 0.5 mg dry mass; lane 3, control bud harvested at 1600 h, day −1; lane 4, control bud harvested at 2300 h, day −1; lane 5, control open flower, day 0; lane 6, control senesced flower, day +1; lane 7, cycloheximide-treated flower, day 0; lane 8, cycloheximide-treated flower, day +1.

Analysis of ubiquitinated petal proteins by Western blotting

Ubiquitinated proteins in the petal extracts were analysed by Western blotting the resolved extracts onto nitrocellulose membrane (0.1 µm pore size) according to the method of Towbin, Staehelin & Gordon (1979). The resulting transblots were developed with affinity-purified antibody raised against SDS-denatured bovine ubiquitin in rabbits using a modification of the procedure of Haas & Bright (1985). This antibody, as opposed to antibody raised against non-denatured ubiquitin, has a high titre for conjugated ubiquitin but reacts less strongly with free ubiquitin. The sensitivity of the immunoblotting procedure was greatly enhanced by hydrating the transblot and autoclaving for 30 min at 121 °C and 76 kPa prior to immunodevelopment (Swerdlow, Finley & Varshavsky, 1986). Immunoreactive bands were visualised by an adaptation of the method of Blake et al. (1984) using alkaline phosphatase-conjugated second antibody in conjunction with the substrates nitroblue tetrazolium and 5-bromo-4-chloro-3-indolylphosphate. The apparent molecular masses of the ubiquitin-immunoreactive proteins were determined relative to the mobility of Rainbow molecular mass marker standard proteins (Amersham International plc).

Immunoblotting of a standard sample of ubiquitin with ubiquitin antibody revealed two immunoreactive bands, one of apparent molecular mass 15.5 kDa and a second which was unresolved from the dye front (Figs 3B and 4B). The bands are the dimer and monomer of ubiquitin, respectively, as demonstrated by SDS–PAGE on 15% gels, whereupon the monomeric ubiquitin is resolved from the dye front (results not shown). Analysis of the petal extracts by Western blotting (Fig. 3B) demonstrated that there are probably up to one hundred ubiquitin-immunoreactive proteins present within the petals at all stages of development, although many are very faint. However, when the immunoblot was developed with preimmune serum, no immunoreactive bands were detected, thereby demonstrating that the reactivity observed with ubiquitin antibody was not a result of non-specific binding nor was it caused by the activity of endogenous alkaline phosphatase(s). In addition, the distribution of ubiquitin-immunoreactive proteins did not reflect the distribution of total protein (compare Figs 3A and 3B). Overall, there is a marked increase in intensity of ubiquitin-immunoreactivity at day −1; this increase corresponds to the onset of protein degradation, which is subsequently observed in the open and senescing daylily flower. Despite the large number of ubiquitin-immunoreactive bands detected, it was possible to detect specific changes in the intensities of these immunoreactive bands during floral development and senescence of the daylily. Two of the most prominent ubiquitin conjugates detected had apparent molecular masses of 160 kDa and 105 kDa, yet neither of these was detected by Coomassie blue-staining. The former band, which was present in the daylily bud at day −1, declined in intensity as the flower opened (day 0) and senesced (day +1). The latter band was detected in the early daylily bud (day −3), reaching its highest intensity in the late bud (day −1) and declining rapidly as the flower opened (day 0) and senesced (day +1). A further intense band of apparent molecular mass 29 kDa, which was apparent in the early bud (day −3), decreased in intensity as the bud developed and opened. At day −1, a band of apparent molecular mass 30 kDa appeared, but disappeared as the flower opened and was absent in the senesced flower. As the flower senesced, a ubiquitin-immunoreactive band of similar mobility to the 29 kDa band observed at day −3 was seen. However, although the bands were of similar mobility, they were not thought to represent identical ubiquitinated proteins. A similar pattern of increased intensity as the flower developed and senesced was also shown by bands of apparent molecular mass 26 kDa and 15.5 kDa. In addition to these progressive changes in the intensity of specific ubiquitinated proteins, bands of apparent molecular mass 36 kDa and 24 kDa were detected only in the late bud (day −1). Ubiquitin-immunoreactive material at the interface between the stacking and separating gels was most intense in the late bud (day −1) and open flower (day 0).

Treatment of daylily buds with cycloheximide effectively froze the ubiquitinated protein pattern at the time of treatment, that is, 2300 h (see Fig. 4B). Thus, the degradation of the 160, 105, 36 and 29 kDa bands was prevented, as was the increase in intensity of the 30, 26 and 15.5 kDa bands. Two ubiquitin-immunoreactive bands of apparent molecular mass 50 kDa and 46 kDa actually increased in intensity in the cycloheximide-treated flowers compared with the corresponding control samples. The latter band may correspond to the abundant 46 kDa band detected on Coomassie blue-stained gels in the late bud (day −1, 1600 h and 2300 h) and cycloheximide-treated flowers (see Fig. 4A).

Discussion

Since the anti-ubiquitin immunoglobulins used in this study were raised against bovine ubiquitin and affinity purified on a column of Sepharose-(yeast)ubiquitin, it is unlikely that they would recognise plant proteins other than those containing the ubiquitin sequence. Importantly, the distribution of ubiquitin-immunoreactive proteins in the petal extracts did not reflect the distribution of major protein bands detected by staining with Coomassie blue R-250, demonstrating that the patterns observed did not result from non-specific binding of the antibody. Similar complex patterns are observed for mammalian proteins ubiquitinated both *in vivo* (Hershko *et al.*, 1982) and *in vitro* (with supplemented ATP) (Ciechanover, Finley & Varshavsky, 1984).

Since floral senescence in daylilies appears to involve the collapse of cellular integrity, there is likely to be an increase in readily extractable structural proteins. To ensure that petal proteins were extracted efficiently by the protocol routinely used, samples at all developmental stages were extracted with the inclusion of detergent (5% SDS) in the initial extraction medium. No significant alteration in the apparent levels of extractable protein (as revealed by Coomassie blue-stained gels) or in the ubiquitination pattern seen on the immunoblots was observed.

The observation that cycloheximide inhibits petal senescence and protein degradation while inhibitors of RNA biosynthesis are ineffective implies that control of petal senescence in daylilies is dependent upon 80S ribosomal protein synthesis. In carnations ethylene-induced petal senescence is also prevented by cycloheximide (Wulster, Sacalis & Janes, 1982). A further species in which cycloheximide has been shown to inhibit senescence is *Oxalis martiana* (Tanaka *et al.*, 1989). The loss of protein (per unit fresh mass) starts before the petals senesce, although this is exaggerated by the increase in fresh mass during flower opening. A similar situation exists in other short-lived flowers (for example, *Ipomoea purpurea*) (Matile & Winkenbach, 1971), although in these species ethylene has been implicated in the control of senescence.

In addition to its function as an inhibitor of protein synthesis, cycloheximide has also been shown to perturb respiration in certain tissues (Ellis & MacDonald, 1970). Inhibition of senescence by cycloheximide in daylily flowers, however, was not caused by uncoupling or inhibition of respiration by cycloheximide (Lay-Yee *et al.*, 1992). Therefore, the inhibition of daylily flower senescence by cycloheximide suggests that senescence requires *de novo* synthesis of specific proteases or the synthesis of proteins which facilitate either protein breakdown or the action of certain proteases.

The physiological significance of ubiquitin in plant physiology and development is not yet clear, but since ubiquitin has a remarkable degree of sequence conservation in eukaryotes and no alternative pathway for protein degradation has been detected in the cytoplasm in plants, it seems likely that ubiquitin conjugation may serve the same functions in plants as it does in animals. Immunoblotting of daylily petal extracts with anti-ubiquitin antibodies reveals that a substantial percentage of plant ubiquitin exists as conjugates *in vivo* and that many petal proteins are modified in this way. Moreover, the pattern of ubiquitination shows considerable change at the onset of protein loss. Similar results were also obtained by immunoblotting of *Digitalis pupurea* (foxglove) and *Petunia* extracts. Moreover, similarities in the intensities of ubiquitinated proteins during floral development and senescence of the three species were noted, despite the flowers showing different patterns of senescence: in foxgloves, ethylene production increases rapidly after pollination and corolla abscission is accelerated (Stead & Moore, 1979), but, for both pollinated and unpollinated flowers, the corolla abscises without any significant loss of fresh mass or corolla constituents (Stead & Moore, 1983); in *Petunia*, pollination induces ethylene production and the corolla wilts (Pech *et al.*, 1987); although the corolla of daylily flowers wilt, they produce very little ethylene either as buds or as the flowers senesce (Lay-Yee *et al.*, 1992). In all three species, the ubiquitinated protein of apparent molecular mass 160 kDa was intense during bud development, but rapidly disappeared upon flower opening. In addition, ubiquitinated proteins of apparent molecular masses 26 kDa and 15.5 kDa accumulated as the *Petunia* and daylily flowers senesced. The 15.5 kDa band corresponded in mobility to the ubiquitin dimer observed in the standard sample of ubiquitin. A ubiquitinated protein band of similar molecular mass is often detected in mammalian cell culture lysates and is attributed to being the dimer of ubiquitin. The discrepancy in molecular mass from the predicted value of 17.2 kDa is probably due to the characteristically anomalous migration of ubiquitin in SDS–polyacrylamide gels (Ciechanover et al., 1980; Wilkinson et al., 1980).

In general, the ubiquitin-immunoreactivity protein band patterns did not correspond to the pattern of proteins observed on Coomassie blue-stained gels. For example, the most intensely stained ubiquitinated proteins had apparent molecular masses of 160 kDa and 105 kDa, whereas the most abundant protein detected had a molecular mass of 46 kDa, and bands of molecular mass greater than approximately 100 kDa were not detected on the protein gel, except for unresolved material at the interface between the stacking and separating gels. However, ubiquitin-immunoreactivity was intense in this region, as shown in Figs 3B and 4B. In no instance was it possible to correlate a change in the intensity of a ubiquitin-immunoreactive protein with a change in abundance of a protein band detectable by staining with Coomassie blue R-250.

Ubiquitination is an important aspect of post-translational modification of proteins, leading to the degradation of the targeted proteins. Although previous studies have demonstrated the ability of plants to conjugate ubiquitin to plant proteins, this study shows that the pattern of protein ubiquitination changes during the normal development and senescence of flowers. In addition, it demonstrates that treatments that delay floral senescence also prevent the degradation of target proteins by the ubiquitin-dependent proteolytic pathway. The aim of these studies was to determine the involvement of ubiquitin, which is known to be involved in selective catabolism, in floral senescence where there is almost total protein degradation occurring. The lack of correlation between the prominent ubiquitinated proteins detected on immunoblots and the abundant proteins visualised on Coomassie blue-stained gels, in addition to the observation that proteins are ubiquitinated at different stages throughout flower development and senescence, clearly demonstrates that selective ubiquitination of petal proteins is occurring against a background of almost total protein catabolism. Therefore, these findings suggest a role for ubiquitin in the degradation of proteins that accompanies floral development and senescence.

Acknowledgements

This work was supported by an award from the AFRC Plant Molecular Biology programme (PGIII/501PMB).

References

Agarwal, M.L. & Cullis, C.A. (1991). The ubiquitin-encoding multigene family of flax, *Linum usitatissimum*. *Gene* **99**, 69–75.

Bachmair, A., Becker, F., Masterson, R. V. & Schell, J. (1990). Perturbation of the ubiquitin system causes leaf curling, vascular tissue alterations and necrotic lesions in a higher plant. *EMBO Journal* **9**, 4543–9.

Bachmair, A., Finley, D. & Varshavsky, A. (1986). *In vivo* half-life of a protein is a function of its amino-terminal residue. *Science* **234**, 179–86.

Bachmair, A. & Varshavsky, A. (1989). The degradation signal in a short-lived protein. *Cell* **56**, 1019–32.

Ball, E., Karlik, C.C., Beall, C.J., Saville, D.L., Sparrow, J.C., Bullard, B. & Fyrberg, E.A. (1987). Arthrin, a myofibrillar protein of insect flight muscle, is an actin-ubiquitin conjugate. *Cell* **51**, 221–8.

Binet, M.N., Steinmetz, A. & Tessier, L.H. (1989). The primary structure of sunflower (*Helianthus annuus*) ubiquitin. *Nucleic Acids Research* **17**, 2119.

Blake, M.S., Johnston, K.H., Russell-Jones, G.J. & Gotschlich, E.C. (1984). A rapid, sensitive method for detection of alkaline phosphatase-conjugated anti-antibody on Western blots. *Analytical Biochemistry* **136**, 175–9.

Bonner, W.M., Hatch, C.L. & Wu, R.S. (1988). Ubiquitinated histones and chromatin. In *Ubiquitin* (ed. M. Rechsteiner,), pp. 157–72. New York: Plenum Press.

Busch, H. (1984). Ubiquitination of proteins. *Methods in Enzymology* **106**, 238–62.

Callis, J., Pollmann, L., Shanklin, J., Wettern, M. & Vierstra, R.D. (1989). Sequence of a cDNA from *Chlamydomonas reinhardii* encoding a ubiquitin 52 amino acid extension protein. *Nucleic Acids Research* **17**, 8377.

Callis, J., Raasch, J.A. & Vierstra, R.D. (1990). Ubiquitin extension proteins of *Arabidopsis thaliana*. Structure, localization, and expression of their promoters in transgenic tobacco. *Journal of Biological Chemistry* **265**, 12486–93.

Chau, V., Tobias, J.W., Bachmair, A., Marriott, D., Ecker, D.J., Gonda, D.K. & Varshavsky, A. (1989). A multiubiquitin chain is confined to specific lysine in a targeted short-lived protein. *Science* **243**, 1576–83.

Chen, K. & Rubenstein, I. (1991). Characterization of the structure and transcription of an ubiquitin fusion gene from maize. *Gene* **107**, 205–12.

Ciechanover, A., Elias, S., Heller, H., Ferber, S. & Hershko, A. (1980). Characterization of the heat-stable polypeptide of the ATP-dependent proteolytic system from reticulocytes. *Journal of Biological Chemistry* **255**, 7525–8.

Ciechanover, A., Finley, D. & Varshavsky, A. (1984). Ubiquitin dependence of selective protein degradation demonstrated in the mammalian cell cycle mutant ts85. *Cell* **37**, 57–66.

Ciechanover, A. & Schwartz, A.L. (1989). Review: How are substrates recognized by the ubiquitin-mediated proteolytic system? *Trends in Biochemical Sciences* **14**, 483–8.

Driscoll, J. & Goldberg, A.L. (1990). The proteasome (multicatalytic protease) is a component of the 1500-kDa proteolytic complex which degrades ubiquitin-conjugated proteins. *Journal of Biological Chemistry* **265**, 4789–92.

Ellis, R.J. & MacDonald, I.R. (1970). Specificity of cycloheximide in higher plant systems. *Plant Physiology* **46**, 227–32.

Ferguson, D.L., Guikema, J.A. & Paulsen, G.M. (1990). Ubiquitin pool modulation & protein degradation in wheat roots during high temperature stress. *Plant Physiology* **92**, 740–6.

Finley, D. & Chau, V. (1991). Ubiquitination. *Annual Review of Cell Biology* **7**, 25–69.

Fortin, M.G., Purohit, S.K. & Verma, D.P. (1988). The primary structure of soybean (*Glycine max*) ubiquitin is identical to other plant ubiquitins. *Nucleic Acids Research* **16**, 11377.

Gausing, K. & Barkardottir, R. (1986). Structure and expression of ubiquitin genes in higher plants. *European Journal of Biochemistry* **158**, 57–62.

Goldknopf, I.L. & Busch, H. (1977). Isopeptide linkage between nonhistone and histone 2A polypeptides of chromosomal conjugate-protein A24. *Proceedings of the National Academy of Sciences USA* **74**, 864–8.

Goldstein, G., Scheid M., Hammerling, U., Boyse, E.A., Schlesinger, D.H. & Niall, H.D. (1975). Isolation of a polypeptide that has lymphocyte-differentiating properties and is probably represented universally in living cells. *Proceedings of the National Academy of Sciences USA* **72**, 11–15.

Haas, A.L. & Bright, P.M. (1985). The immunochemical detection and quantitation of intracellular ubiquitin-protein conjugates. *Journal of Biological Chemistry* **260**, 12464–73.

Haas, A.L., Murphy, K.E. & Bright, P.M. (1985). The inactivation of ubiquitin accounts for the inability to demonstrate ATP, ubiquitin-dependent proteolysis in liver extracts. *Journal of Biological Chemistry* **260**, 4694–703.

Haas, A.L., Warms, J.V.B., Hershko, A. & Rose, I.A. (1982). Ubiquitin-activating enzyme. Mechanism and role in protein-ubiquitin conjugation. *Journal of Biological Chemistry* **257**, 2543–8.

Hatfield, P.M. & Vierstra, R.D. (1989). Ubiquitin-dependent proteolytic pathway in wheat germ: isolation of multiple forms of ubiquitin-activating enzyme E1. *Biochemistry* **28**, 735–42.

Hershko, A. (1988). Minireview: Ubiquitin-mediated protein degradation. *Journal of Biological Chemistry* **263**, 15237–40.

Hershko, A. & Ciechanover, A. (1982). Mechanisms of intracellular protein breakdown. *Annual Review of Biochemistry* **51**, 335–64.

Hershko, A. & Ciechanover, A. (1986). The ubiquitin pathway for the degradation of intracellular proteins. *Progress in Nucleic Acid Research and Molecular Biology* **33**, 19–56.

Hershko, A., Eytan, E., Ciechanover, A. & Haas, A.L. (1982). Immunochemical analysis of the turnover of ubiquitin-protein conjugates in intact cells. Relationship to the breakdown of abnormal proteins. *Journal of Biological Chemistry* **257**, 13964–70.

Hershko, A., Heller, H., Elias, S. & Ciechanover, A. (1983). Components of ubiquitin-protein ligase system. Resolution, affinity purification, and role in protein breakdown. *Journal of Biological Chemistry* **258**, 8206–14.

Hershko, A., Heller, H., Eytan, E., Kaklij, G. & Rose, I.A. (1984). Role of the α-amino group of protein in ubiquitin-mediated protein breakdown. *Proceedings of the National Academy of Sciences USA* **81**, 7021–5.

Hough, R., Pratt, G. & Rechsteiner, M. (1986). Ubiquitin-lysozyme conjugates. Identification and characterization of an ATP-dependent protease from rabbit reticulocyte lysates. *Journal of Biological Chemistry* **261**, 2400–8.

Jabben, M., Shanklin, J. & Vierstra, R.D. (1989). Ubiquitin-phytochrome conjugates. Pool dynamics during *in vivo* phytochrome degradation. *Journal of Biological Chemistry* **264**, 4998–5005.

Jennissen, H.P., Botzet, G., Majetschak, M., Laub, M., Ziegenhagen, R. & Demiroglou, A. (1992). Ca^{2+}-dependent ubiquitination of calmodulin in yeast. *FEBS Letters* **296**, 51–6.

Laemmli, U.K. (1970). Cleavage of structural proteins during the assembly of the head of bacteriophage T4. *Nature (London)* **227**, 680–5.

Laub, M. & Jennissen, H.P. (1991). Ubiquitination of endogenous calmodulin in rabbit tissue extracts. *FEBS Letters* **294**, 229–33.

Lay-Yee, M., Reid, M.S. & Stead, A.D. (1992). Flower senescence in daylily (*Hemerocallis fulva*) – an ethylene-insensitive species. *Physiologia Plantarum* (in press).

Lee, P.L., Midelfort, C.F., Murakami, K. & Hatcher, V.B. (1986). Multiple forms of ubiquitin-protein ligase. Binding of activated ubiquitin to protein substrates. *Biochemistry* **25**, 3134–8.

Lukaszewski, L. & Reid, M.S. (1989). Bulb-type flower senescence. *Acta Horticulturae* **261**, 59–62.

Matile, P. & Winkenbach, F. (1971). Function of lysosomes and lysosomal enzymes in the senescing corolla of the Morning Glory *Ipomoea purpurea*. *Journal of Experimental Botany* **22**, 759–71.

Pech, J.-C., Latché, A., Larrigaudière, C. & Reid, M.S. (1987). Control of early ethylene synthesis in pollinated petunia flowers. *Plant Physiology and Biochemistry* **25**, 431–7.

Pickart, C.M. & Rose, I.A. (1985). Functional heterogeneity of ubiquitin carrier proteins. *Journal of Biological Chemistry* **260**, 1573–81.

Pollmann, L. & Wettern, M. (1989). Review: The ubiquitin system in higher and lower plants – pathways in protein metabolism. *Botanica Acta* **102**, 21–30.

Rechsteiner, M. (1987). Ubiquitin-mediated pathways for intracellular proteolysis. *Annual Review of Cell Biology* **3**, 1–30.

Rechsteiner, M. (1991). Minireview: Natural substrates of the ubiquitin proteolytic pathway. *Cell* **66**, 615–18.

Seufert, W., McGrath, J.P. & Jentsch, S. (1990). UBC1 encodes a novel member of an essential subfamily of yeast ubiquitin-conjugating enzymes involved in protein degradation. *EMBO Journal* **9**, 4535–41.

Shanklin, J., Jabben, M. & Vierstra, R.D. (1987). Red light-induced formation of ubiquitin-phytochrome conjugates: Identification of possible intermediates of phytochrome degradation. *Proceedings of the National Academy of Sciences USA* **84**, 359–63.

Stead, A.D. & Moore, K.G. (1979). Studies on flower longevity in *Digitalis*. Pollination induced corolla abscission in *Digitalis* flowers. *Planta* **146**, 409–14.

Stead, A.D. & Moore, K.G. (1983). Studies on flower longevity in *Digitalis*. The role of ethylene in corolla abscission. *Planta* **157**, 15–21.

Swerdlow, P.S., Finley, D. & Varshavsky, A. (1986). Enhancement of immunoblot sensitivity by heating of hydrated filters. *Analytical Biochemistry* **156**, 147–53.

Tanaka, O., Murakami, H., Wada, H., Tanaka, Y. & Naka, Y. (1989). Flower opening and closing of *Oxalis martiana*. *Botanical Magazine Tokyo* **102**, 245–53.

Towbin, H., Staehelin, T. & Gordon, J. (1979). Electrophoretic transfer of proteins from polyacrylamide gels to nitrocellulose sheets: procedure and some applications. *Proceedings of the National Academy of Sciences USA* **76**, 4350–4.

Veierskov, B. & Ferguson, I.B. (1991). Ubiquitin conjugating activity in leaves and isolated chloroplasts from *Avena sativa* L. during senescence. *Journal of Plant Physiology* **138**, 608–13.

Vierstra, R.D. (1987). Demonstration of ATP-dependent, ubiquitin-conjugating activities in higher plants. *Plant Physiology* **84**, 332–6.

Vierstra, R.D., Langan, S.M. & Haas, A.L. (1985). Purification and initial characterization of ubiquitin from the higher plant, *Avena sativa*. *Journal of Biological Chemistry* **260**, 12015–21.

Vierstra, R.D., Langan, S.M. & Schaller, G.E. (1986). Complete amino acid sequence of ubiquitin from the higher plant *Avena sativa*. *Biochemistry* **25**, 3105–8.

Watts, F.Z. & Moore, A.L. (1989). Nucleotide sequence of a full length cDNA clone encoding a polyubiquitin gene from *Pisum sativum*. *Nucleic Acids Research* **17**, 10100.

Wilkinson, K.D., Urban, M.K. & Haas, A.L. (1980). Ubiquitin is the ATP-dependent proteolysis factor I of rabbit reticulocytes. *Journal of Biological Chemistry* **255**, 7529–32.

Wulster, G., Sacalis, J. & Janes, H.W. (1982). Senescence in isolated carnation petals. Effects of indoleacetic acid and inhibitors of protein synthesis. *Plant Physiology* **70**, 1039–43.

Ziegenhagen, R. & Jennissen, H.P. (1990). Plant and fungus calmodulins are polyubiquitinated at a single site in a Ca^{2+}-dependent manner. *FEBS Letters* **273**, 253–6.

Index